Advance praise for *Technology Transfer for the Ozone Layer*

'Imagine the pride of earning the Nobel Prize for warning that CFCs were destroying the ozone layer. Then imagine that citizens, policymakers, and business executives heeded the warning and transformed markets to protect the earth. This book is the story of why we can all be optimistic about the future if we are willing to be brave and dedicated world citizens.'

MARIO MOLINA, Nobel Laureate in Chemistry and Professor, University of California

'In 2002 I characterized Andersen and Sarma's *Protecting the Ozone Layer* as one of the most impressive environmental books ever written. Now, with Taddonio, they have produced a timely encore that should become one of the most important books for addressing climate change. This authoritative and meticulously researched treatise cuts to the heart of the problem: the crucial issues of technology research, development and diffusion that have been largely lost in the hot air of climate rhetoric. The authors rightfully put them centre-stage, and draw on the highly relevant success of the Montreal Protocol to provide detailed prescriptions for achieving an indispensable global energy technology revolution.'

AMBASSADOR RICHARD BENEDICK, US chief negotiator of the Montreal Protocol and author of *Ozone Diplomacy*

'A major global achievement in the field of scientific understanding and effective policy has been the set of initiatives taken to save the ozone layer, which provides inspiration and a useful model for action in the field of climate change. This book is extremely valuable reading for policymakers and scholars alike particularly in the context of the challenge of climate change being faced globally.'

R. K. PACHAURI, Chairman, Intergovernmental Panel on Climate Change (IPCC) and Director General, The Energy and Resources Institute (TERI)

'2007 is the 20th Anniversary year of the signing of the Montreal Protocol and there is cause for great celebration for the leadership of both developing and developed countries that led to the proper implementation of the Protocol. This book gives an authoritative account of how impossible challenges to the transfer of ozone-friendly technologies were overcome for the good of human society and ecosystems.'

MOSTAFA K. TOLBA, Under-Secretary-General, United Nations, and Executive Director, United Nations Environment Programme, 1976–1992

'The lessons documented in this book show that solutions to climate change are attainable and in the global economic interest – if we accept the challenge and make the commitment to deal with it.'

ALAN MILLER, International Finance Corporation (IFC)

'This book provides a forward-looking, substantive account of how technology transfer, at its best, collaboratively and cost-effectively enables countries to tackle the ozone layer issue – it demonstrates the relevance of lessons learned from the Montreal Protocol to environmental issues faced today.'

MARIA NOLAN, Chief Officer of the Multilateral Fund Secretariat

'The success of the Montreal Protocol has been supported by unprecedented technology development and transfer under the international collaboration. Japanese chemists and engineers are very proud of their voluntary participation in protecting the ozone layer for future generations. We believe the lessons learned during these twenty years help the technological challenge for the climate change. Lift your spirits up by reading how technology transfer can save the Blue Planet again.'

MASAAKI YAMABE, National Institute of Advanced Industrial Science and Technology, and Asahi Glass Company, Japan

'The authors have dedicated their entire personal capacity and lives to the fight for a better environment for all. Without their intelligence and dedication, the Montreal Protocol's unique success would not be there. In this book they extend their wisdom and experience to guide actions on our most serious environmental challenge: climate change. Having negotiated intensively in both of the environment regimes for more than a decade, I can assure you that this guidance is badly needed. I urge everyone to study carefully the valuable lessons from these eminent writers and implement them expeditiously.'

JUKKA UOSUKAINEN, Acting Director General, International Affairs Unit, Ministry of Environment, Finland

'Stephen Andersen and Kristen Taddonio of the EPA and Madhava Sarma of the Ozone Secretariat (retired) do an excellent job showing the many ways that voluntary partnerships speed global environmental protection. Programmes like the Energy Star label on efficient products and initiatives under the Montreal Protocol have produced dramatic results. Imagine what we can accomplish as we continue to transfer technology to protect the climate.'

KATHLEEN HOGAN, Director, EPA Climate Protection Partnerships Division

'A highly informative, well researched compendium of technology transfers effected under the Montreal Protocol, written by authors who have traversed the length and breadth of this successful environment treaty. Stakeholders of current and future global environment treaties will be vastly benefited when they study the whole process of technology transfers effected under the Montreal Protocol to phase out ozone depleting substances.'

ARUN BHARAT RAM, Chairman and Managing Director, SRF Limited (a leading Indian chemical company)

Technology Transfer
for the Ozone Layer

Technology Transfer for the Ozone Layer

Lessons for Climate Change

Stephen O. Andersen, K. Madhava Sarma
and Kristen N. Taddonio

GLOBAL
ENVIRONMENT
FACILITY
INVESTING IN OUR PLANET

Routledge
Taylor & Francis Group

LONDON AND NEW YORK

First published 2007 by Earthscan

2 Park Square, Milton Park, Abingdon, Oxon OX14 4RN
711 Third Avenue, New York, NY 10017, USA

Routledge is an imprint of the Taylor & Francis Group, an informa business

First issued in paperback 2016

ISBN: 978-1-138-98852-1 (pbk)

Typeset by MapSet Ltd, Gateshead, UK
Cover design by Nick Shah

A catalogue record for this book is available from the British Library

Library of Congress Cataloging-in-Publication Data

Andersen, Stephen O.
  Technology transfer for the ozone layer : lessons for climate change / Stephen O. Andersen, K. Madhava Sarma, and Kristen Taddonio.
       p. cm.
  "Sequel and complement to Protecting the Ozone Layer: The United Nations History."
  Includes bibliographical references and index.
  ISBN-13: 978-1-84407-473-0 (hardback)
  ISBN-10: 1-84407-473-0 (hardback)

  1. Ozone-depleting substances. 2. Ozone layer depletion—Prevention—History—20th century. 3. Technology transfer. I. Sarma, K. Madhava, 1938- II. Taddonio, Kristen. III. Andersen, Stephen O. the United Nations history. Protecting the ozone layer : IV. Title.
  TD887.O95A63 2007
  363.738'75526—dc22
                                                            2007021491

Contents

List of Figures, Tables and Boxes

FIGURES

TABLES

BOXES

About the Authors

Stephen O. Andersen

Director of Strategic Climate Projects, Climate Protection Partnerships Division, US Environmental Protection Agency

 Stephen O. Andersen began work on climate and ozone layer protection in 1974 as a member of the Climatic Impact Assessment Project on the effects of supersonic aircraft. With K. Madhava Sarma, he is author of *Protecting the Ozone Layer: The United Nations History* and, with Durwood Zaelke, *Industry Genius: Inventions and People Protecting the Climate and Fragile Ozone Layer.* Prior to joining the US Environmental Protection Agency (EPA), he worked for environmental and consumer non-governmental organizations (NGOs) and was a professor of environmental economics. In 1986, he joined the fledgling EPA Stratospheric Protection team, working his way up to Deputy Director. Since 1988, he has been Co-chair of the Technology and Economic Assessment Panel and has also chaired the Solvents Technical Options Committee, the Methyl Bromide Interim Technology and Economic Assessment and the Task Force on the Implications to the Montreal Protocol of the Inclusion of HFCs and PFCs in the Kyoto Protocol. He was co-editor of the IPCC/TEAP Special Report 'Safeguarding the Stratospheric Ozone Layer and the Global Climate: Issues Relating to Hydrofluorocarbons and Perfluorocarbons'. He pioneered voluntary programmes to phase out CFC food packaging, recycle CFCs from vehicle air conditioning, halt testing and training with halon, and accelerate CFC solvent phaseout in electronics and aerospace. He created the EPA ozone and climate protection awards and helped found the Industry Cooperative for Ozone Layer Protection and the Halons Alternative Research Corporation. He helped negotiate the phaseout of CFC refrigerator manufacturing in Thailand and the corporate pledge to help Vietnam avoid dependence on ozone-depleting substances (ODSs). He served on the team that commercialized no-clean soldering and the team phasing out ODSs from solid rocket motors. He is the recipient of numerous awards, including the 1990 EPA Gold Medal, the 1995 Fitzhugh Green Award, the 1995 UNEP Global Stratospheric Ozone Protection Award, the 1996 Sao Paulo Brazil State Ozone Award, the 1998 US

EPA Stratospheric Ozone Protection Award, the 1998 UNEP Global 500 Roll of Hónour, the 1998 Nikkan Kogyo Shimbun Stratospheric Protection Award, the 1999 Vietnam Ozone Protection Award, the 2000 Mobile Air Conditioning Society Twentieth Century Award for Environmental Leadership, the 2001 US DoD Award for Excellence and the 2007 US EPA Best-of-the-Best Stratospheric Ozone Protection Award. He has a PhD from the University of California, Berkeley.

K. Madhava Sarma

formerly Executive Secretary, Secretariat for the Vienna Convention and the Montreal Protocol, United Nations Environment Programme

 K. Madhava Sarma is currently a consultant on ozone issues and integration of the common aspects of global environmental treaties for greater synergy. With Stephen O. Andersen, he authored *Protecting the Ozone Layer: The United Nations History* (Earthscan, co-published by UNEP, 2002). He was the Executive Secretary of the Secretariat for the Vienna Convention and the Montreal Protocol from 1991 to 2000. During his tenure as Executive Secretary, he served the Parties to the Protocol through the turbulent Meetings of the Parties in Copenhagen, Vienna, Montreal, and Beijing – including three replenishments of the Multilateral Fund for the Implementation of the Montreal Protocol. He streamlined the administration of the institutions of the Protocol, the reporting requirements and other administrative obligations so that Parties could devote their full attention to resolving challenging political issues. Prior to being recruited to head the Secretariat, Madhava Sarma was a senior member of the Indian diplomatic team involved in the Montreal Protocol negotiations between the first and second Meetings of the Parties (1989–1991). During this time, he was often an effective spokesman for the developing country perspective and cosponsored many of the provisions of the London Amendment that satisfied developing countries while creating enforceable obligations to protect the ozone layer. He made other significant contributions as the senior Indian official looking after environmental policy, law, institutions and international cooperation, including responsibility for all global environmental issues. Prior to joining the national Government of India, he served (as a member of the Indian Administrative Service) as Head of District Administration, State Water Supply Board, and as Secretary to the Government, Irrigation and Power. During this state tenure, he was responsible for planning and implementation of many water supply, irrigation and energy projects. He earned the 1996 US EPA Stratospheric Ozone Protection Award and a 1995 award from UNEP 'For Extraordinary Contributions to Ozone Layer Protection', and the 2007 US EPA Best-of-the-Best Stratospheric Ozone Protection Award.

Kristen N. Taddonio

Project Director, Climate Protection Partnerships Division, US Environmental Protection Agency

Kristen N. Taddonio is Manager of Strategic Climate Projects at the US Environmental Protection Agency (US EPA) Climate Protection Partnerships Division. She organizes public–private partnerships for environmental innovation, harmonizes international standards to speed technology market penetration, and directly promotes technology transfer with information, leadership pledges and conferences. In the Climate Protection Partnerships Division she brought together a team of international experts from industry, government, military, and standards organizations who are removing global barriers to climate-friendly refrigerants. The success of her team will allow vehicle manufacturers to market environmentally superior technology worldwide with confidence and safety. Her partners are from Australia, Austria, Belgium, France, Germany, India, Italy, Japan, Netherlands and the United States. At the EPA, she manages an annual budget of more than US$600,000 and organizes the annual Climate Protection Awards, which were established in 1998 to recognize exceptional leadership, outstanding innovation, personal dedication and technical achievements in climate protection. Prior to her latest promotion, she was a technical writer and a marketing associate for the Energy Star programme for new homes. She earned a Masters Degree in International Science and Technology Policy and a Bachelors Degree in International Environmental Resources from the George Washington University's Elliot School of International Affairs, where she graduated *summa cum laude*. She has also earned degrees in Scientific and Technical Communication and Liberal Arts. Her papers have been featured in plenary sessions of conferences and workshops in Austria, France, India, Italy, Japan and the US. She is Co-chair of the United Nations Task Force on the Legacy of the Technology and Economic Assessment Panel (TEAP) of the Montreal Protocol (Report published April 2007, United Nations Environment Programme, and Nairobi, Kenya). In 2007, the Mobile Air Conditioning Society–Worldwide presented Kristen Taddonio with the Government Partner of the Year Award.

Foreword

Throughout the world, the Montreal Protocol is viewed as a great success and a tribute to institutions, countries, and individuals that made it happen. We at the Global Environment Facility (GEF) are proud of our role in supporting countries with economies in transition (CEITs) in their efforts to implement the Montreal Protocol. We are encouraged by these countries' successes and welcome the opportunity to show how technology transfer and financing can solve the many daunting challenges of global environmental protection.

The GEF – the largest funder of environmental protection in developing countries and economies in transition – was created in 1991, at a time when it was clear that Russia and the Newly Independent States and the other countries of Central and Eastern Europe would need the global community's support to meet their obligations to phase out ozone-depleting substances under the Montreal Protocol.

Responding to the appeal of the Parties to the Protocol, the GEF provided financial assistance to them at a crucial juncture and enabled them to implement the Protocol. In 15 years, from 1991 to 2006, these countries have decreased their consumption of ozone-depleting substances from about 296,000 tonnes to 350 tonnes – a reduction of over 99 percent.

Global environmental problems cannot be treated in isolation. At the GEF, we increasingly work with countries to intervene across domains to address climate change, biodiversity conservation, sustainable land management and chemicals management, including pollution of international waters from persistent organic pollutants (POPs).

The GEF strategies for climate change, POPs and ozone layer-depletion are indicative of the flexibility that we exercise. Within each domain of intervention, project developers are encouraged to seek synergies and co-benefits with the other areas: for example, between ozone and POPs, or between climate change and ozone. This ability to work across global environmental issues is one of the greatest strengths of the GEF.

There are two potential ways in which the phaseout of ozone-depleting substances might increase the risk of climate change: using substitutes that have a high global warming potential; and introducing less energy-efficient technologies. Therefore, the focus of GEF's work has been to help the countries transfer from ozone-depleting substances to both ozone-safe and climate-safe options. The GEF funds the conversion to technologies that have the least impact on global warming while being technically feasible, environmentally sound and economically acceptable.

The book also points out another important dimension of GEF's work: bridging the environment and development for sustainable development. It shows that technology conversions in many enterprises were instrumental in helping a number of sectors to modernize and adapt to a market economy.

I am pleased that the authors are recording this vital technology transfer story. A performance study of GEF has praised the Montreal Protocol process for its emphasis on clear goals and for creating an enabling environment for alternatives. The authors of this book have succeeded in bringing out the best from this process. I hope that the stakeholders of climate and other treaties will examine these lessons and adopt those that are suitable for their circumstances.

Monique Barbut
CEO and Chairperson
Global Environment Facility
Washington DC
June 2007

Preface

This is the first authoritative account of how technology was transferred worldwide under the Montreal Protocol. It tells the remarkable story of how governments, industry, consumers and the concerned public can, when faced with an environmental change crisis that threatens the health of the planet, work quickly and creatively to transform markets. As such it holds lessons on how to deal with other mutual and common challenges facing the environment, livelihoods, economic stability and human health across a wide range of spheres.

The story of the Montreal Protocol is worth repeating in all its detail. The Montreal Protocol of 1987 was the first convention based on the precautionary approach and the concept of a 'common but differentiated responsibility'. The preamble of the Protocol says the Parties to the Protocol are:

> Determined *to protect the ozone layer by taking precautionary measures to control equitably total global emissions of substances that deplete it, with the ultimate objective of their elimination on the basis of developments in scientific knowledge, taking into account technical and economic considerations and bearing in mind the developmental needs of developing countries,*

> Acknowledging *that special provision is required to meet the needs of developing countries, including the provision of additional financial resources and access to relevant technologies, bearing in mind that the magnitude of funds necessary is predictable, and the funds can be expected to make a substantial difference in the world's ability to address the scientifically established problem of ozone depletion and its harmful effects,*

These strategies were later made explicit in the 1992 Earth Summit in Rio de Janeiro as Principles 7 and 15 of the Rio Declaration and have been followed by virtually every environmental Convention since.

Piloted by UNEP, the Montreal Protocol allowed developing countries more time than developed countries to implement the control measures so that alternative technology would be mature and affordable. Developing countries had the advantage of 'leapfrogging' over alternatives that entered the market early but were soon made obsolete by technical progress. In 1990, on the urging of developing countries, developed countries agreed to finance the incremental costs of the phaseout in developing countries with its own financial mechanism called the Multilateral Fund. In 1991, the Global Environment Facility (GEF) was created by the Governments to deal with a wide range of global environmental issues, including ozone depletion. The GEF financed the incremental costs of those eligible countries not qualifying for financing under the MLF, including the countries of Eastern Europe and central Asia with economies in transition. The original control measures of the 1987 Protocol were repeatedly strengthened by the Parties to the Protocol on the basis of periodic scientific and technological assessments to provide for the phaseout of nearly a hundred ozone depleting chemicals on a specified time schedule.

The success of the Protocol is now acknowledged by all, even though the phaseout of the ozone-depleting chemicals is by no means complete. 190 governments have ratified the Protocol and are actively committed to phasing out ozone-depleting chemicals. The Fund to date has granted more than US$2.1 billion to the developing countries to switch to ozone-friendly solutions. The GEF has assisted the CEIT to the tune of US$200 million. Technological cooperation over the last 20 years has led to outstanding reductions of over 95 per cent in the consumption of ozone-depleting chemicals. Continuing scientific observations through satellites, balloons and ground-based observation have confirmed this reduction, as elaborated in the periodic reports of the Scientific Assessment Panel.

Protection of the ozone layer involved a large number of stakeholders. Many United Nations organizations did their part, including: the United Nations Development Programme (UNDP); United Nations Environment Programme (UNEP); United Nations Industrial Development Organization (UNIDO); World Health Organization (WHO); World Meteorological Organization (WMO); Food and Agriculture Organization (FAO); and Regional Economic and Social Commissions. International financial institutions, such as the World Bank and the Global Environment Facility, and national financial institutions also played an invaluable part in implementation. Industry and industrial organizations eschewed their usual competitive spirit and shared technologies and techniques to phase out ozone-depleting chemicals. Non-governmental organizations not only kept an alert eye on the issue and sounded the alarm when necessary, but also developed ozone-safe technologies and spread awareness about such technologies. National governments employed many regulatory, economic and policy instruments to achieve the phaseout as planned.

Does the success of the Montreal Protocol process suggest any advice for other global environmental treaties? While the treaties differ from one another, there are many common strands among them. Most, if not all, treaties aim at replacing some of the current environment-unfriendly technologies with environmentally sound technologies. The challenges posed to the Earth's environment by some issues (like climate change) are so serious that the world community has to adopt the new technologies as soon as possible in all the countries.

This is precisely the challenge met by the Montreal Protocol process. It would be sensible for the world community to study the process and adopt its useful features so that time is not lost by reinventing the wheel with every convention. This study will also be relevant to UNEP's Bali Strategic Plan for Technology Support and Capacity-building.

I am grateful to Stephen O. Andersen, who has been a co-chair of the Montreal Protocol's Technology and Economic Assessment Panel (TEAP) since its inception 18 years ago; Madhava Sarma, who served as the Executive Secretary of the Secretariat for the Vienna Convention and the Montreal Protocol for more than nine years; and to Kristen Taddonio, for agreeing to put together this book. It was a labour of love for them. They obtained contributions to this study from many of the people who made it a triumph. It is a timely contribution on the occasion of the twentieth anniversary of the Montreal Protocol.

I hope this history and analysis will please all those who contributed to the success of the ozone agreements, serve as an authentic record of one of the world's great achievements and assist other Conventions in their way forward.

Achim Steiner
Executive Director
United Nations Environment Programme
Under Secretary-General, United Nations

Acknowledgements

Many individuals helped us to compile this book. Foremost, we are grateful to Mr Len Good, the former Chairman, Global Environment Facility (GEF), for his ready support to our research and writing on the topic of this book. His successor, Ms Monique Barbut, was earlier the head of the United Nations Environment Programme (UNEP) Division of Trade, Industry and Economics and was equally encouraging when we explained the idea behind this book to her. Her support continued when she took over as Chief Executive Officer and Chairperson, GEF. We are grateful for her thoughtful Foreword to this book. Dr Kathleen Hogan (Director of the EPA Climate Protection Partnerships Division) inspired the hard work.

We are grateful to Mr Durwood Zaelke and Mr Scott Stone at the International Network for Environment Compliance and Enforcement (INECE) for their substantial contributions of time and logistics. The United States Environmental Protection Agency (US EPA) sponsored Mr Madhava Sarma's investigation of military leadership in ozone and climate protection and allowed Dr Stephen Andersen and Ms Kristen Taddonio to pursue this labour of love. Mr Marco Gonzalez, Executive Secretary of The Ozone Secretariat and his able Administrative and Finance Officer Ms Ruth Batten enthusiastically helped us. Mr Achim Steiner, Executive Director of UNEP, set the stage and Mr Gonzalez put the Protocol in a global context with their authoritative Preface and Introduction. We are honoured by the kind endorsement on the front cover by Nobel Prize winner Dr Mario Molina (University of California) and on the back cover by Ambassador Richard Benedick (US chief negotiator of the Montreal Protocol), Dr Kathleen Hogan (US EPA), Mr Alan Miller (International Finance Corporation), Ms Maria Nolan (Multilateral Fund Secretariat), Dr Rajendra K. Pachauri (Intergovernmental Panel on Climate Change), Dr Mostafa K. Tolba (former Under-Secretary-General, United Nations, and Executive Director, UNEP), Mr Jukka Uosukainen (Finland Ministry of Environment), and Dr Masaaki Yamabe (Japan National Institute of Advanced Industrial Science and Technology).

We have the greatest respect for Ms Lani Sinclair – our experienced editor for this and the previous book (*Protecting the Ozone Layer: The United Nations History*) – who ironed out our English and helped us clarify our thoughts. Earthscan and their editorial team – Mr Hamish Ironside, Mr Robert West, and Ms Alison Kuznets – worked beyond the call of duty with recommendations that strengthened the presentation, accuracy and style. Special thanks to Ms Samira DeGobert for guiding our choice of art and graphics.

We appreciate the work of Mr Rajendra Shende and Mr Tilman Hertz and their colleagues Mr Etienne Gonin, Ms Samira de Gobert, Mr Jim Curlin, Mr Yerzhan Aisabayev and Ms Nirupa Ram of UNEP for their informed chapter on Awareness and Capacity Building. We are grateful to Dr Melanie Miller and Ms Marta Pizano for their chapter on technology transfer for eliminating methyl bromide. Mr János Maté made considerable contributions to the Air Conditioning and Refrigeration chapter and to the Awareness and Capacity Building chapter.

The manuscript was reviewed by Mr Scott Stone and Mr Durwood Zaelke (INECE) and the chapters on technology transfer were reviewed by Ms Maria Nolan (Executive Officer of the MLF), Mr Tony Hetherington (Retired MLF Deputy Chief Officer). Their perceptive comments and amendments to the text strengthen the findings. In addition, each chapter was reviewed for technical accuracy and clarity by experts. We are particularly indebted to Mr Geno Nardini, Mr Jose Pons, Dr Helen Tope and Dr Ashley Woodcock for the Aerosol and Medical Products chapter; to Dr Radhey Agarwal and Mr Stephan Sicars for the Air Conditioning and Refrigeration chapter; to Dr Mike Jeffs and Ms Sally Rand for the Foams chapter; to Mr David Catchpole and Dr Dan Verdonik for the Halons chapter; to Mr William J. Van Houten, Mr Dave Koehler, Dr David J. Liddy, Mr William Nicholls, Mr E. Thomas Morehouse, Mr Peter Mullenhard, and Mr Ronald W. Sibley for the Military chapter; and to Mr Jorge Corona, Mr Brian Ellis, Dr William Kenyon, Mr Darrell Staley, and Dr John Stemniski for the Solvents chapter.

Two dozen of the most respected practitioners of ozone-safe technology transfer contributed text that illustrates how daunting technical and business challenges are overcome. These perspectives are contributed by: Dr Radhey Agarwal (India Institute of Technology), Mr Atul Bagai (UNEP) OzonAction South Asian Network), Mr Natarajan Balaji (UNEP OzonAction South Asian Network) Dr Tom Batchelor (TouchDown Consulting), Mr Ross Bowman (Thiokol Corporation), Mr Gilbert F. Decker (US Army), Mr Yuichi Fujimoto (Japan Electrical Manufacturer's Association), Mr Ramachandran Gopichandran (Centre for Environment Education, Ahmadabad), Mr Kiwohide Hata (Matsushita Electric Company), Dr Richard Helmick (US Naval Sea Systems Command), Mr Bill Holder (US Air Force), Dr Mike Jeffs (European Diisocyanate and Polyol Producers Association), Mr Donald Kaniaru (International Network for Environment Compliance and Enforcement), Mr Osami Kataoka (Daikin Industries), Ms Bella Maranion (US EPA), Mr János Maté (Greenpeace), Mr Alan S. Miller (International Finance Corporation), Mr Toshiyuki Miyajima (Seiko Epson), Mr E. Thomas Morehouse (Institute for Defense Analyses), Mr Hideo Mori, (Otsuka Pharmaceutical Company), Mr Peter Mullenhard (Science Applications International Corporation), Mr Naohiro Yamamura (Japan Ministry of Economy, Trade and Industry), Mr Steve Newman (National Aeronautics and Space Administration), Mr Jeffery J. Norton (US Air Force), Mr Tsutomu Odagiri (Japan Industrial Conference on Cleaning), Mr Akira Okawa (Japan Industrial Conference for Ozone Layer and Climate Protection), Ms Sally Rand (US EPA), Dr Akira Sekiya (National Institute of Advanced Industrial Science and Technology), Dr Shunichi

Samejima (Asahi Glass Foundation), Mr Rajendra Shende (UNEP DTIE), Mr Darrel Staley (Boeing Company), Mr Scott Stone (INECE), Mr Shigehiro Uemura (Japan Industrial Conference for Ozone Layer and Climate Protection), Mr Gary Vest (US Department of Defense, retired), Dr Masaaki Yamabe (National Institute of Advanced Industrial Science and Technology), Mr Robert L. Walker (US Army), and Mr Durwood Zaelke (INECE).

No colleagues are more valuable than those who grant informed access to the files and reports that are primary sources for analysis. Thanks to Mr Laurent Granier at the Global Environment Facility, Mr Rajendra Shende and his colleagues at UNEP, Dr Suely Carvalho and her colleagues of UNDP, Mr Si-Ahmed and his colleagues at UNIDO and Mr Steven Gorman, Mr Viraj Vithoontien, and their colleagues at the World Bank for giving us the Project Completion Reports and for sharing their valuable insights with us.

Also, a special thanks to the Intergovernmental Panel on Climate Change, the Science and Development Network, UNCTAD, the Montreal Protocol Secretariat, and other international organizations for allowing us to quote from their publications and reports and to Dr Carlos Montalvo Corral for granting permission to quote from his book, *Environmental Policy and Technological Innovation: Why Do Firms Adopt or Reject New Technologies?*.

We thank Dr Janet Andersen and Ms K. Ramalakshmi, Dr Stephen Andersen's and Mr K. Madhava Sarma's respective spouses, for their constant support and encouragement and Ms Kristen Taddonio's friends and family for ongoing encouragement and support. Ms Gouthami, one of the daughters of Mr Sarma, deserves his thanks for providing him with the necessary facilities for writing for a considerable time period.

We have benefited substantially from the many contributions and for the review of our drafts by experts, but we are responsible for any errors or omissions in this book.

This book has not been subjected to publication review by the US Environmental Protection Agency (EPA) or the Global Environment Facility (GEF) and therefore does not necessarily reflect the views of the EPA or GEF; no official endorsement should be inferred. The views expressed by individual authors are their own. Mention of trade names, products, or services does not convey official EPA, GEF, or author approval, endorsement, or recommendation.

Stephen O. Andersen, K. Madhava Sarma and Kristen N. Taddonio

Introduction

 The Ozone treaties have been extraordinarily fortunate, born under the right stars as it were. There are thousands of inviduals and institutions connected to the ozone layer issue over the past 33 years and each of them works with missionary zeal to protect the ozone layer. The treaties owe their success to this zeal, which continues to this day.

First came the scientists. Two chemists at the University of California at Irvine – Mario J. Molina and F. Sherwood Rowland – were the first to discover the link between CFCs and ozone depletion. They also made a passionate plea for practical action, and this gave rise to a ban on CFCs in aerosols by many countries. This also led to diplomatic action by UNEP beginning in 1977. Ever since, the scientists, through their many startling discoveries – the 'ozone hole', the 'smoking gun' – and through the four-yearly assessments organized by UNEP and WMO, kept the world community informed and educated on the policy options available. For the first time, scientists played a direct part in diplomatic negotiations and helped the governments not only to understand the phenomenon of ozone depletion and its adverse effects, but also to give concrete policy options, with each option leading to a particular impact on the ozone layer.

The technologists were on hand to analyse the technical and economic feasibility of alternatives, so that governments could make up their minds after weighing all the consequences – environmental, technical and financial.

The industry threw their awesome talent into discovering alternatives to the ozone-depleting substances and spreading these alternative technologies and processes throughout the world.

Many professionals throughout the world joined the effort and contributed their best. The NGOs not only contributed their watchful attention but also helped in bringing to light the hitherto unnoticed ozone-safe technologies.

The depletion of the ozone layer was by far the most serious global environmental problem ever faced by humanity. The objective of the ozone treaties was certainly a difficult one: to persuade the entire world to give up the use of many profitable chemicals that had been praised as wonder chemicals. Those to be persuaded were not only governments, but also the producers of these chemicals, all major multi-national giants of industrialized countries and thousands of industries that used these chemicals considered to be 'irreplaceable'. Behind them were the billions of consumers who wanted and needed the products that contained ozone-depleting chemicals: refrigerators, air conditioning, firefighting equipment and foams.

UNEP had to convince the world that once the depletion was started there would be no place for humanity to hide. The UNEP and its successive Executive Directors, and particularly Dr Mostafa Tolba, shed the image of 'neutral' UN organizations and pushed the diplomats of the world to arrive at the ozone treaties and to strengthen them continuously. The other UN organizations, such as the UNDP, World Bank, UNIDO and WHO, lent their combined might to the goal of saving the ozone layer. The Multilateral Fund created by the Montreal Protocol has worked wonders to enable every developing country in the world to join the effort. When the countries of Eastern Europe and Central Asia faced great economic and political problems, the Global Environmental Facility stepped in to ensure that those countries phased out the ozone-depleting substances, despite their troubles.

The implementation of the Protocol over the last 20 years has led to outstanding reductions of over 95 per cent in the consumption of ozone-depleting chemicals through changing to ozone-safe technologies according to the timetable set by the Montreal Protocol. What was the process that led to such a success throughout the world? Can it be replicated in other situations of threat to global environment? These questions are very relevant in this year of the twentieth anniversary of the Montreal Protocol. The lessons would be useful to the future actions of the Protocol in completing the remaining tasks of the phaseout. Perhaps other multilateral agreements could also gain some advantage by studying the process of the Montreal Protocol.

I am grateful to the authors who have taken the trouble over the past two years to prepare this insightful study into the ways and means of effectively transferring information, knowledge and technology, and supporting national capacity-building within the Ozone Layer Protection Treaties.

Marco Gonzalez
Executive Secretary
Secretariat for the Vienna Convention and its Montreal Protocol
UNEP

Prologue

This book is a sequel and complement to *Protecting the Ozone Layer: The United Nations History*. It is an account of how technology was developed, commercialized and transferred to companies in 190 countries – rich and poor, east and west, and north and south – in order to halt, within a prescribed time schedule, the production and use of chemical substances that destroy the ozone layer.

Throughout its existence, the United Nations has been at the forefront of efforts to protect the global environment. The making of environmental law has been an essential part of that undertaking. Today there are nearly 270 environmental treaties, covering issues such as marine and air pollution, hazardous waste, biodiversity, desertification, and climate change. Development, commercialization and transfer of environmentally sound technologies are the crux of these treaties, and this includes the ozone treaties.

Among the most successful of these treaties are the ozone agreements brokered by the United Nations Environment Programme (UNEP): the Vienna Convention for the Protection of the Ozone Layer (1985) and its Montreal Protocol on Substances that Deplete the Ozone Layer (1987).

In recent years there have been several debates in the international forums of the United Nations and the World Trade Organization on the obstacles to transfer of, and change to, environmentally sound technologies, particularly for the developing countries. These obstacles encompass institutional, social, political, technical and economic factors. The UNEP Bali Declaration is the latest consensus of governments on this issue. Experience regarding ozone-friendly technologies is now more than 20 years old. This book analyses this experience, in the hope that it will both benefit the Montreal Protocol, since there is much more to be done to achieve its objectives, and help the global environmental agreements that protect climate: the United Nations Framework Convention on Climate Change (UNFCCC) of 1992 and the Kyoto Protocol to the UNFCCC of 1998.

There will be some who will protest that the climate treaties do not need advice or that, if they do need advice, the Montreal Protocol experience is not relevant to the climate treaties since the issues are very different. We agree that the issues are different, as indeed any two issues are. However, in this book we hope to convince you and others responsible for protecting the climate that the lessons of the Montreal Protocol process are transferable to many situations. It is a fact that climate change is already occurring. Scientists have noted the existence of change, as well as the adverse effects thereof. They have, through the reports of the Intergovernmental Panel on Climate Change, given dire warnings about the consequences of greenhouse gas emissions from human activities, and have urged the world to take urgent action. It is also a fact that the climate treaties have not achieved much so far. The UNFCCC is now 15 years old, the Kyoto Protocol 9, and the progress made by these treaties is miniscule compared to the progress made by the Montreal Protocol at a similar time stage. The immense complexity of the issues involved may be one reason for the lack of results. However, we will demonstrate that the issues involved in the phase-out of ozone-depleting substances (ODSs) were also complex. Despite this complexity, stakeholders' determined efforts are currently phasing out 96 ozone-depleting substances used by 240 industry sectors in thousands of products, and the phaseout is proceeding ahead of the schedule set by the Montreal Protocol. We include detailed sector-wise accounts of this phaseout to shed light on the lessons of the Montreal Protocol process, but a technically disinclined reader can skip the technical details and still keep track of the creativity displayed by thousands of ozone actors in order to resolve the problems that arose.

The Montreal Protocol experience dispels many myths and reveals many surprises. It will surprise you to learn that technologies to protect the ozone layer came from many parts of the world even before the Protocol entered into force; that concern for future generations motivated unprecedented access to intellectual property (often without charge); that military organizations motivated technical solutions to the most challenging applications; that the cost of financing the incremental costs of technology for countries with economies in transition and developing countries was far less than anyone imagined; and that specialized institutions can outperform large industrial and financial institutions normally charged with carrying out international technology projects.

Governments, scientists, industry, non-governmental organizations and the United Nations system set aside their differences and came together to fight against the catastrophic threat of stratospheric ozone depletion. However, the picture is not perfect, and some mistakes were made. We hope that this book will help others to avoid these mistakes in the future.

Technology change generally came about first in developed countries and only then in other countries, as the Protocol gave a period of grace to developing countries in recognition of the 'common but differential responsibility' principle. The industries that used the ozone-depleting substances are very diverse: the air-conditioning, refrigeration, firefighting, solvent, agriculture, aerosol product and foam sectors were the major ozone-depleting substance consumers, but ODSs were used in thousands of other small applications as well. There were many large enterprises that had the resources to develop new and innovative ozone-friendly technologies, but there were also many small enterprises that needed to be educated on such technologies and how to access them. The governments of developed countries implemented many policies, regulations, awareness and education campaigns, and financial incentives and disincentives to promote ozone-safe technologies. These were later followed by the developing countries, who introduced additional policy innovations. A considerable body of literature has been published by governments as well as by scholars. We have relied on these studies to discern some key features of technology change in the countries involved.

The developing countries are eligible for assistance from the Multilateral Fund, which the Protocol set up in 1991 under Article 10, to meet the incremental costs of implementing ozone-depleting substance control measures. As of March 2007, the Fund has approved 5520 projects, with total funding of over US$2.1 billion in assistance to 143 developing countries. We have analysed the completion reports of about 1000 such projects in order to find answers to the many questions on the various facets of technology change.

When the Protocol was signed in 1987, most Eastern European and Central Asian countries were either part of the USSR or members of the political alliances led thereby. They had communist economic systems. The Meeting of the Parties to the Protocol did not consider these countries to be developing countries. The 1989 political upheavals led to the break-up of the USSR and the collapse of communist governments, and the new governments, more or less, chose the market economy path of the industrialized West. This transition took some time and resulted in the economic collapse of many of these countries, and there was concern that they might be unable to implement the Montreal Protocol control measures. The Global Environment Facility (GEF) here came to their rescue by providing assistance to make the change to ozone-friendly technologies. We have analysed the GEF project completion reports for these countries as well.

In addition to our analyses, we asked many experts on technology change to give their own accounts – from their individual, organization, sector or country point of view – of the problems of technology change and transfer, and the solutions. The accounts of the many that responded to our request have been integrated with our analysis.

We believe that this book is the first of its kind: a comprehensive analysis of environment-friendly technology change that actually occurred throughout the world and in a relatively short period of time. Previous publications on technology transfer for environmental protection have mostly been limited to case studies. We believe that our findings will change many existing perceptions about technology transfer and influence the debate about technology change. We also believe that this book will assist with the implementation of other environmental conventions, particularly those relating to climate change.

Contours of Technology Transfer

INTRODUCTION

International efforts to encourage technology cooperation and transfer for environmental protection began in the 1970s. At the 1972 United Nations Conference on the Human Environment, political leaders called upon the global community to make technologies more available to developing countries.[1] Since the beginning of the 1980s, technology transfer has become an increasingly important issue due to mounting global environmental problems. Developing countries demanded technology transfer as a condition of participation in the control measures of the Montreal Protocol on Substances that Deplete the Ozone Layer, and today technology transfer is included in over 80 regional and international agreements, including Agenda 21, the United Nations Framework Convention on Climate Change (UNFCCC), the Kyoto Protocol, the Organisation for Economic Co-operation and Development (OECD) Environmental Strategy, the Convention on Biological Diversity (CBD) and the United Nations Convention to Combat Desertification (CCD).[2]

This book is among the first to study the success of the Montreal Protocol in technology transfer, with the goal of guiding other technology transfer efforts, particularly for climate protection.

TECHNOLOGY TRANSFER AS A POSITIVE MEASURE

Multilateral environmental treaties have historically used a variety of tools to achieve their objectives. Among these tools are voluntary agreements, trade measures and enforcement measures designed to encourage compliance. Such treaties include 'positive measures' designed to help countries meet their commitments. Technology transfer provisions fall into this category and are included in many multilateral environmental agreements (see Box 2.1).[3]

WHAT IS TECHNOLOGY TRANSFER?

Technology transfer is the intentional 'passing-on' of technology or know-how from one party to another, commonly by purchase, investment or agreements for cooperation (see Box 2.2). There are three distinct components of technology that can be transferred:

Box 2.1 Technology transfer in the Montreal Protocol, UN Framework Convention on Climate Change and Kyoto Protocol

The Montreal Protocol

According to Article 10A of the Montreal Protocol, 'each Party shall take every practicable step, consistent with the programmes supported by the financial mechanism, to ensure that the best available, environmentally safe substitutes and related technologies are expeditiously transferred to Parties operating under paragraph 1 of Article 5 [developing countries]', and that 'transfers [...] occur under fair and most favorable conditions.'

The United Nations Framework Convention on Climate Change

Article 4.5 of the United Nations Framework Convention on Climate Change (UNFCCC) states that developed country Parties should take 'all practicable steps to promote, facilitate and finance, as appropriate, the transfer of, or access to, environmentally sound technologies and know-how to other Parties, particularly developing country Parties', that they should 'support the development and enhancement of endogenous capacities and technologies of developing country Parties', and calls on other Parties and organizations to assist in facilitating the transfer of such technologies.

The Kyoto Protocol

The importance of technology transfer was also recognized by the Kyoto Protocol of 1997 in Article 10c, which asks all Parties to 'cooperate in the promotion of effective modalities for the development, application and diffusion of, and take all possible steps to promote, facilitate and finance, as appropriate, the transfer of, or access to, environmentally sound technologies, know-how, practices and processes pertinent to climate change, in particular to developing countries, including the formulation of policies and programmes for the effective transfer of environmentally sound technologies that are publicly owned or in the public domain and the creation of an enabling environment for the private sector, to promote and enhance the transfer of, and access to, environmentally sound technologies.'

---

1   physical assets, such as industrial plants, machinery, and equipment;
2   information, both technical and commercial, relating to process know-how, choice of technology, engineering design and plant construction, organization and operating methods, quality control, and market characteristics; and
3   human skills, especially those possessed by specialized professionals and engineers.

Some technologies are 'plug-and-play', requiring only minor tailoring to local circumstances. However, most manufacturing and environmental technologies require a significant amount of learning on the part of the technology user. Technology transfer is not achieved until the transferee understands and can use the technology. Although the literature tends to focus on technology transfer from developed countries to developing countries – 'north to south' – technol-

---

BOX 2.2 DEFINITIONS OF TECHNOLOGY TRANSFER

Some authors narrowly define 'technology transfer' as the transfer of technical knowledge from one place to another. Others, such as the Intergovernmental Panel on Climate Change (IPCC) (2001)[4], broadly define it as the transfer of technical knowledge and its utilization, dissemination and diffusion. Indicative definitions of technology transfer include:

the process by which commercial technology is disseminated. This involves the communication, by the transferor, of the relevant knowledge to the recipient. (United Nations Conference on Trade and Development (UNCTAD), 2001)[5]

any process by which one party gains access to a second party's information and successfully learns and absorbs it into his production function. (Maskus, 2004)[6]

most fundamentally a complex process of learning. (Levin, 1993; Kranzberg, 1986)[7]

a broad set of processes covering the flows of know-how, experience and equipment [...] amongst different stakeholders such as governments, private sector entities, financial institutions, non-governmental organizations and research/education institutions. [...] It comprises the process of learning to understand, utilize and replicate the technology, including the capacity to choose it and adapt it to local conditions and integrate it with indigenous technologies. (IPCC, 2001)[8]

---

ogy transfers actually take place within and between all countries. Andersen and Sarma (2002)[9] and IPCC (2001)[10] catalogue technologies invented in developing countries and transferred 'south to north' and 'south to south' for the protection of the stratospheric ozone layer. This overlooked perspective is elaborated in Chapter 13, where the case is made that the technology necessary to protect the climate and other global environment resources is just as likely to come from developing countries as from developed.

Although some sources, such as IPCC (2001),[11] combine the concepts of technology transfer, technology diffusion and technology commercialization, most of the literature distinguishes between the three. In general, 'transfer' of a technology refers to its transmission to another party, 'technology diffusion' refers to the extent to which a technology is utilized (in other words to the number of people or firms that have adopted the technology), and 'technology commercialization' refers to a technology's transition from the developmental stage to the marketplace. Some authors refer to technology commercialization as 'vertical technology transfer'.

EARLY LESSONS IN TECHNOLOGY TRANSFER

Although many international environmental agreements have included provisions for technology transfer, experts observe that these provisions have been

BOX 2.3 FROM TECHNOLOGY TRANSFER TO TECHNOLOGICAL LEARNING AND CAPABILITIES

The initial emphasis in the analysis of international technology transfer, in discussions among policy analysts up to the late 1970s, was on its costs and on whether the choice of technologies was appropriate to the local conditions in developing countries.

Little attention was given in this analysis to the absorptive capacities and domestic technological learning of those who acquired foreign technologies – in other words to the processes involved in assimilating imported technologies and putting them to work efficiently. The underlying assumption seemed to be that once a technology was acquired, its absorption and implementation took place almost automatically and effortlessly. However, it is now widely accepted that this is not the case. The acquisition and absorption of foreign technologies, and their further development, are complex processes that demand significant efforts on the part of those that acquire them.

Several factors contribute to this complexity. First, the acquisition and mastery of technology are both costly and time-consuming. Second, acquired technologies often need to be adapted to local conditions. Third, technologies are not commodities that can be transferred as complete ready-to-use sets; they also contain tacit components that are not easily codified and transmitted in written documents and require extensive learning efforts to be properly understood. In other words, technologies do not consist only of machines and other bundles of hardware; technology is knowledge in different forms, of which hardware is only one element. While machines can be easily transferred, other components cannot; and these also need to be mastered.

This increased understanding of the process of technological development has contributed to shifting the attention of academics and policy researchers from the narrow transfer of technology as such to the associated technological learning efforts and mastery of the acquirers. In this respect, despite its common use, the term 'technology transfer' can be misleading, as it appears to represent the acquirer as a passive receiver of technologies developed elsewhere and technology as something that can be easily transferred in a 'plug-and-play' mode.

In fact, the two processes – the acquisition of foreign technologies via technology transfer on the one hand, and domestic technological learning on the other – are complementary and intertwined. The absorptive capacities of the acquirer are fundamental in laying the foundations for the efficient assimilation of foreign technologies. At the same time, the acquirer of foreign technologies is not a passive receiver, and further technological learning is necessary to make the technologies work efficiently, as well as to develop them further.

Source: Science and Development Network.[12]

historically unsuccessful, with the conspicuous exception of the Montreal Protocol. Both the International Environmental Technology Centre and the United Nations Environment Programme suggest that technology transfer in support of sustainable development 'failed to fulfil expectations and meet clearly evident and pressing needs'.[13] The United Nations Commission on Sustainable Development gives low marks to technology transfer:

> *Hopes for accelerated transfer and diffusion of environmentally sustainable technologies remain largely unfulfilled despite the extensive acceptance that resource-efficient and cleaner technologies will benefit all.*[14]

Lack of attention to local conditions (see Box 2.3) and inattention to market incentives are two reasons commonly offered for this lack of success.

In documenting the history of international approaches to technology transfer, UNCTAD and the World Trade Organization (WTO) report that governments gradually shifted from regulatory approaches to market-based methods to encourage technology transfer. The underlying philosophy of the regulatory approach is that countries should screen technology transfer agreements between foreign and domestic companies and discourage 'unfair' practices, such as export restrictions, by the former. Regulatory approaches encourage transfer of technology from multinational corporations to locally owned businesses and use regulation to remedy any inequalities that might result. The underlying philosophy of the market-based approach, on the other hand, is that 'technology transfer is best achieved in an environment where intellectual property rights are fully protected as private commercial property and in which the market for technology is maintained in as competitive a condition as possible.'[15] Market-based approaches tend to encourage technology transfer from a foreign-owned enterprise to its (often wholly owned) local affiliates, and therefore require no correction for inequalities.

Some academics suggest that regulatory approaches actually discourage technology transfer by putting excessive restrictions on technology owners. According to UNCTAD:

> *The major disadvantage may be that such a regulated approach could be perceived as creating commercial disincentives for trans-national corporations, as the principal owners of technology, against the dissemination of that technology to developing host countries. [In particular] additional costs may arise as a result of intervention in the bargaining process through protective contractual requirements aimed at the promotion of the interests of independent local technology recipients. The imposition of extensive performance requirements could be perceived as limiting the commercial return on the transfer transaction.*[16]

Although market-based approaches have shown promising results for some technologies, some experts warn that such approaches may not work as well for the transfer of environmentally sound ones. For example, the OECD argues that the 'general trend away from public sector to private sector finance for channelling technology' is a concern for the transfer of environmentally sound technologies. This is because there are fundamental differences between the transfer of environmentally sound technologies and the transfer of other technologies (see Table 2.1). Environmental technology transfer is 'more reliant on public funds than on private investment'.[17] Furthermore, companies may not have incentives to develop environmental technology because the benefit they create – enhanced environmental quality – is in the realm of public goods. Therefore firms may need public funds to stimulate research and development on environmental technologies.

In recent years, more emphasis has been placed on the various stakeholders, pathways, stages and barriers that influence the technology transfer process.

Table 2.1 *Similarities and differences between environmental technology and other technologies*[18]

	Environmental technology	Other technologies
Main drivers	Regulation, public policy, multilateral environmental agreements	Market forces: demand, competition, production bottlenecks, etc.
Finance	Public funding important	Largely private funding, including reinvested earnings, venture capital and sales of stocks
Location of research and development (R&D)	More often in universities, public R&D institutes and laboratories	Mainly enterprise-based
Mechanisms for transfer	Transfer to private sector, emerging role of public–private sector partnerships	New structures through inter-firm R&D collaboration as well as partnerships of firms with public R&D
Commercialization	Increasingly private, many small and medium-sized enterprises involved, need for support structures and incentives	Generally private
Application	Often site- or location-specific applications, some environmentally sound technologies could be applied globally (e.g. CFC substitutes)	Increasingly global
Transfer to developing countries and countries with economies in transition	Private commercialization, official development assistance, sometimes funding from multilateral sources (e.g. Multilateral Fund under the Montreal Protocol or funding from the Global Environment Facility)	Almost exclusively through private commercialization

Technical and environmental experts have identified certain stages that occur within most technology transfers (see Box 2.4), but have not identified preset strategies to enhance technology transfer, and generally conclude that technology transfer must be tailored to the specific barriers, interests and influences of specific stakeholders.[19]

BOX 2.4 STAGES OF TECHNOLOGY TRANSFER

The IPCC identifies certain stages in technology transfer, but notes that they do not always occur in a linear fashion. Stages include the identification of needs (assessment), choice of technology, negotiation of conditions of transfer, agreement, implementation, evaluation and adjustment to local conditions, and replication. These stages appear regardless of whether the technology transfer is driven by governments, the private sector or communities.

Government-driven pathways

Assessment = technology assessment by governments involved in the transfer.
Agreement = governmental agreement or government agency mandate.
Implementation = project implementation.
Evaluation/adjustment = project or programme evaluation.
Replication = replication of the project.

Private sector pathways

Assessment = market opportunities or needs assessment,
or persuasion by third parties.
Agreement = contracts and financing.
Implementation = investments or purchases.
Evaluation/adjustment = profitability or service quality.
Replication = reinvestment or repurchase.

Community-driven pathways

Assessment = public pressure.
Agreement = consensus.
Implementation = collective action or joint management.
Evaluation/adjustment = localization.
Replication = diffusion of technology or practice.

Source: Andersen et al (1998)[20]

PATHWAYS AND STAKEHOLDERS

It is important to examine the pathways by which technology is transferred and the stakeholders involved in the process. The IPCC breaks down the technology transfer pathways into three categories: government pathways, private sector pathways and community-driven pathways:

> *Government-driven pathways are technology transfers initiated by governments to fulfil specific policy objectives; private-sector-driven pathways primarily involve transfers between commercially oriented private-sector entities; and community-driven pathways are those technology transfers involving community organizations with a high degree of collective decision-making.*[21]

Other authors use two pathway categories: market (or commercial) channels and non-market (or non-commercial) channels.[22] Market channels are those which involve some form of formal transaction, such as foreign direct investment, joint ventures, licensing, franchising, management contracts, marketing contracts, technical service contracts, turnkey contracts and international subcontracting. Non-market technology transfer channels have no formal arrangements; they occur when information is transferred through technical journals and the like.

Government pathways of technology transfer

Although the vast majority of technology transfer occurs through private sector channels, governments also provide pathways for technology transfer. Many governments support the development of environmentally sound technologies that can be freely distributed to domestic or international firms. This pathway of technology transfer 'can be an important means for governments to catalyse private sector technology transfer'.[23]

Governments also influence private sector pathways of technology transfer by creating policies and regulations that shape the environment in which technology transfer occurs. These policies can encourage technology transfer or inhibit it, depending on how they are designed. For example, strong domestic environmental regulations and financial incentives such as taxes or subsidies increase the demand for environmentally sound technologies. In addition, strict but flexible regulatory regimes, such as those that focus on performance standards and outcomes, can encourage innovation and drive technological development.[24] In extreme circumstances, governments can facilitate technology transfer through 'command and control' measures that compel the use of certain technology and make licensing compulsory.

Private sector pathways of technology transfer

The private sector develops, owns and controls the vast majority of technology and technical innovations; consequently, it plays the most important role in technology transfer.[25] Recognizing this, international environmental agreements frequently call for increased collaboration with the private sector for the purpose of transferring environmentally superior technology. Some of the most important pathways through which the private sector transfers technology are trade in goods and services, foreign direct investment, licensing, joint ventures, and cross-border movement of personnel.

Trade in goods and services (includes purchases, sales, exports and imports)
One important way that firms acquire technology is by buying it. However, the mere acquisition of a technology does not assure its effective transfer; it may need to be adapted to meet the purchasing firm's needs, and depending on the purchasing firm's technical capabilities, after-sales service and training may be necessary.

Foreign direct investment
Foreign direct investment is the most commonly identified channel of market-based international technology transfer. Multinational enterprises introduce new

knowledge, tools, techniques and technologies by directly investing in new productive capacity or acquiring and updating facilities in another country. In the most successful cases, foreign direct investment transfers technology and builds indigenous capacity by creating new infrastructure and training local workers.

Licensing

Licences are agreements that sell the right to use certain proprietary knowledge in a defined way. They typically involve a contract in which a product's manufacturer (or the firm that owns the proprietary rights) grants permission to manufacture that product (or make use of that proprietary material) in return for specified royalties or other payment or access through cross-licensing of other proprietary knowledge.[26] Licensing is often used where there is a barrier to trade or when the owner of the technology considers foreign investment too difficult, too risky or too far from their business plan. Licensing opens the possibility for an enterprise to transfer technology to a foreign market where it would otherwise be unavailable.

Joint ventures

A joint venture is a business in which two or more parties undertake an economic activity together. The joint venture agreement typically specifies the contribution of each partner and the distribution of profits. Joint ventures spread costs and risks, improve access to financial resources, achieve economies of scale, provide access to new technologies and customers, and allow different parties to learn new skills from each other.

Cross-border movement of personnel

When firms relocate or engage in foreign direct investment, they usually send managerial and technical staff to new locations to provide training and supervision. Indeed, 'many technologies cannot be effectively or affordably transferred without the complementary services and know-how of engineers and technicians that must be on-site for some period of time'.[27] In addition, experts familiar with specific technology take those skills to new jobs or start their own companies producing variations of the technology they previously mastered.

Non-market pathways of technology transfer

Not all technology transfer occurs through market interactions. Non-market pathways involve the acquisition of technology without the consent of the provider. Examples of non-market technology transfer pathways include public education, internet searches, industrial espionage, end-user or third-country diversions, imitation, and reverse engineering. Technology transfer also occurs when experts describe their results in technical papers, patent disclosures and other publications, and when employees move between firms or migrate to other regions.[28]

Imitation

Imitation is among the most significant non-market channels of technology transfer. It occurs when 'a rival firm learns the technological or design secrets of another firm's formula or products'.[29] This can be accomplished through simple trial and error, but more often it is achieved through reverse engineering. The legality of imitation products depends on the strength of intellectual property rights legislation and is sometimes a major issue in international trade negotiations.

Data in patent applications and test data

Patent applications and test data contain important technological information. By studying patents, rival firms may be able to sidestep or bypass existing patents or may be able to employ the patented technology without being discovered. Either way, the information contained in patent applications and test data is a productive avenue for the transfer of knowledge and technology.

Employee mobility and migration

When employees change jobs, they take their technological knowledge and skills with them. Their prior knowledge and skills may also help to catalyse new technological developments in their new places of employment, where there may be a cross-fertilization of knowledge.

Technology transfer also occurs when employees migrate temporarily to acquire training or when assigned short-term duties. In addition, private and public sector employees transfer knowledge when they serve on international committees such as the Montreal Protocol Technology and Economic Assessment Panel (TEAP) and its Technical Options Committees (TOCs). The lessons learned constitute a long-lasting form of knowledge transfer. However, clearly the knowledge acquired when experts migrate to a new location will only be transferred back to the original location if they return. Technology transfer is not efficiently accomplished for developing countries if higher foreign standards of living result in a 'brain drain' of the best students and experienced technical experts.

Community pathways of technology transfer

Geographic, sectoral or professional communities can provide a pathway for technology transfer. Community-driven pathways for technology transfer are 'initiated and led by community organizations and entities with a high degree of collective decision making'.[30] Communities can put pressure on the private sector to change practices or utilize different technologies; often this is enough to create change even in the absence of government pressure. Communities can also sponsor meetings, workshops and other public forums where attendees can learn about new technology, creating another possible pathway for transfer. An example in the case of the ozone layer is the Greenfreeze refrigerator, which was promoted by the international environmental non-governmental organization (NGO) community, leading to the widespread adoption of this technology, which is climate-friendly and free of ozone depleting-substance.

Stakeholders

It is just as important to study the stakeholders involved in technology transfer as it is to study the technologies themselves. The actions and interactions of stakeholders determine the rate and direction of technology change. There are many stakeholders involved in technology transfer, including businesses, governments, educational and research institutions, financiers, consumers, the media, and NGOs. When these stakeholders are well established and well connected, knowledge will flow quickly and easily between them, speeding technology transfer (see Table 2.2).

The private sector enterprise is perhaps the most important stakeholder in the technology transfer process, and many theories attempt to explain the factors that lead enterprises to adopt or reject new technologies. One popular theory, introduced by Carlos Corral, hypothesizes that any enterprise contemplating a change of technology is influenced by three factors: risk, social pressure and enterprise capability.[31] One risk is that an enterprise's current operations will have an unacceptable environmental impact. Another risk is that the enterprise will fail to capture significant benefits of the alternative clean technological portfolio. The final risk is the possible negative consequences of action or inaction. This risk depends on the enterprise's market niche, the capability of government regulators, customer willingness to pay, the growth opportunity of adapting the new technology, the costs of new technologies and the financial risks.

The main sources of social pressure for technology change are regulation, industrial standards, market position, market forces, customers' expectations and demands, and public concerns. The most important determinants of change are the firm's current technological capabilities, the availability of technological opportunities, collaboration with research institutions, collabora-

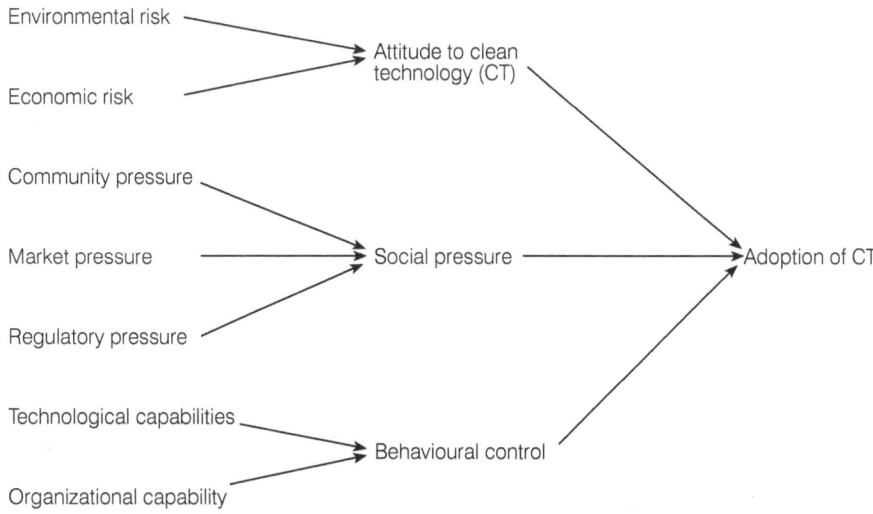

Figure 2.1 *Factors that lead enterprises to adopt or reject new technologies*

Table 2.2 *Stakeholders*

Stakeholders	Motivations	Decisions or policies that influence technology transfer
Governments	• Compliance with international treaties • Development goals • Environmental goals • Competitive advantage • Energy security	• Tax policies (including investment tax policy) • Import/export policies (including bans on manufacture or import of products made with or containing ozone-depleting substances) • Innovation policies • Education and capacity-building policies • Regulations and institutional development • Direct credit provision
Private sector	• Profits • Market share • Return on investment • Sustainable operations • Company reputation and respect • Public service and community health	• Technology R&D/commercialization decisions • Marketing decisions • Capital investment decisions • Skills/capabilities development policies • Structure for acquiring outside information • Decisions to transfer technology • Choice of technology transfer pathway • Lending/credit policies (producers, financiers) • Technology selection (distributors, users)
Donors	• Development goals • Environmental goals • Return on investment	• Project selection and design criteria • Investment decisions • Provision of technology, equipment, technical assistance and training • Procurement requirements • Conditionalities
International institutions	• Development goals • Environmental goals • Policy formulation • International dialogue	• Policy and technology focus • Selection of participants in forums • Choice of modes of information dissemination
Research organizations	• Basic knowledge • Applied research • Teaching • Knowledge transfer • Perceived credibility	• Research agenda • Technology R&D/commercialization decisions • Decision to transfer technology • Choice of pathway to transfer technology
Media	• Information distribution • Education	• Acceptance of advertising • Promotion of selected technologies • Educational curricula
NGOs	• Special interests • Collective welfare	• Promotion of selected technologies • Lobbying for technology-related policies
Individual consumers	• Welfare • Utility • Expense minimization	• Purchase decisions • Decision to learn more about a technology • Selection of learning/information channels • Ratings of information credibility by source

Source: Adapted from Intergovernmental Panel on Climate Change (2001).[32]

tion and influence with suppliers, perceived internal control of the firm, and learning capability. This is summarized in Figure 2.1.

FACILITATING TECHNOLOGY TRANSFER FOR ENVIRONMENTAL PROTECTION: THEORY AND OBSERVATIONS

The environmental and technology transfer literature offers many suggestions about how to facilitate technology transfer and cooperation for environmental protection. Common themes include sensitivity to stakeholder needs, enabling environments and national settings, economic incentives, information-based policies, regulation, capacity-building, intellectual property protection, and financial assistance.

Sensitivity to stakeholder needs

Experts agree that successful publicly sponsored technology transfer projects are sensitive to stakeholder needs. Projects are unsuccessful when managers fail to appreciate the motivations, constraints or limitations of one or more key stakeholders in the technology transfer project. A notable technology transfer failure occurred under the Montreal Protocol in the transfer of refrigerant recovery and recycling technology from developed countries to the Czech Republic. Project planners failed to realize that poorly paid technicians would prefer to recover and re-sell refrigerant on the street rather than bring the refrigerant to the reclamation facility for purification. The failure to bring in the refrigerant made the reclamation facility uneconomic, and the sale of unpurified refrigerant resulted in large-scale damage to refrigeration and air-conditioning equipment from the use of unpurified refrigerant.[33]

The IPCC and others[34] recommend analysis prior to the start of a project to identify the motivations, desires and constraints of all parties involved in the technology transfer process. To illustrate this point, the IPCC summarized key stakeholders and their motivations in its special report on technology transfer. This summary is given in Table 2.2.[35]

Enabling environments and national settings

Technology transfer is more likely to succeed in a country with a well-developed national system of innovation, combined with public or regulatory pressure to stimulate change. Conversely, technology transfer is less likely to succeed when national innovation systems are weak and there is little pressure to drive technology change. National systems of innovation are generally defined as a network of institutions that influence the rate and direction of technical progress within a country. The OECD elaborates on this topic in its influential report 'National innovation systems'.[36]

The national system of innovation depends on business, government and educational institutions. Their actions and interactions determine the rate and direction of technology change. In a country where these institutions are well established and well connected, knowledge will flow quickly and easily between them. This is important because 'the smooth operation of innovation systems depends on the fluidity of knowledge flows – among enterprises, universities and research institutions'.[37] The technology transfer literature suggests that by strengthening national systems of innovation, countries can improve their capacity to develop and absorb environmentally sound technologies.[38]

Economic incentives

Economic incentives, including taxes, subsidies and other market interventions, provide a powerful stimulus for technology adoption and are commonly recommended to spur technology transfer. In order of effectiveness, the OECD and the International Energy Agency (IEA) rank taxes first, followed by cap-and-trade systems and then subsidies. Subsidies have sometimes spoiled markets for a new technology by perpetuating dependency, stifling competition, promoting inferior products or otherwise blocking market forces from operating properly.[39] The impact on technology transfer and diffusion will depend on 'the strength of economic incentives for technology adoption'.[40]

Information-based policies

According to many authors, information-based policies address market failures that arise from imperfect information and can be extremely helpful in motivating firms to adopt environmentally sound technology.[41] Information-based policies include publicity campaigns, demonstration projects, technology certification, eco-labelling, industry codes, product standards and subsidies to technological consulting services. All of these policies spread awareness, reduce uncertainty and encourage technology adoption.

The OECD and the IEA (2003) also recommend technical and environmental performance standards and voluntary agreements as effective tools.[42] They contend that firms and consumers do not always have the knowledge or information necessary to select the most efficient technology; therefore standard development corrects a key market failure and facilitates the diffusion of the socially optimal technology. The importance of information-based policies is discussed at length in Chapter 13 on awareness and capacity-building.

Regulation

Well-designed and well-enforced regulations speed up the transfer and diffusion of environmentally sound technologies.[43] Regulation comes in the form of product bans, technology standards, environmental quality requirements and many other categories. In the absence of formal regulation, 'informal regulation' applied by private sector groups such as neighbourhood organizations, non-governmental organizations and trade unions can be effective.[44]

Sound regulatory frameworks are particularly important for the effectiveness of international technology transfer projects.[45] According to the IPCC, regulatory uncertainty poses a risk to investors and discourages the transfer of environmentally sound technologies. 'Well-developed administrative law', on the other hand, encourages transfer by 'assuring private actors relatively prompt and articulated regulatory decisions, an absence of excessive corruption, and coordination between multiple agencies with regulatory responsibilities'.[46]

Capacity-building

Capacity-building is vital for the successful integration of environmentally sound technologies in developing countries.[47] Without adequate infrastructure and trained labour, new technologies may fail to diffuse.[48] The subject of capacity-building is further discussed in Chapter 13.

Intellectual property protection

Intellectual property protection can sometimes encourage and sometimes stifle technology cooperation. On the one hand, a country with strong intellectual property rights legislation is more attractive to firms who are considering foreign direct investment (FDI), and FDI is a major means by which technologies spread across borders.[49] Thus strong intellectual property rights laws could in theory encourage the transfer and diffusion of environmentally sound technologies. On the other hand, 'intellectual property restrictions attach significant costs to adoption of new technologies', and this can slow the diffusion of those technologies.[50] The issue of intellectual property rights in the context of the Montreal Protocol is discussed at length in Chapter 13.

Financial assistance

The lack of financing is a major barrier to the transfer and diffusion of new technologies. Lending institutions are understandably risk-averse, and capital markets usually look for short-term payback. It is particularly difficult to secure financing in risky projects with long-term pay-offs.[51] Given that lack of credit is a critical barrier to technology adoption in certain situations, subsidized credit can provide a means by which to speed technology transfer and diffusion.[52] Financial assistance is especially important to facilitate technology adoption in developing countries.[53] The Intergovernmental Panel on Climate Change (2001)[54] highlighted the importance of financing from commercial banks, equity markets, the Global Environment Facility and the multilateral development banks. Subsidized credit programmes – both public and private – have had, on the other hand, very mixed results.[55] Chronic problems include the diversion of loans by borrowers to non-targeted activities, low repayment rates, the creation of financially unsustainable lending institutions and the undermining of existing credit markets.

Common sources of financial assistance include official development assistance, export credit, the Global Environment Facility (GEF), the Multilateral Fund (MLF) and multilateral development banks.

Official development assistance

Many industrialized countries provide official development assistance (ODA) to developing countries to support their economic development. Although ODA usually does not finance the development and diffusion of technology, it can be used to finance capacity-building programmes crucial to the success of technology transfer. One drawback of ODA is that it has often been 'tied', such that aid recipients are obligated to enter into contracts only with companies from donor countries, possibly limiting technical choice and increasing costs. Total global spending on ODA was US$78.568 billion in 2004, with a small but increasing proportion for environmental protection.[56]

Under terms of the Montreal Protocol, nations that donate to the MLF are allowed to contribute up to 20 per cent of their assessment through ODA projects. Before the MLF was up and running, each donor nation was allowed to conduct projects of its own design, but later the MLF required that projects be approved in advance by its Executive Committee.

Export credit

Many industrialized countries have export credit agencies that promote their business and economic interests abroad. These provide assistance to domestic businesses that are trying to establish a presence in emerging markets.

The Global Environment Facility

The Global Environment Facility 'is an independent financial organization that provides grants to developing countries for projects that benefit the global environment and promote sustainable livelihoods in local communities'.[57] It was established in 1991 and currently over 30 countries contribute funds for GEF projects. The GEF funds a variety of projects, from its small grants programme (US$50,000 or less) to enabling activities, medium-sized projects and full-sized projects (more than US$1 million).[58] Any person or group can propose a project, but the project has to meet two key criteria: first, it 'must reflect national or regional priorities and have the support of the country or countries involved' and second, 'it must improve the global environment or advance the prospect of reducing risks to it'.[59] GEF-funded projects are managed by the United Nations Development Programme, the World Bank and the United Nations Environment Programme.

The GEF assisted in financing technology transfer to countries with economies in transition (CEITs) for the phaseout of ozone-depleting substances (ODSs). After the collapse of communism in Eastern Europe and the break-up of the USSR in 1991, the new states were in a state of social and economic turmoil. Despite this turmoil, they were both determined and required to meet the rigorous phaseout requirements for ODSs established under the Montreal Protocol. These countries did not meet the qualifications for financial assistance from the MLF because they had not been designated as 'developing countries' by the Meetings of the Parties to the Montreal Protocol. A political resolution was achieved when the GEF stepped in and helped the CEITs to acquire the technologies needed to comply with the ozone-depleting substances phaseout schedule.

The Multilateral Fund

The Multilateral Fund (MLF) for the Implementation of the Montreal Protocol on Substances that Deplete the Ozone Layer was established in 1990 to help developing countries phase out the production and consumption of ODSs. Countries designated by the Meeting of the Parties to the Montreal Protocol as developing and that consume less than 0.3 ODS kilograms per capita are eligible for assistance. The fund covers the incremental costs associated with technology transfer, including the costs of on-site engineering, equipment purchase and installation, training, and start-up. Capacity-building projects, such as the establishment of national ozone offices and regional ozone network offices, are also eligible for funding. The MLF finances and audits the projects which are managed by implementing agencies. The implementing agencies of the MLF are the United Nations Environment Programme, which acts as an information clearinghouse, conducts awareness training and helps countries establish country programmes for ODS phaseout; the United Nations Industrial Development Organization; the World Bank; and the United Nations Development Programme, which implements investment, technical assistance and demonstration projects.

Multilateral development banks

Multilateral development banks (MDBs) 'are institutions that provide financial support and professional advice for economic and social development activities in developing countries'.[60] They receive their funding from developed country donors and make grants or loans to developing country governments. The term 'multilateral development banks' typically refers to the World Bank Group and four Regional Development Banks: the African Development Bank, the Asian Development Bank, the European Bank for Reconstruction and Development and the Inter-American Development Bank Group.

MDBs provide financing through:

- long-term loans at market interest rates from funds borrowed on the international capital markets and provided to governments in developing countries;
- very long-term loans (often termed credits) at well below market interest rates funded through direct contributions from governments in donor countries; and
- grant financing, which is offered for technical assistance, advisory services or project preparation.[61]

CONCLUSION

The international community began promoting technology transfer as a means to protect the environment in the 1970s. Since that time, experience has provided valuable insights. It is now clear that technology transfer requires more than mere access to new technologies; domestic technological capacity and local skills are equally important. Publicly sponsored technology transfer projects that recognize these factors have been most successful.

The majority of technology transfers take place in the private sector; however, governments and communities are also important sources of and catalysts for technology transfer. Governments and communities influence the rate and direction of technology transfer through policies, regulations, social pressure and spending practices. Governments wishing to promote the transfer and diffusion of technologies for environmental protection have used various methods to achieve their objectives. Economic incentives, information-based policies, regulations, capacity-building activities and financial assistance are all methods that have proven successful.

The following chapters of this book examine the technology transfer experience of the Montreal Protocol in detail, paying special attention to the results of the financial assistance provided by the Multilateral Fund and the Global Environment Facility to help developing countries phase out the use of ozone-depleting substances.

Background of the Ozone and Climate Agreements

THE SCIENCE OF OZONE DEPLETION[1]

Without a protective ozone layer in the atmosphere, animals and plants could not exist, at least upon land. It is therefore of the greatest importance to understand the processes that regulate the atmosphere's ozone content. (The Royal Academy of Sciences, Sweden, announcing the Nobel Prize for Chemistry, 1995, for Paul Crutzen, Mario Molina and F. Sherwood Rowland)

Introduction

The ozone layer forms a thin shield in the stratosphere, approximately 20–40km above the Earth's surface, protecting life below from the sun's ultraviolet (UV) radiation. It absorbs the lower wavelengths (UV-C) completely and transmits only a small fraction of the middle wavelengths (UV-B). Nearly all of the higher wavelengths (UV-A) are transmitted to the Earth, where they cause skin aging and degrading of outdoor plastics and paint. Of the two types of UV radiation reaching ground level, UV-B is the more harmful to humans and other life forms.

Manufactured chemicals transported by the wind to the stratosphere are broken down by UV-B, releasing chlorine and bromine atoms that destroy ozone. As ozone is depleted, increased transmission of UV-B radiation endangers human health and the environment by increasing skin cancer and cataracts, weakening human immune systems, and damaging crops and natural ecosystems. Furthermore, most ozone-depleting substances (ODSs) are also greenhouse gases that contribute to climate change, causing sea level rise, intense storms and changes in precipitation and temperature.[2]

Thomas Midgley invents CFCs

In 1928 Thomas Midgley, an industrial chemist working at General Motors, invented a chlorofluorocarbon (CFC) as a non-flammable, non-toxic compound to replace the hazardous materials, such as sulphur dioxide and ammonia, then being used in home refrigerators. CFCs and other similar halocarbons were non-reactive, long-lasting materials that did not directly harm humans. CFCs and other ODSs were considered 'wonder chemicals' and were implemented in thousands of applications over the next 60 years. They were used as refrigerants in automobile air-conditioners, as propellants in aerosol sprays, as feedstocks in manufacturing plastics, as extinguishing agents in firefighting, as solvents for electronic components and in many other applications.

Early warnings about damage to the ozone layer

In 1970 Paul Crutzen, from The Netherlands, demonstrated the importance of catalytic loss of ozone by the reaction of nitrogen oxides and theorized that chemical processes that affect atmospheric ozone can begin on the surface of the Earth. He showed that nitric oxide (NO) and nitrogen dioxide (NO_2) react in a catalytic cycle that destroys ozone, without being consumed, thus lowering the steady-state amount of ozone. These nitrogen oxides are formed in the atmosphere through chemical reactions involving nitrous oxide (N_2O) which originates from microbiological transformations on the ground. He therefore theorized that increasing atmospheric concentration of nitrous oxide, which could result from the use of agricultural fertilizers, might lead to reduced ozone levels. His hypothesis was that:

> *NO and NO_2 concentrations have a direct controlling effect on the ozone distributions in a large part of the stratosphere, and consequently on the atmospheric ozone production rates.*[3]

At the same time, James Lovelock, from the UK, developed the electron-capture detector, a device for measuring extremely low organic gas contents in the atmosphere. Using this device aboard a research vessel in 1971, Lovelock measured air samples in the North and South Atlantic, and in 1973 he reported that he had detected CFCs in every one of his samples, 'wherever and whenever they were sought'.[4] He concluded that CFC gases had already spread globally throughout the atmosphere.

In another article, published in 1970, Halstead Harrison of the Boeing Scientific Research Laboratories in the US hypothesized that:

> *with added water from the exhausts of projected fleets of stratospheric aircraft, the ozone column may diminish by 3.8 per cent, the transmitted solar power increase by 0.07 per cent and the surface temperature rise by 0.04 degrees K in the northern hemisphere.*[5]

Harrison added that:

> *several authors have expressed concern that exhausts from fleets of strato-spheric aircrafts may build up to levels sufficient to perturb weather both in the stratosphere and on the surface. Indeed, calculations indicate that the quantity of added water vapour may become comparable to that naturally present.*

At the time, the projected fleets of supersonic transport aircraft (SSTs) were estimated at 500 in the US and 350 in other countries.

In the US in 1971, Harold Johnston, who had carried out extensive studies of the chemistry of nitrogen compounds, showed that the nitrogen oxides produced in the high-temperature exhaust of the proposed fleet of SSTs could contribute significantly to ozone loss by releasing the nitrogen oxides directly into the stratospheric ozone layer.[6] In 1972 Crutzen elaborated on this theory in a paper that explained the process by which ozone is destroyed in the stratosphere and presented estimates of the ozone reduction that could result from the operation of the supersonic aircraft.[7]

Responding to James Lovelock's measurements of CFCs accumulating in the atmosphere, in 1972 the DuPont Company arranged a panel on 'The Ecology of Fluorocarbons' for the world's CFC producers.[8] As a result, 19 companies formed the Chemical Manufacturers Association's Fluorocarbon Program Panel, a research group that eventually provided at least US$20 million in funding for research at academic and government facilities worldwide.

The Molina–Rowland hypothesis: CFCs linked to ozone depletion

In a paper published in the 28 June 1974 issue of *Nature*, Mario J. Molina and F. Sherwood Rowland hypothesized that when CFCs reach the stratosphere, ultraviolet radiation causes them to decompose and release chlorine atoms, which in turn become part of a chain reaction as a result of which a single chlorine atom could destroy as many as 100,000 molecules of ozone.[9] 'The chemical inertness and high volatility which make these materials suitable for technological use also mean that they remain in the atmosphere for a long time,' Drs Molina and Rowland wrote. They concluded:

> *Chlorofluoromethanes are being added to the environment in steadily increasing amounts. These compounds are chemically inert and may remain in the atmosphere for 40–150 years, and concentrations can be expected to reach 10 to 30 times present levels. Photo-dissociation of the chlorofluoromethanes in the stratosphere produces significant amounts of chlorine atoms, and leads to the destruction of atmospheric ozone. [...] It seems quite clear that the atmosphere has only a finite capacity for absorbing Cl atoms produced in the stratosphere, and that important consequences may result. This capacity is probably not sufficient in steady state even for the present rate of introduction of chlorofluoromethanes. More accurate estimates of this absorptive capacity need to be made in the immediate future in order to ascertain the levels of possible onset of environmental problems.*

Rowland and Molina called for a ban on aerosol CFCs when in September 1974 they told the American Chemical Society:

> *If nothing was done in the next decade to prevent further release of chlorofluorocarbons, the vast reservoir of the gases that would have built up in the meantime would provide enough chlorine atoms to ensure continuing destruction of the ozone layer for much of the twenty-first century.*[10]

Many governments, particularly in the US and some Scandinavian countries, immediately responded with bans on use of CFCs in cosmetic aerosols. This led to some reduction of CFC use and emissions. However, the consumption of CFCs and other ODSs in non-cosmetic sectors increased rapidly, and total ODS emissions continued to rise.

The Antarctic ozone hole

Scientists continued their atmospheric observations and analysis, and their findings increasingly validated the findings of Rowland and Molina. As early as October 1981, Dobson instrument measurements from Japanese, British and other Antarctic research stations recorded a dramatic 20 per cent reduction in ozone levels above Antarctica.[11] None of the Antarctic scientists published their results or consulted other stations to confirm their observations, however. Joseph Farman, head of the Geophysical Unit of the British Antarctic Survey:

> *could only assume that something had gone wrong with his Halley Bay apparatus. He knew, of course, about the Molina–Rowland theory and the scientific debate over the relationship between man-made chemicals and ozone depletion, but the Dobson reading was simply too low to suggest anything but an instrument malfunction.*[12]

The next year, during the 1982 Antarctic spring in October, readings from a new Dobson instrument registered similar low ozone levels. At the same time, the ozone-measuring devices aboard the Nimbus 7 satellite had also registered low ozone levels, but the computers that logged the devices' measurements had been programmed to identify extremely low ozone measurements as erroneous, and therefore to ignore them.

In 1984 the first published results of research on ozone depletion over Antarctica appeared when Shigeru Chubachi of the Japanese Meteorological Research Institute in Ibaraki reported his findings reflecting extensive ozone observations carried out at Syowa Station. Chubachi reported the annual variation of total ozone and that the smallest value of total ozone since 1966 was observed in September to October 1982, when readings showed ozone levels of under 250 Dobson units.[13]

In May 1985 Joseph S. Farman, B. G. Gardiner and J. D. Shanklin of the British Antarctic Survey published in *Nature*[14] their findings confirming that ozone levels above Antarctica had been significantly depleted every Antarctic spring since at least 1981. Their paper attributed the ozone depletion to CFCs, yet scientists would not be confident in this conclusion for several more years. The phenomenon of ozone depletion over Antarctica became known as the 'ozone hole'.

Environmental scientists describe bleak future with ozone depletion

Scientists' reports that ozone depletion was sure to increase the incidence of skin cancer attracted significant attention in the US and other countries. However, they also described other impacts with far greater global consequences, particularly the potential reduction in global food supply and proliferation of diseases caused by the effects of ultraviolet radiation on the human immune system.[15] Scientists testified that the potential of ultraviolet radiation to damage crops and plants was indisputable.[16]

THE INCREASING USE OF OZONE-DEPLETING SUBSTANCES

Chemical emissions responsible for ozone layer depletion came from various sources. In about 1900 methyl bromide became the first ODS to be commercialized.[17] Carbon tetrachloride was the second. Chlorofluorocarbons (CFCs) are the most familiar ozone-depleting substances, but the first CFCs (CFC-11 and CFC-12) were not commercialized until the 1930s. Halon-1211 and 1301 were commercialized in the 1940s. CFC-113 solvents were commercialized in the 1960s.

The most toxic ODSs – methyl bromide and carbon tetrachloride – were initially marketed for uses where high levels of exposure resulted in death and injury (both immediate poisoning and long-term health effects) and subsequently suffered market and regulatory rejection for consumer uses. Other, safer ODSs – CFCs, halons and methyl chloroform – enjoyed growth mostly unconstrained by toxicity concerns, entering through aggressive marketing into a vast number of utilitarian and luxury uses.

Table 3.1 presents an overview of the many uses of controlled ODSs.

Table 3.1 *Ozone-depleting substances and sectors where used*

Commercial Designation(s)	Refrigerant	Fire Extinguishant	Solvent	Propellant	Foam-Blowing	Process Agent or Feedstock	Pesticide	Miscellaneous
Carbon Tetrachloride (Halon-104)	H	H	S			S/S	S	H[18]
Methyl Chloroform	H	H	S			S/S		M[19]
Methyl Bromide (Halon-1001)		H				S/S[20]	S	H[21]
CFC-11	S		S[22]		S	S/S		S[23]
CFC-12	S	M[24]		S	S			S[25]
CFC-113			S			M, N		
CFC-114	S			M				
CFC-115	S			M				
HCFC-22	S	M			M	S		
HCFC-123	N							
HCFC-141b			N		N			
HCFC-142b					M, N	S		
HCFC-225			N					
Halon-1211		S						
Halon-1301	M	S						
Halon-2402		S						

Note: H = historic use; S = significant use; M = minor use; N = negligible use.

ODSs marketed for critical, cosmetic, and convenience products

By the late 1980s more than 250 separate product categories were made with, or contained, ozone-depleting substances. Some products were unnecessarily used for cosmetic and convenience products where alternatives were readily available. However, many other ODS products had become vital to society. The more critical uses included:

* medical applications, such as metered-dose medicine inhalers, sterilization, cleaning of heart pacemakers and artificial limbs, and blood substitutes;
* refrigeration for meat and fish processing, vegetable storage, and frozen food, blood and medicines;
* air-conditioning in buildings and vehicles;
* foam insulation for refrigerated appliances, building insulation and industrial applications;
* cleaning of critical electronic and mechanical components, including for weapons detection and guidance, safety systems for nuclear and hazardous chemical facilities, and aircraft flight control;
* fumigation for quarantine and pest control;
* fire protection and explosion suppression in telecommunications, naval and commercial shipping, aircraft, oil and gas processing, and transport;
* safety foams used in vehicles as padding and structure;
* industrial processes, including nuclear fuel processing and aluminium manufacturing; and
* laboratory and analytical uses, including leak testing and as a dielectric medium in scientific and medical equipment.

ODSs were also used in wineglass chillers, tyre inflators, dust blowers, toys, noise-making horns and tobacco puffing to reduce tar.

GOVERNMENTS AND UNEP RESPOND
TO THE SCIENTIFIC FINDINGS ON ODSs

UNEP World Plan of Action, 1977

The first international meeting on ozone layer protection occurred in March 1977, just three years after Molina and Rowland published their paper linking CFCs to ozone depletion and years before the discovery of the 'ozone hole'. On instructions from its Governing Council, UNEP organized a meeting of experts from many countries, in Washington, DC. This meeting resulted in a World Plan of Action on the Ozone Layer. As part of this plan, UNEP established a Coordinating Committee on the Ozone Layer (CCOL), in which all interested countries shared the results of their studies on the ozone layer. In 1978 UNEP initiated the diplomatic process of intergovernmental meetings to negotiate a solution to the problem of ozone depletion.

The Vienna Convention for the Protection of the Ozone Layer, 1985

The 1985 Vienna Convention for the Protection of the Ozone Layer was the first global treaty action to address the problem of ozone depletion. The Convention was only a framework because industry, supported by many European governments, had unresolved doubts about the threat to the ozone layer. It requested that the Parties to the Convention study, research and report on various aspects of ozone depletion. It contained no steps to curb the consumption of ODSs, but provided for further protocols as needed to deal with ozone depletion.

THE MONTREAL PROTOCOL ON SUBSTANCES THAT DEPLETE THE OZONE LAYER, 1987

After many diplomatic discussions arranged by UNEP, on 16 September 1987 governments agreed on the Montreal Protocol to reduce the consumption of ozone-depleting substances. It was signed by 24 countries and the European Economic Community. The UNEP Secretariat for the Vienna Convention and the Montreal Protocol, called the Ozone Secretariat, was created to provide the requisite support to the Meetings of the Parties in strengthening the Protocol, monitoring implementation and reporting problems to Parties when necessary. The Protocol of 1987 listed eight ODSs and, due to the continuing doubts of some countries, prescribed only mild measures to freeze and reduce production and consumption of these. Even as it was signed, scientists, proactive diplomats and non-governmental organizations knew that the Protocol was wholly inadequate to protect the ozone layer. However, the Protocol could be strengthened in the future because it provided for periodic assessment of the control measures based on available scientific, environmental, technical and economic information through scientific, environmental, technology and economic panels of experts (Article 6).

The Scientific, Environmental, Technology and Economic Assessment Panels were informally initiated in 1988 at a UNEP Conference held in The Hague in The Netherlands. On directions from the Parties to the Protocol, the Panels conducted a full assessment of ozone depletion, the environmental impacts, and the technical and economic feasibility of alternatives. In 1989, 1991, 1994, 1998, 2002 and 2006 the Assessment Panels integrated their findings in synthesis reports that presented options to the governments on the measures needed to better protect the ozone layer. Based on these reports, the Parties adjusted and amended the Protocol many times in order to strengthen its provisions. The current control measures phase out the production and consumption of 96 ozone-depleting chemicals according to the timetable in Appendix 1 of this book.

The control measures for the phaseout of ODS production and consumption are specified on the basis of 'ODP-weighted' quantities for 'groups' of chemically similar gases. The quantities are calculated by multiplying the quantity of each controlled substance by its ozone depletion potential (ODP), totalling within each group. This flexibility allows each Party to choose to produce and consume the combination of controlled chemicals that best suits its needs. The Protocol encourages the practices of recovery and recycling by not counting those quantities recovered and recycled, and allows Parties to credit quantities destroyed against total production and consumption. There are three classes of exceptions to the phaseout: feedstocks to other industries in which ODSs are almost entirely consumed; process agents, in which emissions are tightly controlled; and applications authorized by Meetings of Parties for essential use (such as metered-dose inhalers for asthma and chronic obstructive pulmonary disease).

The original 1987 Montreal Protocol entered into force on 1 January 1989. Each country had a different internal ratification process. In the US, for example, Congress had to vote in favour of the US becoming a Party. The control measures of the 1987 Montreal Protocol were made stricter with the adjustments and amendments to the Protocol approved by the Parties in 1990, 1992, 1995, 1997 and 1999; each time the Protocol was strengthened through amendments, Parties had to ratify them. Each adjustment entered into force at a specified date. At any point in time, some countries that had ratified the Montreal Protocol may not have ratified all of its amendments. The Ozone Secretariat website (http://ozone.unep.org) contains the status of ratification and is updated periodically.

Structure of the Montreal Protocol and Party obligations

All Parties to the Montreal Protocol have the obligation to implement the Articles thereof. Certain key points are as follows:

- Article 2 mandates the phaseout of ODSs by the Parties according to a prescribed timetable;
- Article 4 obliges all Parties to ban trade in ODSs with non-Parties. The Article provides for identification of products containing ODSs and products made using, but not containing, ODSs. Article 4A specifies the obligations for control of trade between the Parties;
- Article 5 permits developing countries which consume ODSs in quantities less than specified limits to delay their implementation of the control measures by a specified number of years. Hence qualifying developing countries are called 'Parties operating under Article 5' or, in short, 'Article 5 Parties'. The first Meeting of the Parties decided on which countries could be classified in this category; succeeding Meetings of the Parties revised these decisions. All other countries are called 'non-Article 5 Parties'. In this text we sometimes use the terms 'Article 5' and 'non-Article 5 Parties' to refer to developing and developed countries respectively. Almost all the national governments of the world (190 governments) and the European Union have ratified the Protocol and are Parties thereto. There are 145 Parties currently classified as Article 5 Parties. The Ozone Secretariat website contains an updated list of Article 5 Parties.
- Article 7 mandates baseline and annual reporting by Parties of both production and consumption of all ozone-depleting substances;

- Article 8 forms the basis for action in the case of non-compliance with any of the obligations;
- Article 9 requires Parties to conduct research and development, to exchange information on alternatives to ODSs and control strategies, and to cooperate in promoting public awareness of the environmental effects of the emissions of controlled substances. It also requires each Party to submit to the Secretariat a summary of the activities it has conducted pursuant to Article 9 within two years of entry into force and every two years thereafter; and
- Article 10 established a Financial Mechanism, including the Multilateral Fund (MLF) for the Implementation of the Montreal Protocol, to help meet the agreed incremental costs of the ODS phaseout by Article 5 Parties. Non-Article 5 Parties contribute to this fund. Article 10A also mandates that every Party take every practicable step to transfer the best available substitutes and technologies to Article 5 Parties.

THE ROLE OF THE MULTILATERAL FUND

The Multilateral Fund (MLF) plays the major role in financing technology transfer and cooperation for developing countries operating under Article 5. The developed country Parties contribute to the Fund according to the ratio of their contributions to the United Nations and as decided by the Meetings of the Parties. Other governmental, intergovernmental and non-governmental sources were also welcome to contribute to the Fund. Bilateral aid agencies of developed countries could spend up to 20 per cent of their contributions to the Fund directly in the recipient countries or regions, subject to the approval of their programmes by the Executive Committee.

The Fund started out with US$200 million for the period 1991–1993. Every three years, the Fund was replenished after a needs assessment made by the Meetings of the Parties. The replenishments were in the following amounts:

1994–1996	US$455.0 million
1997–1999	US$466.0 million
2000–2002	US$440.0 million
2003–2005	US$474.0 million
2006–2008	US$400.4 million

As of March 2007, the Fund has approved 5520 projects, with total funding of over US$2.1 billion in assistance to 143 developing countries. This level of funding is expected to result in the permanent, annual phaseout of about 236,000 ODP tonnes of ODS consumption and nearly 157,000 tonnes of production, with an ODP tonne representing a tonne multiplied by the ODP of an ozone-depleting substance. Projects approved by the Executive Committee have thus far resulted in the permanent annual phaseout of 191,000 ODP tonnes of ODS consumption and 116,000 tonnes of production.

The MLF Secretariat

The MLF Secretariat is based in Montreal, Canada, and consists of a small number of internationally recruited professional staff and local support staff. The Secretariat is headed by the Chief Officer.[26] The Secretariat develops and manages plans and budgets, reviews all activities submitted for funding, makes recommendations, disburses funds, monitors the implementing agencies, prepares policy papers, and undertakes other activities as directed by the Executive Committee.

The MLF Executive Committee

The responsibility for overseeing the operations of the MLF rests with an Executive Committee, supported by the Fund Secretariat. The Executive Committee develops and monitors the implementation of specific operational policies, guidelines and administrative arrangements, including the disbursement of resources; develops the three-year plan and budget for the MLF, including allocation of resources among the implementing agencies; develops the criteria for project eligibility and guidelines for the implementation of activities supported by the MLF and reviews the performance reports and expenditures of those activities; reports annually to the Meeting of the Parties on the activities exercised; and makes recommendations as appropriate.[27]

The Executive Committee consists of 14 members, 7 from Article 5 Parties and 7 from non-Article 5 Parties. The members are selected by each group of Parties and formally endorsed by the Meeting of the Parties. The Chair and the Vice-Chair of the Executive Committee are subject to annual rotation between the Article 5 and non-Article 5 Parties, while the group not providing the Chair selects the Vice-Chair.

Some decisions of the Executive Committee on investment

Over time, the Executive Committee has developed rules for investments to phase out ozone-depleting substances in Article 5 Parties. The Executive Committee has:

- requested implementing agencies and countries to negotiate technology transfer fees with vendors to cover groups of projects;
- decided not to finance the phaseout by transnational corporations or enterprises permitted to operate in 'free zones' with output for export only; funding proportional to the percentage of local ownership was provided for local enterprises partly owned by a transnational corporation;
- decided not to finance taxes, duties or other such transfer payments, the loss of economic subsidies, or rates of return in excess of cost of capital that might incorporate non-economic financial effects such as administered prices or interest rates;
- implemented guidelines for enterprises that exported part of their production to non-Article 5 Parties to ensure that consumers there were not subsidized and that companies based there were not subject to unfair competition;

- required ODS equipment to be destroyed or rendered unusable after project completion;
- granted concessions to low-volume ODS-consuming countries (annual ODS consumption of 360 tonnes or less), including separate allocations, higher cost-effectiveness thresholds, and special projects such as refrigerant management plans to suit their needs; and
- adopted sector and sub-sector cost-effectiveness threshold values, capping the maximum amount that could be financed per ODP tonne reduced.

Implementing agencies of the MLF

The United Nations Development Programme (UNDP), the United Nations Environment Programme (UNEP), the United Nations Industrial Development Organization (UNIDO) and the World Bank implement the projects funded by the MLF. In addition, a number of developed countries provide bilateral assistance to Article 5 Parties, with costs counted as a contribution to the Fund.[28]

The developing countries can choose to collaborate with any of the four implementing agencies or with Parties offering bilateral assistance in preparing their country programmes or project proposals.

The UNDP, UNIDO and the World Bank assist in feasibility and pre-investment studies and in investment projects. UNEP's Division of Technology, Industry and Economics (UNEP/DTIE) carries out no investment projects, but rather helps to establish the infrastructure within which projects can proceed. This includes carrying out institutional strengthening activities (such as establishing National Ozone Units within each country), facilitating regional networks, and helping to prepare country programmes, especially for low-volume-consuming countries. UNEP/DTIE also provides clearinghouse functions and produces a range of training materials.[29]

All the agencies report to the Fund Secretariat on the status of activities related to country programmes, prepare periodic progress reports on projects, prepare an annual report on income and expenditures of previous years, and prepare a final report after completion and/or termination of each project. They coordinate among themselves when preparing activities for phaseout of ODSs.

THE ROLE OF THE GLOBAL ENVIRONMENT FACILITY

When the Montreal Protocol was approved in 1987, the countries of Eastern Europe and republics of the former USSR were not classified as developing countries under the Protocol and therefore had to fulfil the same phaseout schedule as industrialized countries. In 1989 and 1990, with the collapse of communism, these countries broke up and introduced market mechanisms. Many of them were in the political and economic chaos of transition and did not have the resources to implement the Protocol, but they were not eligible for financing from the MLF. The Global Environment Facility (GEF) stepped in to assist these so-called countries with economies in transition (CEITs). The GEF financed the phaseout projects of these countries in accordance with a programme of compliance drawn up for each of these countries by the Meetings of the Parties to the Protocol. UNEP, the UNDP and the World Bank are their implementing agencies.

Good results from GEF assistance

According to data reports under Article 7 of the Protocol, total consumption of CFCs in the countries funded by the GEF decreased from about 224,000 tonnes in the base year to 1300 tonnes by 2004, a drop of more than 99 per cent. Production has been reduced accordingly. Of the four original producers of ODSs, only Russia sustained a considerable production capacity, which was phased out at the end of 2000.

Between 1991 and 2004, the GEF allocated more than US$177 million to projects to phase out ODSs and co-financed US$182 million.

RESULTS OF PROTOCOL IMPLEMENTATION

The results of implementation have shown excellent compliance with the Protocol so far. The reported figures for 2006 show that the non-Article 5 Parties have phased out all consumption of CFCs, halons, carbon tetrachloride, methyl chloroform and methyl bromide (a total of 1.1 million ODP tonnes), with the exception of the 2000 ODP tonnes for essential uses and 6160 ODP tonnes for critical methyl bromide use they were allowed. Article 5 Parties reduced their consumption from a baseline of 274,400 tonnes to 90,300 tonnes and are well on their way to phasing out according to the time schedule of the Protocol.

The hydrochlorofluorocarbons (HCFCs) are not due for phaseout by non-Article 5 Parties until 2030, with Article 5 Parties permitted an additional ten years beyond that.

STATUS OF THE OZONE LAYER TODAY

The ozone hole in the Antarctic spring of 2006 was among the largest ever recorded – covering almost 30 million square kilometres – partly because of super-cold temperatures that may paradoxically be a further indicator of global warming. The Antarctic ozone exhibits widespread and massive local depletion in the heart of the ozone hole, frequently exceeding 90 per cent, and the total integrated column ozone in spring is considerably below that observed in earlier decades. The depth of the ozone losses in the Arctic is considerably smaller, and their occurrence is far less frequent. Thus the widespread and deep ozone depletion that characterizes the Antarctic ozone hole is a unique feature on the planet.[30]

The Scientific Assessment of Ozone Depletion[31] presents a clear picture of how close humanity came to destroying life on Earth. It also estimates how long it might take to restore the ozone layer, first to its 1980 condition, before the

Antarctic ozone hole was discovered, and ultimately to a state in which there is no significant chlorine and bromine loading contributed by the production and consumption of substances phased out by the Montreal Protocol. The Montreal Protocol stopped chemical companies from producing millions of tonnes of long-life ozone-depleting substances, many of which are also greenhouse gases, and eliminated ODS consumption in almost all sectors. The Scientific Assessment finds that although the Montreal Protocol is working, it will take decades for the stratospheric ozone layer to recover. Most significant to this book is the Assessment's conclusive finding that:

> *failure to comply with the Montreal Protocol would delay, or could even prevent, recovery of the ozone layer.* (Scientific Assessment Panel September, 2006)

Other significant findings of the Scientific Assessment Panel are as follows:

- there is clear evidence of a decrease in the atmospheric burden of ODSs, and some early signs of stratospheric ozone recovery;
- Antarctic ozone abundances are projected to return to pre-1980 levels around 2060–2075, roughly 10–25 years later than estimated in the 2002 Assessment;
- changes in climate will influence if, when and to what extent ozone will return to pre-1980 values in different regions;
- large ozone losses will probably continue to occur in cold Arctic winters during the next 15 years; and
- elimination of all emissions of ozone-depleting substances after 2006 would advance ozone layer recovery by about 15 years, from 2049 to 2034.

THE SCIENCE OF CLIMATE CHANGE

Introduction

Evidence presented on the website of the United Nations Framework Convention on Climate Change (UNFCCC, www.unfccc.int) predicts a bleak future if greenhouse gas emissions are not controlled and efforts are not begun to adapt to the climate change that is inevitable from the pollution already released into the atmosphere. The average temperature of the Earth's surface has risen by approximately 0.74°C since the late 1800s and is expected to increase by another 1.8°C to 4°C by 2100. Should the necessary action by governments, organizations and consumers not be taken, the temperature rise – a rapid and profound change – will be greater than any century-long trend in the last 10,000 years.

The principal reason for the rising global thermometer is a century and a half of industrialization: the burning of ever-greater quantities of oil, gasoline and coal, the cutting of forests, and the practice of certain farming methods which increase the emission of natural 'greenhouse gases' (GHGs) such as carbon dioxide, methane and nitrous oxide. Certain man-made chemicals – Hydrofluorocarbons (HFCs), Perfluorocarbons (PFCs) and Sulphur hexafluoride (SF_6), in addition to the chemicals controlled by the Montreal Protocol – also cause global warming and climate change.

The natural balance of GHGs is critical for life on Earth: they keep some of the Sun's warmth from reflecting back into space, and without them the world would be an unstably cold, barren and agriculturally unproductive place. But in unnatural and increasing quantities GHGs are pushing the global temperature to artificially high levels and altering the climate.

The current warming trend is expected to cause extinctions of species important to the prosperity of humans and natural ecosystems. Numerous plant and animal species, already weakened by pollution and the loss of habitat, are not expected to survive the next 100 years. Human beings, while not threatened in this way, are likely to face mounting difficulties. Recent severe storms, floods and droughts are evidence that computer models predicting more frequent 'extreme weather events' are on target.

The average sea level rose by 10 to 20cm during the 20th century, and an additional increase of 18 to 59cm is expected by 2100: higher temperatures cause ocean volume to expand, and melting glaciers and ice caps add more water. If the higher end of expected sea level rise is reached, the oceans will overflow the heavily populated coastlines of such countries as Bangladesh and the US (Florida, the States adjoining the Gulf of Mexico and the Eastern Seaboard) and island nations, causing the disappearance of some nations entirely (such as the island state of the Maldives); foul freshwater supplies for billions of people; and spur mass migrations and other hardship.

Agricultural yields are expected to drop in most tropical and subtropical regions, and in temperate regions, too, if the temperature increase is more than a few degrees Celsius. Drying of continental interiors, such as central Asia, the African Sahel and the Great Plains of the US, is also forecast. These changes would cause, at the minimum, disruptions in land use and food supply, and at the maximum, catastrophic starvation. And the range of diseases such as malaria may also expand or the diseases become more virulent.

Over a decade ago, most countries joined an international treaty – the United Nations Framework Convention on Climate Change – to begin to consider what can be done to reduce global warming and to cope with whatever temperature increases are inevitable. The Kyoto Protocol of 1997 has more powerful (and legally binding) measures, but fewer countries have ratified it and some who have are struggling to comply. In addition, since 1988 the Intergovernmental Panel on Climate Change has reviewed scientific research and provided governments with summaries and advice on climate problems.

What can be done to protect the climate?

Getting more electricity, transport and industrial output for less coal, oil or gasoline is a no-lose situation: more profit, less pollution, less global warming. 'Combined-cycle' turbines, in which the heat from burning fuel drives steam turbines while the thermal expansion of the exhaust gases drives gas turbines, can boost the efficiency of electricity generation by 70 per cent. In the longer term, new technologies could double the efficiency of power plants. Gasoline fuel cells and other advanced automotive technologies can cut carbon dioxide emissions from transport roughly in half, as can 'hybrid' petroleum/electricity vehicles, some of which are already on the market. Improved vehicle air-conditioners can reduce global fuel use by about 2 per cent and reductions in vehicle refrigerant greenhouse gas emissions can reduce equivalent emissions by a comparable amount. Natural gas releases less carbon dioxide per unit of energy than coal or oil – hence switching to natural gas is a quick way to cut emissions.

Industry, which accounts for over 40 per cent of global carbon dioxide emissions, can benefit from combined heat and power co-generation, other uses of waste heat, improved energy management and more efficient manufacturing processes. Installing more efficient lighting and appliances in buildings can significantly cut electricity use. Improving building insulation can greatly reduce the amount of fuel needed for heating or air-conditioning.

Solar energy and wind-generated electricity – at current levels of efficiency and cost – can replace some fossil fuel use, and are increasingly being used. Expansion of hydro-electric power, where appropriate, could make a major contribution to lowering greenhouse gas emissions. Biomass sources of energy – such as fuel wood, alcohol fermented from sugar, combustible oils extracted from soybeans and methane gas emitted by waste dumps – can help cut green-house gas emissions, but only if vegetation used for the purpose is replaced by equal amounts of replanted vegetation (so that the carbon dioxide released by biomass combustion is recaptured through photosynthesis). Nuclear energy produces virtually no greenhouse gases, but there are concerns over safety, transport and disposal of radioactive wastes, not to mention weapons proliferation.

New technologies have become available for 'capturing' the carbon dioxide emitted by fossil fuel power plants before it reaches the atmosphere. The carbon dioxide is then stored underground in empty oil or gas reservoirs, unused coal beds, or the deep ocean. While not exactly 'renewable,' this approach, which is already in limited use, is being scrutinized for possible risks and environmental impacts.

Actions under the Montreal Protocol can also protect climate. The phaseout of hydrochlorofluorocarbons (HCFCs) can be accelerated, ODSs can be collected from building insulation, refrigeration and air-conditioning equipment, and other products for destruction. Parties and enterprises can avoid the use of HFCs as substitutes for most ODS uses and can minimize emissions with smaller refrigerant charge, more efficient foam blowing, better containment, and taxes or deposits that provide an incentive to minimize use and to finance collection and disposal at the end of useful life.

The role of 'carbon sinks' in climate protection

Trees and other green plants, using only sunlight for energy, take carbon dioxide out of the atmosphere, releasing oxygen and storing carbon in a safe and useful way. Forests, which provide many benefits for the environment, can be major allies in the battle against climate change and global warming. Deforestation, which is occurring all over the world, has a doubly damaging effect: it reduces the number of trees that can recover the carbon dioxide produced by human activities and it releases into the atmosphere the carbon contained in the trees that are cut down.

Carbon stored in agricultural soils can often be preserved or enhanced by switching to 'no-tillage' or 'low-tillage' techniques, which slow the rate at which organic soil matter decomposes. In rice fields, emissions of methane, a powerful greenhouse gas, can be suppressed to some extent through tillage practices, water management and crop rotation. Using nitrogen fertilizers more efficiently can reduce emissions of nitrous oxide, another potent greenhouse gas.

Adaptation to climate change

Since major effects from climate change now appear inevitable, it is vital for countries and communities to take practical steps to protect themselves from the likely disruption and damage that will result.

THE UNITED NATIONS FRAMEWORK CONVENTION ON CLIMATE CHANGE OF 1992

The first attempt to address the causes of climate change at the international level was with the United Nations Framework Convention on Climate Change (UNFCCC). The UNFCCC was signed at the 1992 United Nations Conference on Environment and Development, also known as the Rio Earth Summit, and calls on its Parties to prevent 'dangerous anthropogenic interference' with the Earth's climate system. The UNFCCC does not place any controls on greenhouse gas (GHG) emissions.

The UNFCCC places the heaviest burden for fighting climate change on 27 industrialized nations and 12 'economies in transition' (countries in Central and Eastern Europe, including some states formerly belonging to the Soviet Union, called 'Annex I' countries because they are listed in the first annex to the treaty), since they are the source of most past and current greenhouse gas emissions.

Industrialized nations agree under the Convention to support climate change activities in developing countries by providing financial support above and beyond any financial assistance they already provide to these countries. A system of grants and loans has been set up through the Convention and is managed by the Global Environment Facility. Industrialized countries also agree to share technology with less advanced nations.

Because economic development is vital for the world's poorer countries – and because such progress is difficult to achieve even without the complications added by climate change – the Convention accepts that the share of greenhouse gas emissions produced by developing nations will grow in the coming years. It nonetheless seeks to help such countries limit emissions in ways that will not hinder their economic progress. The Convention acknowledges the vulnerability of developing countries to climate change and calls for special efforts to ease the consequences. There are currently 190 Parties to the UNFCCC.

Bodies of the Convention and partner agencies

The Conference of the Parties (COP) is the prime authority of the Convention. A Subsidiary Body for Scientific and Technological Advice (SBSTA) counsels the COP on matters of climate, the environment, technology and method. It meets twice a year. A Subsidiary Body for Implementation (SBI) helps review how the Convention is being applied, for example by analysing the national communications submitted by member countries, and deals with financial and administrative matters. The SBI also meets twice a year.

Partner agencies include the Global Environment Facility (GEF), which has existed since 1991 to fund projects in developing countries that will have global environmental benefits. The job of channelling grants and loans to poor countries to help them address climate change, as called for by the Convention, has been delegated to the GEF because of its established expertise.

The Intergovernmental Panel on Climate Change (IPCC) provides services to the Convention, although it is not part of it. The IPCC was established by the World Meteorological Organization (WMO) and the United Nations Environment Programme (UNEP) in 1988. It is open to all members of the UN and WMO. The declared role of the IPCC is to assess, on a comprehensive, objective, open and transparent basis, the scientific, technical and socio-economic information relevant to understanding the scientific basis of risk of human-induced climate change, its potential impacts, and options for adaptation and mitigation. The IPCC does not carry out research, nor does it monitor climate-related data or other relevant parameters. It bases its assessment mainly on peer-reviewed and published scientific/technical literature, and operates out of three Working Groups and a Task Force.

The UNFCCC Secretariat

A secretariat staffed by international civil servants supports the Convention and its supporting bodies. It makes practical arrangements for meetings, compiles and distributes statistics and information, and assists member countries in meeting their commitments under the Convention. The Secretariat is based in Bonn, Germany.

THE KYOTO PROTOCOL

The Kyoto Protocol to the UNFCCC was negotiated in 1997 and requires the 39 industrialized and countries-in-transition parties to reduce GHG emissions (excluding the GHGs controlled by the Montreal Protocol) by an average of 5.8 per cent of 1990 levels for its first commitment period, which runs from 2008 to 2012. The Kyoto Protocol does not require developing countries to reduce GHG emissions, although it mandates every Party (Article 10) to 'formulate, implement, publish and regularly update national and, where appropriate, regional programmes containing measures to mitigate climate change and measures to facilitate adequate adaptation to climate change'. It entered into force on 16 February 2005. Future mandatory targets are expected to be established for 'commitment periods' after 2012. These are now being negotiated.

The Parties with mandatory targets for reduction of GHGs may reach their targets through three market-based mechanisms – the International Emissions Trading System, the Clean Development Mechanism (CDM) and Joint implementation (JI).

Emissions trading in the 'carbon market'

The Kyoto Protocol allows countries that have emissions units to spare (in other words emissions permitted them but not 'used') to sell this excess capacity to countries that are over their targets. This so-called 'carbon market' – so named because carbon dioxide is the most widely produced greenhouse gas, and because emissions of other greenhouse gases will be recorded and counted in terms of their 'carbon dioxide equivalents' – is both flexible and realistic. Countries not meeting their commitments will be able to 'buy' compliance. A global 'stock market' where emissions units are bought and sold is currently developing.

The Clean Development Mechanism: Getting developing countries involved

The Kyoto Protocol does not set limits on the GHG emissions of developing nations; instead, it includes a Clean Development Mechanism for reductions to be 'sponsored' in countries not bound by emissions targets. Developed countries pay for projects that cut or avoid emissions in poorer nations and are awarded credits that can be applied to meeting their own emissions targets.

The Clean Development Mechanism is overseen by an Executive Board. To be certified by the Clean Development Mechanism Executive Board, a project must be approved by all involved parties, demonstrate a measurable and long-term ability to reduce emissions, and promise reductions that would be additional to any that would otherwise occur.

Joint implementation: Mutual help for countries with emissions targets

Joint implementation (JI) is a programme under the Kyoto Protocol that allows industrialized countries to meet part of their required cuts in GHG emissions by paying for projects that reduce emissions in other industrialized countries. In practice, this will probably mean facilities built in the countries of Eastern Europe and the former Soviet Union – the so-called transition economies – paid for by Western European and North American countries.

The sponsoring governments will receive credits that may be applied to their emissions targets; the recipient nations will gain foreign investment and advanced technology (but not credit toward meeting their own emissions caps, which they have to do themselves). The system has advantages of flexibility and efficiency. It is often cheaper to carry out energy-efficiency work in the transition countries, and to realize greater cuts in emissions by so doing. The atmosphere benefits wherever these reductions occur. The operation of the JI mechanism is similar to that of the CDM.

The Kyoto Protocol entered into force in February 2005 and has so far been ratified by 164 countries, including Brazil, China and India. The US and Australia have not ratified Kyoto.

CLIMATE PROTECTION FROM THE ODS PHASEOUT

The Montreal Protocol is the world's most successful environmental treaty, having phased out about 95 per cent of ODSs. By phasing out the production of chemicals that destroy the ozone layer, it has protected the Earth from harmful ultraviolet radiation, which causes skin cancer and cataracts, suppresses the human immune system, damages natural ecosystems, and decreases agricultural production.

In addition to destroying the ozone layer, ODSs are also greenhouse gases, with global warming potential (GWP) thousands and even tens of thousands of times greater than carbon dioxide (CO_2). For example, CFC-12 has a 100-year GWP of 10,720 relative to CO_2 (GWP=1).

According to a new scientific study published in the *Proceedings of the National Academy of Sciences*,[32] the greenhouse gas emission reductions achieved by the Montreal Protocol by 2010 will be equivalent to about 11.7 gigatonnes of carbon dioxide per year. This is about five to six times greater than the emissions reductions that the Kyoto Protocol, assuming full compliance, will achieve from 2008 to 2012.

The study shows that the discovery that ODSs were destroying the ozone layer in 1974 provided an 'early warning' that altered what otherwise would have been a steady annual increase in ODS production and use. This early warning delayed climate change by 35 to 41 years (CFC emissions were growing at 7 per cent annually). Furthermore, the Montreal Protocol provided up to a 12-year delay by eliminating the uses that persisted after the early warning.

Most important, the study also shows that strengthening the Montreal Protocol will further protect climate, in addition to advancing the restoration of the ozone layer to its pre-1980s state.

The accelerated phaseout of HCFCs and other measures could avoid emissions of the equivalent of 1.2 gigatonnes of carbon dioxide per year by 2015. The study notes that the greatest climate benefits of strengthening the Montreal Protocol are achieved if low-GWP substitutes delivering the highest life-cycle climate performance (LCCP) are selected, and if energy efficiency standards and refrigerant containment requirements are simultaneously strengthened worldwide.

BOX 3.1 THE DUAL BENEFIT OF THE MONTREAL PROTOCOL: OZONE AND CLIMATE PROTECTION

The 1987 Montreal Protocol on Substances that Deplete the Ozone Layer is a landmark agreement that has successfully reduced the global production, consumption and emissions of ozone-depleting substances (ODSs). ODSs are also greenhouse gases that contribute to the radiative forcing of climate change.

Nobel Prize-winners Mario Molina and Sherwood Roland warned the world about ozone depletion in 1974. Citizens and governments reacted by boycotting and banning CFC aerosol products and ultimately phased out ODSs under the Montreal Protocol. These actions put the ozone layer on the path to recovery and delayed climate change. The reduction in ODSs since 1990 under the Montreal Protocol is equivalent to about 7–12 years of growth in radiative forcing of CO_2 from human activities. Additional climate benefits that are significant compared with the Kyoto Protocol reduction target could be achieved by accelerating the phaseout under the Montreal Protocol by managing the emissions of substitute fluorocarbon gases and/or implementing alternative gases with lower global warming potentials (GWPs). The figures below illustrate and quantify the extraordinary benefits of the Montreal Protocol.

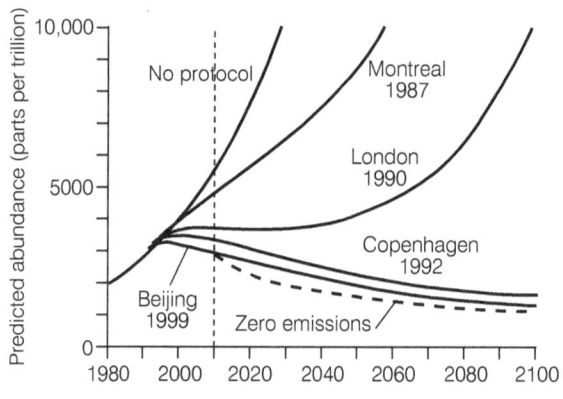

Figure 3.1 *Effect of each Montreal Protocol control measure*

Note: The Montreal Protocol and its subsequent amendments have slowed and reversed the accumulation of stratospheric chlorine and bromine from ODSs.

Figure 3.2 *Effective stratospheric chlorine*

Note: Effective stratospheric chlorine values summed over the principal ODSs are in decline since reaching a peak in the mid to late 1990s.

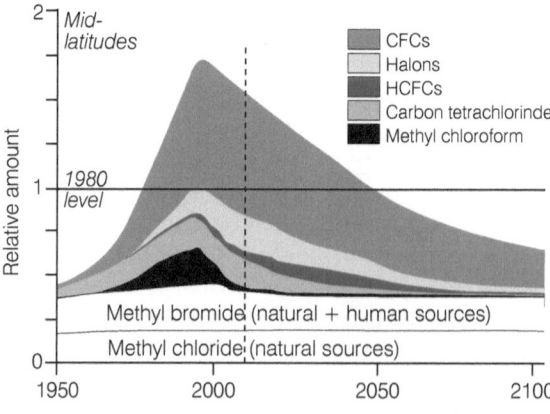

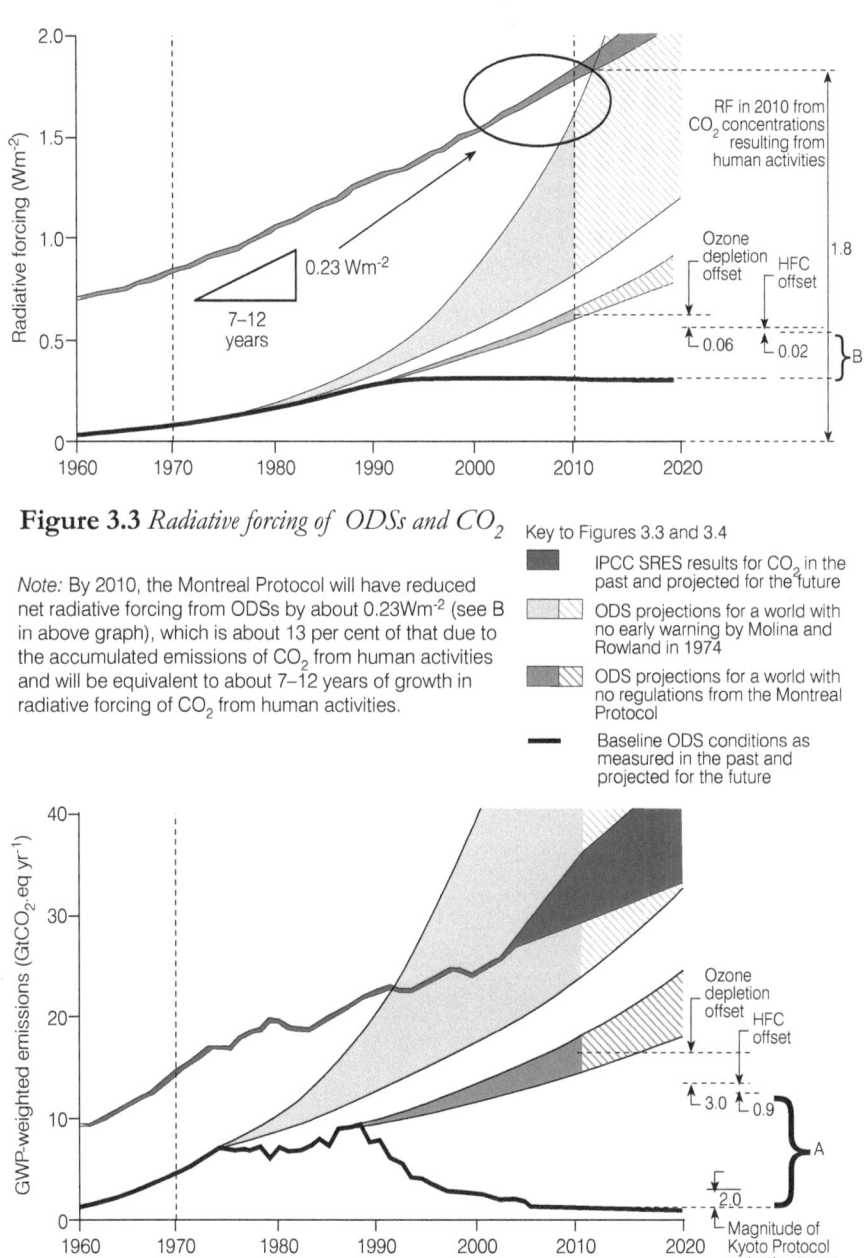

Figure 3.3 *Radiative forcing of ODSs and CO₂*

Key to Figures 3.3 and 3.4

▪ IPCC SRES results for CO₂ in the past and projected for the future

▨ ODS projections for a world with no early warning by Molina and Rowland in 1974

▨ ODS projections for a world with no regulations from the Montreal Protocol

— Baseline ODS conditions as measured in the past and projected for the future

Note: By 2010, the Montreal Protocol will have reduced net radiative forcing from ODSs by about $0.23 Wm^{-2}$ (see B in above graph), which is about 13 per cent of that due to the accumulated emissions of CO_2 from human activities and will be equivalent to about 7–12 years of growth in radiative forcing of CO_2 from human activities.

Figure 3.4 *Emissions of ODSs and CO₂*

Note: By 2010, the Montreal Protocol will have reduced net GWP-weighted emissions from ODSs by about $11 GtCO_2$-eq yr^{-1} (see A in above graph), which is 5–6 times the reduction target of the first commitment period (2008–2012) of the Kyoto Protocol.

Source: Velders et al (2007)[33]

4

Technology Change
in Developed Countries

INTRODUCTION

Actions by the developed countries under the Montreal Protocol demonstrate that technology transfer of environmental technology can be rapid and cost-effective. This is true for the enterprises developing the technology, for the governments sponsoring the environmental protection, for the enterprises receiving the technology, and for the citizens who purchase new products that are responsibly manufactured and owned. Technology selection and transfer under the Montreal Protocol is unprecedented in terms of environmental scrutiny, validation of technical performance, and disclosure of which technology was selected by the most knowledgeable customers.

This chapter describes how and why enterprises that developed technology to avoid ozone-depleting substances (ODSs) chose to share that technology on favourable terms worldwide. It is an extraordinary deviation from the situation reported in other case studies of technology transfer, and many readers may find the truth too good to believe. It is possible that the Montreal Protocol experience is the only occasion so far when public and private stakeholders considered technology cooperation a matter of human survival, stepped out of their narrow self-interests and promoted actions that allowed humanity to survive on Earth.

This chapter explains how more than 240 sectors with products dependent on ODSs halted most uses within ten years and ended up more satisfied with the performance of the replacement products than they had been with the ODSs.

COOPERATION IN INDUSTRIALIZED COUNTRIES SET THE STAGE FOR WORLDWIDE TECHNOLOGY COOPERATION UNDER THE MONTREAL PROTOCOL

By the mid-1970s, it was known that chlorofluorocarbons (CFCs) were accumulating in the atmosphere.[1] However, CFC industry stakeholders and some scientists contended there was no need for an urgent response. They argued that ozone depletion was only a hypothesis based on chemical reactions observed in the laboratory and through computer models; that there had been no atmospheric observations of the proposed ozone destruction cycle; and that there were questions about the relative magnitude of CFCs as a source of stratospheric chlorine. CFC manufacturers and their customers argued for delay in a regulatory response for a few years until scientific research answered these outstanding questions.[2] Early on, industry estimated that hundreds of thousands of jobs and billions of dollars of income depended on CFCs.

Despite industry arguments to wait for more scientific evidence, public concern for the ozone layer was very high, particularly in the US, Canada, Denmark, Finland, Norway and Sweden. US aerosol product sales slumped. The S. C. Johnson Company earned the historic distinction in June 1975 of being the first enterprise to abandon CFCs. Within weeks, Sherwin-Williams, Bristol Meyers and Mennen joined S. C. Johnson in positively advertising their CFC-free products and advertising against the use of competitors' CFC products. In 1976 the US Government announced plans to ban 'non-essential' aerosol products. Canada, Sweden and Norway also announced bans, and many nations discussed further action.

CFC manufacturers declare chemical substitutes too expensive

In early 1986 representatives of DuPont, Allied and ICI separately reported that between 1975 and 1980 they had identified compounds meeting environmental, safety and performance criteria for some CFC applications, but had terminated research and development of these environmentally superior products when they concluded that none would be as inexpensive to customers or as profitable to manufacturers as CFCs.[3] DuPont, the world's largest producer of CFCs, ended a five-year US$15 million research programme in 1980, largely because preliminary results indicated substitutes and alternatives would be uneconomical to make, require difficult changes in manufacturing plants, take a minimum of ten years to develop and market, and involve controversial toxicity-testing for adverse health effects.[4]

At a United Nations Environment Programme (UNEP) workshop in Rome on 26–30 May 1986, DuPont stated that the costs of chemical substitutes for CFCs, mainly hydrofluorocarbons (HFCs), would be three to six times the price of CFCs, while ICI estimated costs at eight to ten times the CFC price. DuPont justified its policy of not commercializing substitutes by explaining that:

> *Neither the marketplace nor regulatory policy, however, has provided the needed incentives to make these equipment changes or to support commercialization of the other potential substitutes. If the necessary incentives were provided, we believe alternatives could be introduced in volume in a time frame of roughly five years.*[5]

Thus after a decade of industry opposition to regulation, industry claimed that it was the lack of regulation that prevented it from introducing products to protect the ozone layer.

Box 4.1 Development of HCFC-225

Shunichi Samejima, Asahi Glass Foundation[6]

Asahi Glass (AGC) announced in 1989 its success in developing the manufacturing process for HCFC-225, which was the first new chemical alternative to CFC-113 solvent invented after the signing of the Montreal Protocol. This was one of the most important outcomes of the project team in AGC which had been involved in the development of various alternatives to CFCs.

AGC has been a leading manufacturer in Japan of various chlorinated chemicals since the 1950s and started its production of a series of CFCs, CFC-11, -12, -113, and so on in 1964 by converting chlorinated methanes to chlorofluoromethanes (CFMs) under fluorination reaction by hydrogen fluoride in the presence of specified catalysts.

In 1987, before the Montreal Protocol was first signed in September, the US Environmental Protection Agency (EPA) invited a team of world-class experts in fluorine chemistry to advise the agency about plausible chemical alternatives to CFCs. The experts made a list of likely alternatives that was so comprehensive that it included virtually all important chemicals which have been commercially produced as alternatives for various ODSs since the signing of the Montreal Protocol. One of the expert members invited to the panel, Masaaki Yamabe (director of Fluorochemicals R&D in AGC at that time)[6], recalled how Thomas Midgley had methodically considered a wide range of chemicals as possible refrigerants. He applied this creative strategy to imagine every possible fluorinated substitute for CFC-113. During the meeting, Dr Yamabe surprised the other experts by announcing possible new hydrochlorofluorocarbon (HCFC) chemicals with a propane (three-carbon) structure as alternatives to CFC-113, for which no alternatives had been proposed. Soon after coming back to Japan from this expert panel, Dr Yamabe started up a corporate project team for the timely development of alternatives to CFCs. One of the important targets of the team was to discover alternatives to CFC-113, which represented almost 50 per cent of total CFC consumption in Japan as a cleaning solvent, mainly for electronics industries. The research strategy was to develop a new chemical suitable as a drop-in replacement for CFC-113.

Most CFCs have been developed by a 'trial-and-error' process. Our strategy was very unique in predicting a most appropriate chemical structure by the use of computational chemistry. In order to meet the requirement of chemical and physical properties for a drop-in replacement, the computational molecular design strongly indicated the choice of a HCFC structure, but researchers could not find appropriate compounds in the survey of about 70 chemicals with one or two carbon numbers. Then they turned to focus on a new series of compounds with three carbons, and finally developed HCFC-225 with many desirable properties and a technically and economically feasible manufacturing process.

Between 1990 and 1994 HCFC-225 was produced with a pilot plant to supply samples for both market evaluation and toxicological studies, and full commercialization started in 1994. The important point of success was to have internationally recognized organizations such as the Programme for Alternative Fluorocarbon Toxicity Testing (PAFT) and the Alternative Fluorocarbons Environmental Acceptability Study (AFEAS) conduct the safety and environmental assessment.

With this line-up of an alternative to CFC-113 from Japan in addition to the other alternatives to CFCs, the developed countries were encouraged about the complete phaseout of all the CFCs by the end of 1995 as required by the amended Montreal Protocol.

Regulation stimulates and promotes alternatives

Prior to 1987 the majority of enterprises depending on CFCs and other ODSs ultimately controlled by the Montreal Protocol had done little to investigate alternatives. In January 1988 the US EPA, Environment Canada and the Conservation Foundation organized the first annual technology conference on alternatives to ODSs, with more than 1000 participants. The scientific evidence for depletion of the ozone layer was no longer questioned, governments offered to collaborate on new technology, ODS-using enterprises sought alternatives and potential suppliers were motivated by the market opportunity. At the conference:

- AT&T held a press conference to announce that a semi-aqueous solvent made from orange peel could clean electronics as well as CFC-113; the new technology was from an innovative small enterprise, Petroferm;
- solvent equipment suppliers announced methods to cost-effectively reduce CFC emissions by a third to a half; and
- executives of the Mobile Air Conditioning Society, General Motors and the EPA announced plans to commercialize CFC-12 recycling for motor vehicles. Dozens of experts stayed after the conference to begin work on the plan.

Industry leadership, technical innovation and cooperation marked 1988 and 1989, as ODS producers abandoned CFCs and focused on chemical alternatives. The Ozone Trends Panel of the World Meteorological Organization released its report on 15 March 1988 confirming ozone layer depletion. Within ten days of the release, DuPont committed to an orderly global transition to a total phase-out of fully halogenated CFCs, signalling drastic changes in markets and technology. The Pennwalt Corporation almost immediately seconded DuPont's position and urged that CFC production be halted as soon as practical. By October 1988 ICI also announced support for a CFC phaseout. CFC prices were uncertain, and DuPont began to notify customers that supplies might be rationed.[7]

The announcements by CFC producers supporting international agreements to limit global CFCs and the rapid rejection of those chemicals by their industrial customers had financial implications that further stimulated the quest for substitutes and the necessity of cooperation.[8]

Enterprises depending on ODSs organized quickly to welcome new ozone-safe technology. The combination of an international treaty that could only become more stringent and some organizations pledging to move beyond regulation gave suppliers confidence that a totally ozone-safe solution would enjoy large markets.[9]

BOX 4.2 CRISIS AND OPPORTUNITY FOR DAIKIN

Osami Kataoka, Daikin Industries[10]

Because Daikin produces both fluorocarbon refrigerants and air-conditioners that use these refrigerants, the ozone depletion issue was a big crisis, but it was also a great opportunity. The chemical division of Daikin established its alternative refrigerant development team in 1989, and in 1992 the air-conditioning division established the alternative refrigerant equipment project, anticipating the Copenhagen Amendment that accelerated the phaseout of CFCs.

The first alternative for CFC-12 was HCFC-142b, which was commercially produced by Daikin in 1988. Daikin was the first company in Japan to commercially produce HFC-134a in 1991, which helped rapidly eliminate CFC-12 in many applications such as refrigerators and vehicle air-conditioning systems.

Phasing out CFCs and HCFCs from air-conditioning and refrigeration systems required great efforts to resolve reliability problems because the higher polarity of HFCs almost completely eliminates miscibility of refrigerant to oil. Daikin commercialized the swing compressor, which eliminated the heavy metal contact of the rotary compressor, resulting in higher reliability as well as higher efficiency.

Daikin is very proud to have almost finished its work on the ozone depletion issue, but climate change is next and is a more difficult task. However, due to the close relationship of its chemical and air-conditioning divisions, Daikin will surely take a leading role in solving climate change and other environmental issues that may follow.

INDUSTRY AND MILITARY MOTIVATIONS FOR LEADERSHIP ON OZONE PROTECTION

Industry leadership was an important ingredient in the accelerated and cost-effective phaseout of ODSs. Experts from government and NGOs who worked directly with enterprises on phaseout projects have documented that in many cases the overwhelming motivation was to protect the ozone layer.[11] In many other cases, however, enterprises undertook voluntary efforts and leadership projects in response to a wide range of factors.[12] At least two influential industry organizations that originally organized to fight against restrictions on CFCs transformed themselves to support regulations to protect the ozone layer: the Alliance for Responsible Atmosphere Policy (ARAP) and the Association of Fluorocarbon Consumers and Manufacturers (AFCAM). Several dozen other existing organizations created substantial internal subcommittees on ozone layer protection.

By 1989 US industry, environmental NGOs and the US EPA could announce the world's first voluntary national CFC phaseout in food packaging, including a pledge by US suppliers to transfer the technology licence-free worldwide. Canada's Nortel and Japan's Seiko Epson announced corporate goals of a complete CFC-113 phaseout on accelerated schedules, and a global agreement had been reached to recycle CFC-12 from automobile air-conditioners.[13]

By 1990 dozens of organizations had announced that they would phase out ODSs, rather than merely reduce emissions, giving confidence to Protocol

Table 4.1 *Industry motivation to speed protection of the ozone layer*

Motivation	Underlying forces
Social	Some industry sectors or organizations had become increasingly powerful members of the global diplomatic and governance community, choosing to play the role of champion, of exemplar, in helping to forge environmental policy.[14]
Reputation	The reputation of a company today is more than product, goodwill, performance and reliability. Customers care about human rights, the environment and other aspects of sustainability. Because reputation affects sales, it also affects financing.
Regulatory	Early action initiated by industry avoids 'command and control' actions initiated by governments that rigidly prescribe particular technical solutions, and allows industry to search and choose the solution that satisfies the environmental criteria at the least cost while maintaining or enhancing product performance and appeal. Head-starts avoid desperate technology choices.
Economic	'Pollution prevention pays' with less waste, more product and less disposal. Prompt response to new science can avoid liability for health and environmental damage. Phaseout can impress customers and stimulate invention of new technology (for use, sharing or licensing). Sector leadership allows experts to collaborate and share the expense of development. Supplier competition and economies of scale reduce costs.
Public relations	As pressure from environmental NGOs increases, protests can embarrass enterprises and damage reputations with consequences to market share. ODS emissions data for enterprises and facilities were publicly available in many countries.[15]
Recognition	Awards and other recognition from government and NGOs are free advertising, and environmental awards build reputation. Environmental leadership attracts the best applicants seeking careers; a sense of public benefit can be a powerful team-building and motivating force for innovation and performance.[16]

negotiators to move from the original Protocol, which required merely a freeze on halons and a 50 per cent reduction in CFCs, to a total phaseout, and giving confidence to suppliers that a total ozone-safe solution would enjoy large markets.[17]

COMMERCIALIZATION OF ALTERNATIVES

Commercialization of technology requires at least seven separate steps; these are elaborated in the sections below.

1) Motivate action, including research and development

The Montreal Protocol motivated action. The 1987 controls were considered daunting, and there was every expectation that controls would be strengthened

---

BOX 4.3 CREATIVITY AND CHALLENGE: THE SEIKO EPSON SAGA FOR PROTECTION OF THE OZONE LAYER

Toshiyuki Miyajima, General Manager, Trust-based Management Department, Seiko Epson Corporation

In December 1988 the Seiko Epson Group resolved to eliminate the use of CFC-113, which we were using at a rate of 1400 tons per year, and began taking actions towards that end. The enterprise's top management established a clear policy, specifically that:

> *We cannot continue to use substances once we know them to be environmentally harmful; we must therefore stop all operations that make use of CFC-113.*

In response to this policy, all employees began working toward the elimination of CFC-113, and thorough analyses were made of all manufacturing processes that used CFC-113. Finally, after eliminating or replacing certain cleaning processes with aqueous cleaning, we achieved our goal of eliminating CFC-113 in October 1992. Also in 1992 we pledged to eliminate another ODS solvent, trichloroethane (TCA), and accomplished that goal in 1993.

This four-year effort to eliminate CFC-113 and trichloroethane was Epson's first serious attempt to address a global environmental problem. Its corporate culture, rooted in 'Creativity and Challenge', made its employees fit to respond to global environmental problems. The President of Seiko Epson at that time, Tsuneya Nakamura, had the vision that business or even individuals should behave as though global problems were their own. So Seiko Epson disclosed the proprietary technology the enterprise used to solve its own problems. By publishing handbooks on precision cleaning without ODSs, Seiko Epson shared technology at conferences in many developed and developing countries and invited many experts to see Seiko Epson's centralized cleaning facilities to allow small and medium-sized suppliers to phase out ODSs at affordable cost.

In the 2006 financial year, Seiko Epson established an 'Action 2010 General Environmental Policy' and began working toward the reduction of environmental burdens throughout a product's life cycle. This included three major themes: global warming prevention, resource recycling and savings, and chemical substance control.

Powered by the technology cooperation experience gained in the course of the ozone layer protection activities, Seiko Epson continues to drive advances in its environmental performance, one of the missions stated in the corporate philosophy.

---

as scientists more firmly linked ODSs to the Antarctic ozone hole and worldwide increases in ultraviolet light.

2) Identify and invent plausible alternatives and substitutes
Plausible alternatives and substitutes had been catalogued by governments and government authorities and public interest research organizations.[18] For example, the US EPA had published a study of chemical alternatives to ODSs and had amassed thousands of pages listing almost every ODS use and its alternatives and substitutes.[19]

Chemical enterprises were among the first to respond to the new markets created by the Montreal Protocol. They dusted off the patents for the products they had identified in the 1970s in response to public concern over the Molina–Rowland hypothesis and undertook a methodical search for new chemicals that would be ozone-safe, otherwise environmentally acceptable, and technically and economically suitable.[20]

ODS customers took a wider view and considered chemicals that had been replaced by ODSs or had been developed after ODSs had captured markets.[21] One measure of the success of former ODS customers and their current suppliers is that today only 20 per cent of the ODSs that would have been used if the question of ozone depletion had not arisen have been replaced using in-kind chemical substitutes, and an increasingly small portion of that world supply is coming from the enterprises originally responsible for the manufacture of chemicals that destroy the ozone layer.[22]

3) Test technical performance and economic feasibility

To test technical performance and economic feasibility, dozens of existing national and international organizations made ODS replacements a high priority, and at least six industry associations were started with the express goals of speeding the elimination of ODSs: the Alternative Fluorocarbons Environmental Acceptability Study (AFEAS), the Halon Alternatives Research Corporation (HARC), Halon Users National Consortium (HUNC), the Industry Cooperative for Ozone Layer Protection (ICOLP), the Japan Industrial Conference for Ozone Layer Protection (JICOP) and the Programme for Alternative Fluorocarbon Toxicity Testing (PAFT).

Military organizations in Australia, Canada, Sweden, the UK and the US agreed to replace prescriptive standards with performance standards and to publicly disclose test results and technology choice.

The American Refrigeration Institute (ARI), the American Society of Heating, Refrigerating and Air-Conditioning Engineers (ASHRAE), the International Institute of Refrigeration (IIR) and Oak Ridge National Laboratories agreed on a full spectrum of refrigerant property and performance testing and made the information publicly available.

The US Motor Vehicle Manufacturers Association (MVMA), the Society of Automotive Engineers (SAE) and the Mobile Air Conditioning Society (MACS) guided and coordinated testing of materials' compatibility and lubricity testing for refrigerant candidates to replace CFC-12 in vehicle air-conditioning.

The new organizations side-stepped anti-trust barriers to cooperation by taking advantage of laws like the US Cooperative Research Act, which encourages cooperation among independent entities, even competitors, when necessary to maximize the well-being of consumers. The US EPA, frequently a partner with industry, pushed its authority and rules to publish case studies of ODS phaseouts, describing in detail the technology selection criteria, the technology screening results and the choices made by enterprises respected worldwide.[23]

The Industry Cooperative for Ozone Layer Protection (ICOLP) was a partnership of the US EPA and electronic and aerospace enterprises from Canada, Japan, the UK and the US, organized to allow the pooling of cooperative resources for the purpose of researching and developing alternative technologies to the use of CFCs and other ODSs.

The Halon Alternatives Research Corporation (HARC) pooled global resources from military, petroleum, chemical and other stakeholders to develop alternatives to halons used in fire protection.

The US Department of Defense (DoD)/National Aeronautics and Space Administration (NASA)/Thiokol Rocket Tiger Team negotiated cooperation among global rocket designers and manufacturers and published a how-to handbook on avoiding ODSs in the manufacture of solid rocket motors.

Technologies developed for the most demanding applications more than satisfied other uses. For example, alternatives to military, aerospace and automotive safety applications, where circuit failure is not tolerated, can be easily applied to entertainment, appliance and other high-volume products.

4) Validate environmental acceptability and secure government approval
Environmental authorities validated the environmental acceptability of alternatives and substitutes and published their results.

In 1900, when the first ODSs were commercialized, there were very few laws for worker safety, public health and environmental protection. Highly toxic chemicals were commercialized without much testing, and there was little appreciation of long-term health or environmental consequences such as stratospheric ozone depletion.

When the Montreal Protocol was signed in 1987, scientists, diplomats, environmental NGOs and the public urged fast action to phase out ODSs but were reluctant to compromise health and safety. Very low-toxicity ODSs were used in emissive applications with high human exposure; environmental and safety authorities were unsure whether more toxic chemicals or more flammable substitutes could be safely used as alternatives. CFC and halon systems were frequently discharged into occupied areas during use and service. Workers were exposed to particularly high concentrations of propellants, solvents and foam-blowing agents.

A remarkable sense of global community and purpose, and the desperate pace of regulation, caused governments and industry to lower their guard and truly work together to identify and promote technology. Under the influence of the Montreal Protocol, environmental NGOs, government negotiators and regulators, UNEP and its assessment panels, chemical manufacturers and their customers, engineers, and academics consolidated and perfected criteria.[24]

Environmental authorities, conspicuously the US EPA, required every new alternative and substitute to ODSs to be explicitly judged environmentally acceptable (though not necessarily technically suitable). The EPA called its review of replacements for ODSs the Significant New Alternatives Policy (SNAP) programme. Because the US represented a significant portion of the global market for replacements for ODSs, most enterprises worldwide submitted their products for EPA review. Because the SNAP programme was highly financed, environmentally sophisticated and credible, enterprises earning the SNAP listing usually satisfied other environmental authorities or could secure rapid approval by submitting the same proof.[25] The EPA publishes lists of acceptable options, often with use qualifications, and unacceptable options prohibited from use.

Chemical enterprises agreed to jointly finance all the studies of toxicity and atmospheric fate necessary to satisfy the demands of the most stringent environmental authorities, and to publicly disclose the results.

5) Commercialize support infrastructure for the replacement technology
Commercialization of any new technology requires confidence in its technical and economic performance relative to the ODS technology it will replace and other technology that will compete for new market share.

Some markets for ODS substitutes and alternatives were quickly determined and rapidly transformed. For example, hydrocarbons were already used as aerosol propellants in many countries and were the clear economic and environmental choice for aerosol products and foam packaging. Flexible-cushioning foam production was quickly converted to already available substitutes, including water and methylene chloride. These changes rapidly decreased demand for CFCs in Europe, where CFC aerosol products were prevalent, by over 50 per cent from 1990 to 1992, with smaller reductions in demand in Japan, North America and other regions.

Partnerships of industry and government accelerated market transformation by literally and confidently 'picking winners' and announcing the schedule for market transformation, and promoted the new technology by removing barriers and mandating deadlines.

In 1988 the Foodservice and Packaging Institute (FPI), the EPA and NGOs agreed to replace CFC-12 foam packaging blowing agents with either HCFC-22 or hydrocarbons within one year and to replace HCFC-22 as soon as technically feasible, even though HCFC was not controlled at the time by the Montreal Protocol.

In 1989 the global mobile air-conditioning community, working with the US EPA, agreed to replace CFC-12 with HFC-134a worldwide. This assured a market large enough for chemical suppliers to begin construction of full-scale HFC-134a chemical production facilities even before regulatory approval was final. The agreement also gave the mobile air-conditioning parts suppliers confidence to invest hundreds of millions of dollars in retooling for future designs. Nissan and Volvo announced plans to begin transition to HFC-134a immediately.[26] The automobile industry choice of HFC-134a gave confidence to other refrigerant customers that HFC-134a would be available from more than one enterprise at competitive prices, speeding their choice and implementation of ozone-friendly technology.

6) Remove barriers to the new technology
One serious complication to the ODS phaseout was that both government regulations and private standards had become barriers to change. Enterprises were uncertain about the criteria governments and customers would apply for acceptable alternatives, and in many countries it was unclear how rapidly governments could act. There were also concerns that new chemicals would need to satisfy multiple authorities, including governments, insurers and technical standards organizations. Fire safety rules for weapons systems, aircraft, ships,

racing cars, and other uses with flammable or explosive risks often mandated the use of halon fire protection technologies. Military soldering standards requiring cleaning with CFC-113 had become the global badge of honour for high reliability products and had been embedded in most national manufacturing standards for electronics products. Bellcore (the global telecommunications manufacturing standard) also required cleaning with CFC-113. Quarantine standards also encourage or require the use of methyl bromide for certain pests and products.

In these cases, enterprises and governments seeking change organized strategic projects: first, to authorize the use of ODS replacements (most frequently changing from prescriptive standards compelling ODSs to performance standards satisfying technical criteria); second, to endorse the use of alternatives and substitutes satisfying the performance criteria; and finally, to prohibit the use of ODSs, regardless of technical performance. For example, the US Department of Defense (DoD) formed a committee that published soldering performance criteria, prescribed a test method and supervised the test using an industry–government test validation team to certify the cleaning results.[27] Nortel and AT&T used their influence to harmonize the Bellcore soldering standard with the DoD performance standard.

Today, there are no significant standards organizations requiring ODS use, with the exception of methyl bromide for quarantine.

7) Communicate and compete against the ODSs allowed by the Montreal Protocol

Enterprises offering the best of the replacements for ODSs had significant help from the UNEP organizations that support the Montreal Protocol and from the government and industry organizations in their efforts to communicate and compete for markets where ODSs had dominated.

The Technology and Economic Assessment Panel (TEAP) attracted some of the most capable experts in the world, including experts from enterprises and military organizations with the most ambitious ozone protection goals. These experts went well beyond any previous or subsequent United Nations technical committee in:

- cataloguing alternatives and substitutes;
- reporting technical performance and environmental acceptability;
- disclosing which technology was being selected by the most sophisticated and demanding enterprises and military organizations;
- reporting on the evolving satisfaction and dissatisfaction with technologies selected; and
- self-effecting solutions to regulatory and standards barriers that would otherwise slow protection of the ozone layer.[28]

The US EPA often published reports written with its industry partners presenting selection criteria, results of options screening according to those criteria, examples of corporate choice among the screened options, and case studies of implementation and satisfaction with the selected technology. These reports often listed the names and contact information for the enterprises featured in

BOX 4.4 JAPANESE SUPPORT FOR PHASEOUT OF ODSs IN SMALL AND MEDIUM-SIZED ENTERPRISES

Akira Okawa, Japan Industrial Conference for Ozone Layer and Climate Protection (JICOP)[29]

One of the most difficult problems in the phaseout of ODSs in Japan concerned the phaseout of ODSs used for cleaning in smaller enterprises, in particular the conversion of 1,1,1-trichloroethane, a universal industrial solvent.

The Montreal Protocol of 1987 was adjusted in 1992 to phase out 1,1,1-trichloroethane by the end of 1995, nine years earlier than the original target of the end of 2004 set in 1990. The biggest consumer of this ODS was cleaning. Peak annual consumption of ODSs in Japan was about 180,000 tons for 1,1,1-trichloroethane and about 80,000 tons for CFC-113. The industrial associations in each sector worried about their phaseout. Moreover, throughout the country there were many unorganized smaller enterprises which did not belong to the industrial associations; it was difficult to inform such enterprises, and Parliament discussed whether, if the control actions were forced according to the Montreal Protocol schedule, the smaller enterprises might have to stop their production process and be panic-stricken by the sudden difficult availability of ODSs.

The enterprises providing substitute technologies in the field of cleaning are classified as manufacturers of substitute solvents, manufacturers of new cleaning equipment using such solvents, and manufacturers providing accessory equipment (in particular environmentally friendly equipment) necessary for using such solvents and such cleaning equipment. At that time, the enterprises belonged to different industrial associations and had no common stage for exchange.

Some enterprises providing substitute cleaning technologies for phaseout of ODSs became anxious about the difficulties in the cleaning sector and started to consider the necessity of cooperation between different sectors during the summer of 1992. The best approach was seen to be for the manufacturers concerned to unite in the common interests to diffuse the substitute cleaning technologies for ozone protection. In April 1994, after a half-year preparation period, the Japan Industrial Conference on Cleaning (JICC) was established.

The JICC was supported indirectly by the JICOP, an organization controlling the ozone protecting actions in the whole industry. The JICC also systematically supported the technological conversion at the cleaning sites of smaller enterprises after the phaseout of CFC-113 and methyl chloroform (1,1,1-trichloroethane) by participating actively in actions for smaller enterprises at the state and local government levels. As a result, the phaseout of ODSs at the sites of smaller enterprises could be realized without as much trouble as the control schedule had been expected to cause.

The JICC carried out the following activities to assist the smaller enterprises in the phaseout of ODSs at the cleaning sites:

- it sent about 70 experts of the member enterprises directly to the cleaning sites of the smaller enterprises to instruct them about the application of substitute technologies according to an 'instruction system at the cleaning sites of smaller enterprises';
- it cooperated with the 'cleaning evaluation and examination system' to give a guide for selection of substitute technologies;
- it screened projects for tax credit eligibility to promote actions for the phaseout of ODSs; and

---

- it sent experts organized by the Smaller Enterprises Information Center (an affiliated association of the local government in each prefecture) to present 'non-CFC/ethane seminars' and to visit the sites of the smaller enterprises.

Providing support to other industrial associations is another of the JICC's activities. The member enterprises sent 53 experts to be members of the '1,1,1-trichloroethane cleaning technical committee' organized by JICOP from 1990 to 1994 before establishing the JICC. The JICC cooperated with other associations in investigating the technical requirements for many cleaning processes and the suitability of commercially available ozone-safe alternatives.

The JICC presented the plans and results of these activities each year at the International Conference on Ozone Protection Technologies held in the US from 1994 to 1997. The JICC attracted the world's attention as an industrial association independently established by manufacturers of different sectors to provide technologies which could solve a global environmental problem at a wide range of product-manufacturing facilities that had previously used ozone-depleting solvents.

---

the EPA report so that other enterprises facing the same challenge could ask whether the technology had provided lasting satisfaction.

Accomplishing the seven first steps of technology commercialization proved to be of high value to technology transfer.

Open access to detailed information allowed enterprises in developing countries to be fully informed when they selected their own technology. Better still, enterprises worldwide had detailed information on what the most capable enterprises had selected and whether that first choice had evolved into even better choices. Furthermore, choices by multinational enterprises implemented globally often created the local supplier infrastructure that allowed local enterprises to work with experts familiar with the new technology.

Command-and-control and economic incentives

There were other actions in industrialized countries that accelerated technology cooperation, reduced the cost of global compliance with the Montreal Protocol, and ultimately reduced the portion of the cost of global compliance paid by developed countries as contributions to the Multilateral Fund. Many of these strategies were designed by governments to encourage domestic compliance with the control measures of the Protocol. But they were also strong incentives for developed and developing trading partners to reduce their own ODS use.[30] The strategies that had global influence are described in the following sections.

Bans on production or import of ODS products
In 1978 the US became the first country to ban the manufacture and import of most cosmetic and convenience CFC aerosol products. Several other countries enacted similar bans by the end of 1990 (Sweden in 1979, Canada in 1980, Norway in 1981, Taiwan in 1983, Australia in 1989, Austria in 1989, Brazil in 1989, Indonesia in 1990 and Thailand in 1990).

In Thailand, Japanese and Thai manufacturers and the US EPA negotiated the first national phaseout of the production and import of CFCs in refrigera-

Box 4.5 European Union leadership on HCFCs

Tom Batchelor, Director, TouchDown Consulting, formerly European Commission specialist on ozone layer protection

A community of European nations existed through a series of predecessor relationships, dating back to 1951. The predecessor relations were officially consolidated and strengthened in 1957 with an agreement signed by six countries. The community became known as the 'European Union' (EU) in 1992, following the Maastricht Treaty. Since that time, the EU has grown into a supranational and intergovernmental union of 27 democratic member states in Europe. It is the world's biggest trader, accounting for 20 per cent of global imports and exports, with an estimated GDP of US$13.4 trillion. The EU has the world's third-largest population (after Asia and Africa), some 493 million citizens.

Prior to 2000, when developing the current ozone protection Regulation (EC) No 2037/2000, the European Commission (the executive arm of the EU) conducted surveys and held meetings with stakeholders. Small and medium-sized enterprises (SMEs) were given special consideration as they could have faced higher costs than larger enterprises and therefore might be more financially vulnerable to a change in legislation.

These consultations identified cost-effective alternatives for almost all uses of HCFCs, yet they were not being commercialized. Some stakeholders held the view that the HCFC phaseout would have adverse economic consequences for them. For example, some claimed that hydrocarbons (HCs) were suitable only for domestic refrigerators, ammonia only for industrial uses. Many alternatives required additional costs in their manufacture and use to take account of servicing difficulties and safety. Refrigeration associations were reluctant to increase their use of hydrofluorocarbons (HFCs), fearing that further legislative measures might curtail their use.

The 'driver' for the EU at that time was to put in place procedures that contributed towards protection of the environment for the benefit of current and future generations. The challenge was to combine that concept with continuing economic growth in a way that would be sustainable over the long term. The EU environment policy then and today is based on the belief that stringent environmental standards stimulate innovation and business opportunities, and that economic, social and environment policies are closely integrated.

From 1994 until 2000, Regulation (EC) No 3093/94 permitted the use of HCFCs as solvents and for foam production, but banned HCFCs for servicing several types of refrigeration and air-conditioning systems. However, after October 2000 Regulation (EC) No 2037/2000 prohibited all uses of HCFCs and prohibited imports of equipment containing HCFCs manufactured after the date of the 'use ban', as set out below.

Banned from 1 January 1996

- as refrigerants in non-confined direct-evaporation systems, domestic refrigerators and freezers, and vehicle air-conditioning systems (except for military uses).

Banned from 1 January 1998

- as refrigerants in rail transport air-conditioning; and
- as a carrier gas for sterilization.

Banned from 1 January 2000

- as refrigerants in public-transport air-conditioning, public and distribution cold stores and warehouses, and all equipment with shaft input rated at 150kW and over.

Banned from 1 October 2000

- in aerosol products;
- as solvents in unrefrigerated open-top degreasers, in adhesives and mould-release agents, and for drain cleaning;
- in foam safety applications and for rigid insulation; and
- in all uses in other sectors, unless specifically allowed.

Banned from 1 January 2001

- in all refrigeration and air-conditioning equipment except stationary air-conditioning with a cooling capacity of less than 100kW.

Banned from 1 January 2002

- for the production of extruded polystyrene rigid insulating foams, except in insulated transport.

Banned from 1 July 2002

- in all refrigeration and air-conditioning equipment except reversible air-conditioning/ heat pump systems.

Banned from 1 January 2003

- for the production of polyurethane foams for appliances, of polyurethane flexible-faced laminate foams and of polyurethane sandwich panels, except where these last two are used for insulated transport.

Banned from 1 January 2004

- in reversible air-conditioning/heat pump systems; and
- in all foams.

Thus the EC set a schedule for the reduction and phaseout of HCFCs that was faster than required by the Montreal Protocol, and at the same time imposed its own trade measures. The accelerated schedule provided an incentive for companies to be more innovative in developing alternatives and to transfer the HCFC-free technology. The import prohibition had an extraordinary influence on discouraging the use of HCFCs by enterprises in developing countries wanting to retain or expand exports to the EU that relied on HCFC technology.

The EC crafted HCFC restrictions to:

- halt their use within the EC and to avoid future dependence on them for servicing equipment;
- prohibit imports from both developed and developing countries; and
- allow the continued export of HCFC products from the EU.

These restrictions effectively protected EU enterprises from any cost disadvantage they otherwise might have faced from imported products or in export markets. The prohibition on imports acted as a powerful incentive to enterprises wanting to market products in the EU. Any enterprise manufacturing non-HCFC products would have an incentive to market those products worldwide to avoid the complications of inventory, parts supply and service of equipment using another refrigerant, but would have the choice of exporting HCFC equipment if it had a sales advantage in any market. Furthermore, enterprises serving only domestic markets in Eastern European countries were required to avoid HCFCs so that their country could satisfy the environmental compliance requirements of potential EC membership.

The extraordinary influence of the EC HCFC ban in countries with economies in transition (CEITs) and developing countries is described in the chapters on refrigeration and foam.

Within the EU, the use of HCFCs increased from 1989 to 1998 as they were used to replace CFCs; this was followed by step-wise reductions in their use in response to the 'use caps' that rolled out over a four-year period from the date that Regulation (EC) No 2037/2000 came into force (see Figure 4.1). Notice that the actual consumption was dramatically below the Montreal Protocol control schedule and also below the limits set by EC legislation. The reduction 'beyond compliance' is a result of the use bans, the market transformation by customers seeking to avoid the purchase of equipment that could not be serviced in the future, and corporate and customer leadership seeking to avoid any use that would jeopardize human heath and the sustainability of ecosystems.

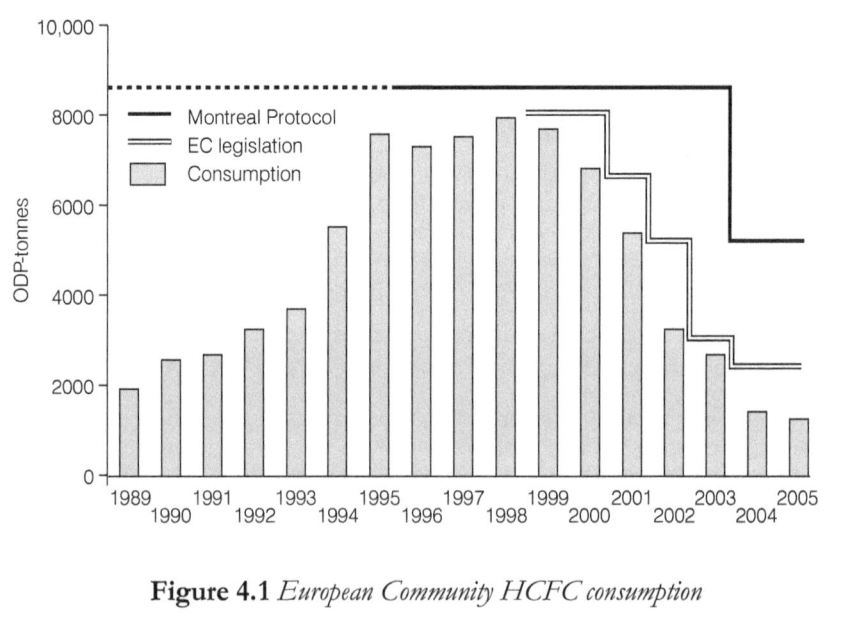

Figure 4.1 *European Community HCFC consumption*

tors. Thailand became the first developing country to enact an environmental trade restriction.

Taxes and fees
Many countries[31] have taxes or fees on ODSs that are intended to discourage use and raise revenue. In some cases, the revenue is spent on programmes to

Box 4.6 Governmental support for ODS phaseout in Japan

Naohiro Yamamura, Japan Ministry of Economy, Trade and Industry[32]

In Japan, the Ozone Layer Protection Act has been effective since May 1988 and the Fluorinated Gases Recovery and Destruction Act since June 2001. As for the emission reductions of greenhouse gases, an industry voluntary action programme was introduced in 2002. The 'Responsible Use Principle for HFCs' was introduced in 2002 in cooperation with the US EPA, the Ministry of Economy, Trade and Industry (METI), and UNEP.

In addition, METI provides other incentives to promote the introduction of equipment using no ozone-depleting substances. These include tax incentives for equipment replacement and/or installation, with special depreciation allowances (16 per cent in the first year), reduced property tax, and financial incentives (lower interest-rate loans).

METI has been supporting the activities of JICOP and other industrial associations to protect the ozone layer and encouraging industries to contribute to protecting the global environment.

encourage ozone layer protection, while in other cases the revenue is not targeted to environmental activities.

The US has one of the most substantial and therefore influential taxes on ODSs, imported products containing ODSs and 'floor stocks' of ODSs held for future use. When the US tax was first imposed, customers reacted strongly to dramatic increases in ODS prices. Price signals, such as increased cost of ODS due to tax on ODS, encouraged early transition and improved the cost-effectiveness of the transition.

The Republic of (South) Korea charged an ODS tax and dedicated it to research, education and technical assistance to enterprises making the transition away from ODSs.

Singapore enforced a reduction of ODS supplies through a bidding process designed to capture the 'monopoly rent' that chemical suppliers would have charged customers as ODS supplies became more scarce under the national phaseout. The bid price was set at a level that kept the total ODS use within the Protocol limits. Revenue from the permits financed research and technology assistance to enterprises seeking alternatives and substitutes.

Voluntary action with government tax incentives

Most governments, including the US, require that any taxes collected be paid to the national treasury and that any spending be authorized by the legislature. These rules made it challenging to have a comprehensive programme to tax ODSs, make alternatives relatively more economic, and encourage containment and recycling. However, Japan has long used revolving tax-or-social-change programmes to transform markets, protect the environment, and encourage and protect employment.

Industry leadership pledges for developing countries

Enterprises working on the phaseout under the Montreal Protocol were incredibly agile and clever in taking advantage of 'environmental targets-of-opportunity'. Environmental targets-of-opportunity are situations that are discovered by accident and are out of character with normal business practices. Typically, these are situations that have the potential to be a scandal if discovered or at least an embarrassment to enterprises, governments or both. The strategy is to find a way to resolve the situation quickly, honourably and affordably, while never assigning any blame. Prior to the discovery that ODSs were destroying the ozone layer, it was desirable to promote these chemicals over more toxic or flammable alternatives. Some forward-looking enterprises had taken action before the Montreal Protocol was signed, but for many, it was the global agreement that made it clear that ODSs were unacceptable.

Several technology transfer pledges stand out:

- the 1988 pledge by the automotive community to recycle CFC-12;
- the 1988 voluntary phaseout of CFC foam in food packaging;
- the 1990 pledge by the automotive community to phase out CFC-12 by 1995;
- the 1990 pledge by Japanese enterprises to phase out ODS use at their facilities in developing countries within one year of the phaseout at facilities in Japan;

BOX 4.7 THE FIRST VOLUNTARY NATIONAL SECTOR PHASEOUT

Alan S. Miller, International Finance Corporation

In 1986–1987 children and grass roots organizations in the US campaigned for McDonald's to stop using CFC packaging for its fast-food products. Friends of the Earth called it 'styro wars' and activists called McDonald's the 'arch enemy'. McDonald's hired an opinion polling organization to interview children to determine their level of concern. The findings were that children would rather ask their parents to take them somewhere else than destroy the ozone layer. As a result, McDonald's pledged to phase out its own CFC use and to require its suppliers to eliminate CFC use in foam packaging within 18 months. This corporate leadership shocked the food-packaging industry and its customers into action.

In 1988 I was the Director of the Center for Global Change (CGC) at the University of Maryland. CGC, the US EPA, Friends of the Earth (FOE USA), the Natural Resources Defense Council (NRDC) and the Environmental Defense Fund (EDF) formed a partnership with the food-packaging industry to phase out CFCs voluntarily in the manufacture of foam packaging. The foam manufacturers agreed to end the use of CFCs within one year (by December 1988) and agreed to use HCFC-22 only as an interim alternative. The enterprises committed themselves to developing a safer and more environmentally sustainable alternative to HCFCs, and a working group was formed to address this issue. The environmental NGOs publicly praised the industry's efforts to protect the ozone layer and urged other industries to follow suit. This agreement led to similar voluntary phaseout agreements in other countries, including Canada, The Netherlands and the UK, and promoted the development of environmentally sound foam-blowing agents.

Box 4.8 Technology transfer for household refrigerators in Thailand

Yuichi Fujimoto, Consultant, and Former President, JICOP [33]

Japan is very proud of its leadership and cooperation in transforming markets to phase-out ozone-depleting substances (ODSs). Japan accelerated technical development, encouraged the selection of environmentally superior technical options, and reduced the costs in both developed and developing countries. This remarkable history is illustrated by the 1996 phaseout of ODSs for household refrigerators in Thailand. This phaseout was only one year later than in the developed countries, despite the fact that developing countries had until 2010 to achieve the phaseout according to the Montreal Protocol. Here is the story.

In September 1990 The Singapore–US Seminar for Ozone Layer Protection was held in Singapore. This seminar was organized by the Government of Singapore, the Association of Southeast Asian Nations (ASEAN) and the US EPA. The seminar brought together about 500 people from ASEAN countries to find ways to reduce ODSs in the developing countries and discuss technology cooperation. Dr Stephen O. Andersen (US EPA) invited me and two experts from Hitachi and Matsushita to represent Japan on behalf of the Japan Electrical Manufacturers Association (JEMA) where I was the environmental executive. This was when I met Dr Andersen for first time, and since then we have worked together for ozone and climate protection.

At the general session of the meeting, there were presentations by the representatives of Southeast Asian countries. The Singapore Government said that the consumption of ODSs was decreasing with the national tendering system that auctioned the rights to use CFCs, and used the revenue to support implementation of alternatives. Malaysia was also expecting to decrease its ODS consumption in 1991. Only Thailand reported increasing consumption of ODSs, with 50 per cent of national use by Japanese enterprises and 25 per cent by US corporations, mainly for products exported to markets in developed countries. Hence responsibility for the ODSs increase in Thailand was attributed mainly to Japan, and conference participants seemed to imply that, as the representative of Japanese industry, I was at fault. I was asked about the policy of Japanese corporations regarding the phaseout of their factories in developing countries. I answered that Japanese enterprises would conduct the phaseout in Thailand after their changeover in Japan, even though I had no official confirmation of this. I reported this to our government after coming home. I also reported back to the member enterprises of JEMA, asking them to improve their image on this issue. I wondered if I had been too brave, but I had complete confidence that Japan had an obligation to future generations to protect the global environment.

In 1991, at the meeting of the Economic Options Committee of the Technology and Economic Assessment Panel (TEAP) held in Bangkok, a representative of the Bangkok office of the United Nations Environment Programme (UNEP) criticized Japan for the delay and poor management by Japanese corporations in the efforts for ozone layer protection. The local Japanese enterprises at that time had poor understanding of ozone layer protection and responded poorly to UNEP's accusations. The audience adopted a very strong position in saying that enterprises from the developed countries should always take care of environmental issues, but they were often apt to consider only economic merits.

Under these difficult circumstances, a Japan–US–Thailand ozone layer protection meeting was held in Bangkok in March 1992. There were 10 experts from the US, including Dr Andersen, and more than 40 from Japan, including Mr Yoshihiko Sumi, chief of the Ozone Layer Protection Office of the Ministry of Economy, Trade and Industry, and Mr Yasunobu Kishimoto, Chairman of the Japan Industrial Conference for Ozone Layer Protection (JICOP) and the chairman of one of the biggest chemical companies, Showa Denko. The meeting decided that Japanese, American and Canadian enterprises operating in Thailand would cease to use ODSs in 1995–1997. ODS use in refrigerators would be stopped in 1996, just one year after being stopped in Japan and in the same year required by the Montreal Protocol for developed countries. This was a very proud day for Japan and for me. It was the first time that private enterprises decided to voluntarily transform manufacturing in a developing country. But we knew that it would not be easy.

Six Japanese enterprises (Hitachi, Matsushita, Mitsubishi, Sanyo, Sharp and Toshiba) were manufacturing household refrigerators in Thailand. Four enterprises (Hitachi, Matsushita, Mitsubishi and Toshiba) were using compressors from a local supplier, Kulthorn Kirby, according to the Thai Government's industrial policy, which required local product content and joint venture. Kulthorn Kirby had developed compressors to use the new ozone-safe refrigerant HFC-134a, but these new compressors miserably failed durability tests conducted by the Japanese refrigerator enterprises.

At an important project meeting in Bangkok, Japanese engineers reported that the moving surfaces of the compressor were contaminated with scrap, that the pipes in closed refrigerant circuits would clog and that the necessary reliability of the refrigerators would not be assured. Japan proposed to improve the compressors in order to secure the total reliability of the refrigerators. Kulthorn Kirby, which was manufacturing with technology under licence from a US enterprise, was doubtful whether it could manufacture compressors with the high reliability required by the Japanese in the set time. They insisted that the same type of compressors were exported to Australia and successfully used and that the reliability demanded by Japan was too severe. Japan investigated whether it could use compressors from countries other than Thailand, but this solution was found to be difficult because it violated the agreement to use local compressors and because it would be difficult and expensive to supply the huge quantity required at short notice. At that point, the changeover in 1996 seemed almost impossible.

At the follow-up meeting of the Japan–US–Thailand partnership, Japan proposed to delay the voluntary phaseout date, but the Thai Government Department of Industry and Works (DIW) persisted in keeping the schedule because this had been decided by the cabinet and made public as the UNEP programme. Japan insisted that the time delay was caused by the Thai corporation Kulthorn Kirby and pledged that it would never compromise the reliability of its refrigerators. The Thai Government did not accept the delay. It said that the phaseout schedule was Japan's responsibility and told Kulthorn Kirby it would authorize imports of foreign compressors if Kulthorn Kirby could not satisfy the Japanese customers. A positive solution was found when JEMA and the four enterprises decided to help and support Kulthorn Kirby technology in order to keep the schedule on time. They had several meetings with Kulthorn Kirby and provided millions of dollars' worth of know-how free of charge to secure reliability. Their efforts made the changeover successful in 1996 for the four enterprises; the two other enterprises (Sanyo and Sharp) also completed the task successfully.

The expenses for this path-breaking technology transfer were borne by the Japanese enterprises voluntarily, and even the bilateral funds available from the Japanese Government were not used. It was very much appreciated when, at the request of the local manufacturers, Thailand became the first developing country in the world to enact an environmental trade barrier, which required that refrigerators imported to Thailand or manufactured in Thailand be CFC-free. The ban on the import of the lower-priced ODS refrigerators saved money for customers in Thailand and its export markets because the HFC refrigerators were more reliable and had high energy efficiency. It proved to be a very wise move for the Thai Government to protect the ODS-free products.

After the Thailand success, Japan–US leadership meetings were held in Malaysia in 1993, in Indonesia in 1994 and 1996, in Vietnam in 1994 and 1995, and in the Philippines in 1996. Each of these efforts contributed successfully to ozone layer protection, and the Vietnam meetings produced a pledge by multinational enterprises to not increase dependence on ODSs and to choose replacements wisely.

It is a great source of satisfaction to us that Japanese industry contributes so much to the resolution of global environmental issues, especially since stopping ODS use in Thailand was a very ambitious proposal. The Japan Electrical Manufacturers Association (JEMA), representing six Japanese enterprises, was given Ozone Protection Awards by UNEP and the US EPA in 1997 and by the Japan Ministry of Economy, Trade and Industry in 1999.

• the 1992 pledge by Coca-Cola to phase out CFC-12 refrigerants worldwide;
• the 1995 pledge by more than 40 multinational enterprises from seven countries to help the Government of Vietnam protect the ozone layer by investing only in modern, environmentally acceptable technology in their Vietnam projects; and
• the 2000 pledge by Coca-Cola to cease purchase of cold-drink equipment with HFC refrigerants or insulation materials where cost-efficient alternatives were commercially available.

Product labelling

Some countries require by law that products containing ODSs be specially labelled; some also require products made using ODSs to be labelled.[34] Product environmental labelling educates consumers about the extent of ODSs in products and the environmental consequences of ozone depletion, empowers consumers to avoid and/or boycott ODS products, and encourages product manufacturers to halt ODS use to satisfy customers and avoid administrative expenses and penalties.

The US label mandated by the Clean Air Act says 'WARNING! Contains (Manufactured with)…, a substance which harms public health and environment by destroying ozone in the upper atmosphere'. Marketing experts in the US predicted that consumers would reject labelled products, particularly electronics products, where brand competition is fierce, and toys, where environment and safety are a priority. Realizing that it would be administratively impossible to determine all products made using but not containing ODSs, however, the EPA suspended its plan in exchange for an agreement by members of the Industry Cooperative for Ozone Layer Protection (ICOLP) to enforce the intent of the labelling law. The dozen multinational enterprises in the Cooperative from Canada, Japan, the US and the UK wrote a joint letter to their thousands of suppliers worldwide informing them that henceforth they would buy only from suppliers that certified ODS-free facilities. Few enterprises in developing countries could afford to lose such important customers and chose instead to completely transform their factories. ICOLP and its enterprises provided technical assistance to their developing country suppliers, including advice on technology choice, price discounts on ODS-free technology, on-the-job training, and assistance in implementing the technology to satisfy product performance requirements. For example, Nortel financed an ODS solvent phaseout expert to assist Mexican enterprises; Motorola managers coordinated Cooperative experts through the Motorola manufacturing centre in Kuala Lumpur; Nortel and Seiko Epson offered factory tours and training in China; and the Cooperative published handbooks with the US EPA. In effect, enterprises devoted time and money to the phaseout, rather than devoting time and money to paperwork to comply with the labelling regulations. Thus the US national labelling law accelerated the ODS phaseout worldwide, particularly the ozone-depleting solvent phaseout.

Public domain technology

Intellectual property was not much of a problem under the Montreal Protocol, although there were some exceptions to this, as will be elaborated in Chapter 12. First and foremost, this was because the best technology generally had more than one supplier, but also because many technologies were cooperatively developed and administratively delivered to the public domain for unrestricted global use.

Chemical alternatives like HFC-134a, which replaced CFC-12 in many refrigeration and air-conditioning applications, had been disclosed as a possible refrigerant more than 50 years earlier by CFCs' inventor Thomas Midgley; many enterprises had patented unique chemical manufacturing pathways in the 1970s. Because HFC-134a is a small part of the total cost of refrigerant and air-conditioning products and because each of the patented HFC-134a manufacturing processes was comparatively efficient, no single enterprise had a monopoly advantage. CFC manufacturers rushed HFC-134a to market under the false impression that there were few not-in-kind substitutes, and with the sceptical view that customers could significantly reduce emissions through housekeeping and recovery/recycling. As a consequence of this management failure, HFCs were drastically overproduced and sold at near-cost for the first 15 years of the Protocol.

The portion of intellectual property that is traditionally protected by secrecy was offered worldwide by enterprises and associations anxious to demonstrate corporate citizenship and mindful that failure to reduce global emissions would result in even faster ODS phaseout.

Significant technology disclosure included:

- publication by the Foodservice and Packaging Institute (FPI) of a handbook for producing CFC-free foam food packaging using hydrocarbons and HCFCs;
- US Navy development of recovery and recycling equipment for halons; and
- handbooks on cleaning without ODSs published by ICOLP, the US military and other industry associations.

More than a dozen multinational enterprises developed new technology and allowed unrestricted global use. These environmental teams included:

- members of the automotive industry and their suppliers, who developed CFC-12 recycling equipment and agreed to not enforce component and application patents;
- ICOLP members AT&T, EPA, Ford, IBM, Motorola, Nortel, Texas Instruments and others, who developed and promoted no-clean soldering;
- Minebea Japan, who developed aqueous and hydrocarbon bearing cleaning and shared the technology worldwide;
- Nortel, who developed and patented no-clean manufacturing quality control equipment and donated the patent to the public domain;
- Digital Equipment Corporation for its computer-controlled aqueous cleaning innovation; and
- Seiko Epson for comprehensive disclosure of ODS-free precision cleaning, published in English, Japanese and Chinese.

Box 4.9 Industrial cooperation on phasing out ODSs in Japan

Akira Okawa, JICOP

The Japan Electrical Manufacturers' Association (JEMA) was important in domestic and international efforts to implement the Montreal Protocol. In Japan, JEMA created awareness for protecting the ozone layer, diffused technologies, trained manufacturing engineers and service technicians, and prepared the route for the Japanese total phaseout of 1,1,1-trichloroethane (TCA). Internationally, JEMA was a significant force for totally phasing out CFC refrigerant and foam-blowing agents for refrigerators and supported the developing countries in reducing their consumption and emissions of CFCs.

CFCs had been used for domestic refrigerators as refrigerant and as blowing agents for thermal insulating foam. Although their consumption was relatively small in the total consumption of the industry (below 2.4 per cent in the base year of CFCs (1986): refrigerants about 700 metric tonnes and blowing agents about 3000 metric tonnes), JEMA started quickly to reduce and phaseout CFCs within the international framework of ozone protection. The total phaseout of CFCs for refrigerators was achieved at the end of 1995.

In February 1990 JEMA organized nine refrigerator manufacturers into the Action and Study Committee to choose the best technology to eliminate CFCs. The biggest technical problem was to maintain refrigerator energy efficiency and reliability for the relatively long product lifetime (an average of about 10 years). This problem was solved by cooperation on selection of the refrigerant, compressor, lubricant and design. JEMA has taken the world lead in converting to substitutes since 1993 and realized the total phaseout of use of CFCs for refrigerators by the end of 1995.

JEMA has also organized a comprehensive national plan to recover CFCs at servicing and when refrigerators are destroyed at the end of their useful product life.

Picking winners among technology choices

One of the most effective ways to protect the global environment is for industry and government authorities to systematically pick winners among technology choices.

Today, business participants in climate negotiations often use the slogan 'Don't Pick Winners' to make the case that government should set the goal, create performance measures, and then stand back and let markets work. Part of the case they make is that government has sometimes chosen badly. This may be true, but industry also has been known to choose badly. The evidence overlooked by the slogan is that industry/government partnerships often make the best choices. A list of successfully picked winners to protect the ozone layer includes HFC-134a to replace CFC-12 in vehicle air-conditioning; hydrocarbons to replace CFC-12 in refrigerators; no-clean to replace CFC-113 in electronics manufacture; HCFC-123 to replace CFC-11 in building air-conditioning chillers; hydrocarbons and HCFC-22 to replace CFC-11 in food service packaging; and integrated pest management to replace methyl bromide.

Box 4.10 Japanese support the total ODS phaseout by CEITs and developing countries

Tsutomu Odagiri, Japan Industrial Conference on Cleaning (JICC)[35]

The Japan Industrial Conference on Cleaning (JICC) made the best use of actions in Japan for protecting the ozone layer in the field of cleaning to support developing countries.

In 1992 Japanese industry started organized activities for supporting developing countries through the Japan Industrial Conference for Ozone Layer Protection (JICOP). The JICC's members participated in all the JICOP conferences (the first being the Japan–US–Thailand Trilateral Ozone Protection Conference in Bangkok in 1992) to support developing countries. In addition, and with the support of the Japanese Government, JICOP has organized 'ozone-protecting actions seminars' for government executives of developing countries every year since 1990, with the JICC responsible for lectures on substitute cleaning technologies and organizing visits to cleaning sites.

The JICC has independently carried out the following activities:

- in Korea the JICC started instruction of substitute cleaning technologies according to the technology contract with the Korea Testing and Research Institute in 1994 and now continues the instruction of environmentally friendly technologies;
- in China the JICC sent instructors to local substitute cleaning technologies seminars, organized its own seminars and exchanged technologies with the China Cleaning Process Technology Cooperation Association; and
- in Nigeria the JICC sent experts to the ODS Substitute Cleaning Technology Workshop and sent a Japanese technological research team.

Box 4.11 Japanese leadership in training CEITs and developing country experts

Shigehiro Uemura, Japan Industrial Conference for Ozone Layer and Climate Protection (JICOP)[36]

Every January and February since 1990, Japan has hosted 15–20 mid-level government officials from 15–17 developing countries who are engaged in ozone protecting actions. So far, 326 trainees from 42 countries have participated; many remain active in implementing the Montreal Protocol as representatives of their governments. The Ministry of Foreign Affairs provides the budget, the Ozone Layer Protection Policy Office of the Ministry of Economy, Trade and Industry (METI) plans the course, the Japan International Cooperation Agency (JICA) carries out the project and the Japan Industrial Conference for Ozone Layer and Climate Protection (JICOP) has been charged with all practical procedures such as training curricula, planning, selection of instructors and negotiations for local training.

The course topics include the international development of ozone-protecting actions, ozone depletion science, the environmental impacts of ozone depletion, and the plans and problems for reduction and phaseout of ODSs for each application.

For about 30 days from mid-January to mid-February each year, participants attend lectures by experts of industry, the Administration and universities and visit manufacturing facilities where ODSs have been phased out.

The training is interactive. Many hours of presentation and discussion are provided so that trainees can master the diplomatic and technical challenges faced by their countries and plan the policies to phase out ODSs. As a result, we can see more active exchange between trainees. But the needs are different for each of the developing countries and are changing year by year. Japan now tailors the classes to find solutions to specific problems rather than providing information on general topics of ozone protection. The curriculum changes with the needs.

The role of conferences and workshops

The character and importance of conferences and workshops changed dramatically once the Montreal Protocol was signed. Businesses dependent on ODSs became more interested in technical options, and suppliers of new technology aggressively pursued the new customers. The flagship conference, 'Substitutes and Alternatives to CFCs and Halons', was held on 13–15 January 1988, organized by the US EPA and co-sponsored by Environment Canada and the Conservation Foundation, which was then headed by William K. Reilly, later appointed Administrator of the US EPA. This conference became an annual tradition, most recently expanding its focus to include climate protection in addition to ozone layer protection. It is the platform for political and technology announcements and the organizing forum for partnerships and other collaboration. It was informally integrated into the Montreal Protocol network through co-sponsorship and international agenda planning teams, with TEAP and technical options committee members often setting the agenda and presenting summaries of the most appealing new technologies.

BOX 4.12 THE MONTREAL PROTOCOL PARTNERSHIP STRATEGY IS ALIVE AND WELL IN CLIMATE PROTECTION

Sally Rand, US EPA

The US Semiconductor Partnership was started by a team of EPA experts who had no experience in ozone layer protection, but who were coached by those who had. The founding industry experts, however, included AT&T and other senior executives and technical managers who had experienced the advantages of corporate leadership on the CFC-113 phaseout.

That US partnership became international with the help of TEAP's Stephen O. Andersen, Yuichi Fujimoto and colleagues, who organized the conference where Japanese industry signed on and pledged to recruit enterprises worldwide.

The PFC Reduction/Climate Partnership for the Semiconductor Industry become a catalyst for enterprises in Europe, Asia and North America to set the first global target for reducing perfluorocarbon (PFC) greenhouse gas emissions under the World Semiconductor Council. Emissions from this sector – which would have otherwise grown 10-fold over the decade from 2000 to 2010 – are not increasing, thanks to early voluntary action.

The EPA Voluntary Aluminum Industry Partnership was created by an EPA team that was not as experienced as, or coached by, ozone champions, but was soon transferred to my management as work evolved from technical cooperation to technology implementation. Since the launch in 1994 enormous progress has been made in reducing PFC emissions and simultaneously increasing process efficiency from primary aluminium production in the US and worldwide, with members accounting for over 70 per cent of the world's primary aluminium production.

The SF_6 Emission Reduction Partnership for the Magnesium Industry was created by an EPA team that was both experienced and coached by ozone champions and was also guided by the experts from the EPA Voluntary Aluminum Industry Partnership. In 2003 partners and their International Magnesium Association established a goal to eliminate SF_6 emissions from magnesium production and casting operations by the end of 2010.

Looking back, it is obvious that the Montreal Protocol trained and paved the way for cooperation for climate protection. All these programmes are now fully international and they all have strong elements of technical cooperation. And like the industry leadership under the Montreal Protocol, enterprises are sharing proprietary technology and know-how, either without charge or on a clear and favourable basis.

Scholars and policy strategists will want to ask why industry leadership of partnerships was successfully transferred from ozone to climate, yet industry participation in climate assessments has been less successful.

All this good work had a profound impact on the selection of alternatives and substitutes and on the financing of technology transfer.

Organized technology transfer to CEIT and developing countries

Many countries, including Australia, Canada, Germany, Japan and The Netherlands, have undertaken bilateral and multilateral projects to train for, promote and implement ozone protection technology. This was financed as overseas development assistance (ODA), as ministerial projects and as an authorized activity counting toward the contribution to the Multilateral Fund.

COST OF THE ODS PHASEOUT WAS LESS THAN PREDICTED

Today, industry, government and non-government organizations all agree that the costs of eliminating ODSs are far less than the consequences of ozone depletion. Prior to 1986, however, there was no agreement on the economics of taking action. In some cases, the costs were overstated because some of the investments would have occurred without the Montreal Protocol and because the benefits of lower operating costs, less maintenance, and higher product quality and reliability were not accounted for in economic models.[37] For example, fuel savings, lower maintenance paid for new refrigeration and air-conditioning equipment, and the increased reliability and performance of electronic equipment paid returns beyond their transition costs. Aerosol manufacturers recovered their cost of investment, and customers received better value for money with environmental protection as a no-cost bonus.

CONCLUSION

More than 240 sectors with products dependent on ODSs halted most uses within ten years and often ended up more satisfied with the performance of their products than when they used ODSs. What is more, they demonstrated that technology transfer for environmental protection can be cost-effective. Consumer concern stimulated early action, regulation provided incentives for technology development and commercialization, and industry and military leadership hastened the transition to ozone-friendly technology. Corporate and military leaders energized markets by publicly accepting scientific warnings, by promoting ozone-safe alternatives, and by declaring ambitious goals to phase out both production and consumption of ODSs on schedules far faster than any Protocol or government schedule. Public–private partnerships greatly assisted with the development, identification and diffusion of environmentally superior technologies. Environmental authorities did their part by using market incentives, such as ODS taxes and labelling of products made with or containing ODSs, to level the competitive playing field and to jump-start new technology. Military organizations joined with environmental ministries and public–private partnerships to eliminate barriers to new technology – shifting from prescriptive standards compelling the use of ODSs to performance standards encouraging the use of the most environmentally acceptable processes and products.

Costs associated with ozone layer protection turned out to be far less than the exaggerated forecasts, and the savings are incalculable. Today, industry, government and non-government organizations agree: protecting the ozone layer was positively the best decision they could have made.

Technology transfer was hastened by corporate decisions to release patented technology and know-how and by proactive measures such as globalizing their national and corporate phaseout schedules and compelling their suppliers to be ODS-free. The leadership, partnerships, pledges and cooperation that featured in phasing out ODSs are unprecedented in global environmental protection and show promise for climate and other challenges.

Military and Space Agency Leadership to Protect the Ozone Layer[1]

The Montreal Protocol was an extraordinary technical and policy challenge to modern military organizations. They realized that every weapon system depended on ozone-depleting substances (ODSs) and that their continued use would jeopardize the very objectives they fought to achieve. Military leadership on the Montreal Protocol has been successful in phasing out most use of ODSs and has forever changed military thinking. In this new world, threats to the climate and ozone layer are matters of 'environmental security'; technical choices by government/industry partnerships are 'environmental technology winners'; procurement is an incentive for 'green technology'; and military standards are shifting from 'prescriptive' to 'performance' in order to consider environmental protection on a par with other criteria.

Military actions under the Montreal Protocol rewarded and encouraged military organizations throughout the world to become increasingly aware of the impact of their operations on the local, regional and global environment. Environmental management has been integrated into the operations and policies of armed services worldwide, and in many countries the armed forces have assumed a leadership role in specific areas of environmental protection. There are many reasons for this 'greening' of the armed forces: improving the health, safety and well-being of military personnel and the civilian communities among whom they live; saving costs by using energy and materials more efficiently; reducing waste-management burdens; complying with national, regional, or international regulations and policies; and improving the public image of the armed forces. But perhaps the most important factor for US and some other military organizations has been that environmental conditions affect military readiness, and hence national security.[2]

THE HISTORY OF THE USE OF ODSs BY MILITARY AND SPACE AGENCIES

All ozone-depleting substances controlled by the Montreal Protocol contain common chemical ingredients – chlorine and bromine – but they vary widely in environmental properties such as atmospheric lifetime and toxicity. The most toxic ozone-depleting substances, methyl bromide and carbon tetrachloride, were restricted on the basis of safety alone for civilian uses in developed countries long before the Montreal Protocol. However, these toxic chemicals were allowed for military fire extinguishing uses because the consequences of not extinguishing a fire during combat are far worse than the long-term impacts on human health. Other ODSs like chloroflourocarbons (CFCs) and hydrochlorofluorocarbons (HCFCs) were uninhibited by regulations since they did not appear to have adverse effects. Indeed, they were often thought of as 'wonder chemicals' because they had many desirable qualities and could be used in many applications.

As a result of their toxicity differences and disparate application requirements, the ODSs controlled under the Montreal Protocol had very different commercial histories. Three of these histories are particularly relevant to both military and civilian uses: fire protection, solvents, and air-conditioning and refrigeration.

Ozone-depleting halons in military firefighting applications

The halons that deplete stratospheric ozone have had two distinguishable applications in fire protection: as 'streaming agents' and as 'flooding agents'. Streaming agents are typically discharged from a nozzle directly into the fire, typically from hand-held or wheeled fire extinguishers. Flooding agents are typically discharged from multiple nozzles to flood an entire room with chemicals at a concentration that extinguishes fire.

Carbon tetrachloride (halon-104) was introduced as a streaming fire extinguishing agent in around 1900. By 1910 carbon tetrachloride was widely used to extinguish gasoline fires from automobiles and other equipment with internal combustion engines. The military was quick to adopt this technology in combat vehicles, where the need for effective fire extinguishing agents is critical. While evacuation from a burning aircraft, ship, submarine or land vehicle is normally the best survival strategy, evacuation in a combat situation may put occupants in even greater peril or may be technically impossible.

Unfortunately, carbon tetrachloride, though a remarkable extinguishing agent, turned out to do harm as well as good. By 1917 carbon tetrachloride was suspected to harm human health, and in 1919 the first recorded deaths attributed to carbon tetrachloride occurred. Two men were working on the construction of a submarine when one man's clothing caught fire. His friend tried to put the fire out with a hand-held carbon tetrachloride fire extinguisher, and both died from inhaling the fumes.[3] Another commonly used multipurpose (streaming and flooding) fire extinguishing agent, methyl bromide (halon-1001), was even more toxic than carbon tetrachloride. It was used in the 1920s to 1940s in Europe, notably in British and German aircraft and ships during World War II, but was never popular in other parts of the world or for use in hand-held extinguishers.[4] During World War II, Germany developed bromochloromethane (halon-1011) to replace methyl bromide. It was a more effective fire extinguishing agent, but still had a relatively high toxicity, similar to that of carbon tetrachloride.[5] Clearly, less toxic firefighting alternatives were needed.

In 1947 the Purdue University Research Foundation, financed by military organizations and others, evaluated the extinguishing performance of more than 60 new candidate agents. The US Army Corps of Engineers simultaneously conducted toxicological studies. This collaboration identified four candidates – halon-1202, halon-1211, halon-1301 and halon-2402. Halon-1301 and halon-1211 proved to be the least toxic of the four candidates and less toxic than the halons they replaced (halon-1001, halon-1011 and halon-104).

One of the halons identified, halon-1202, was an extremely effective fire extinguishant, particularly when it came to extinguishing electrical fires.[6] Unfortunately, it was also the most toxic. Nonetheless, the US Air Force selected halon-1202 for military aircraft engine protection due to its superior fire suppression performance in that application. Halon-1301 ranked second in fire extinguishing effectiveness and was the least toxic. It was best suited for total flooding fire protection systems because it could be discharged rapidly and was able to extinguish most fires at concentrations of only 5 per cent by volume, which is a chemical concentration generally considered safe for humans. Carbon dioxide, by comparison, requires concentrations in excess of 34 per cent to extinguish fires, which is more than enough to kill a human in minutes.[7] The US Federal Aviation Administration approved halon-1301 for commercial aircraft engine fire protection, and the US Army soon developed a portable halon-1301 fire extinguisher for use inside armoured vehicles. Halon-1211 and halon-2402 were best suited for portable fire extinguishers.[8]

These breakthrough fire suppression technologies were also quickly adopted by the civilian sector. Halon-1211 was widely marketed in portable fire extinguishers. When the first commercial computer rooms began to appear in the 1960s, halon-1301 was used to protect the equipment. Over the next 20 years, it was widely marketed as a total flooding extinguishant for protecting

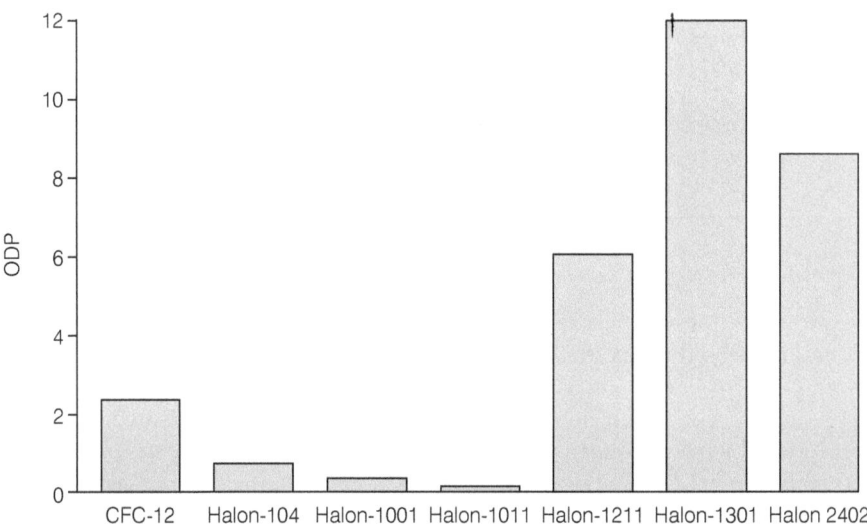

Source: World Meteorological Organization

Figure 5.1 *Comparison of the ozone depletion potential (ODP) of halons and CFC-12*

BOX 5.1 SPECIFIC APPLICATIONS FOR SOME HALONS

- Halon-1211 is used in hand-held systems and in small fixed systems protecting unoccupied spaces.
- Halon-1301 is used as a total flooding agent, in hand-held extinguishers on board some aircraft, ships and tactical vehicles.
- Halon-2402 is used in certain missile systems. Used more widely in a few countries, notably Russia and Italy.

---

computer and communications equipment, repositories of cultural heritage, shipboard machinery spaces, and petroleum pipeline pumping stations.[9] The automobile racing community also adopted halons to extinguish vehicle engine and fuel spill fires; by the early 1990s this industry alone placed in service over 1 million halon-1301 fire extinguishers for driver, crew and engine compartment protection.[10]

Growing concerns about the toxicity of carbon tetrachloride (halon-104), methyl bromide (halon-1001), bromochloromethane (halon-1011) and bromochlorodifluoromethane (halon-2402) led to their official commercial death as fire extinguishing agents in developed countries. By the end of the 1960s the newly developed halons had all but replaced the old ones.[11] Limited use of halon-2402 continued in what was then the Soviet Union and countries dependent on Soviet weapons systems and commercial aircraft and ships.

BOX 5.2 HALON USE IN AIRCRAFT, SHIPS
AND OTHER VEHICLES

- Engine nacelles and auxiliary power units.
- Cabin and cockpit.
- Cargo compartments.
- Fuel tank inerting.
- Dry bays.
- Lavatories.

- Machinery spaces.
- Engine compartments.
- Control rooms or electrical compartments.
- Crew compartments.
- Engine compartments.
- Portable extinguishers.

Ozone-depleting solvents in military applications

Ozone-depleting solvents became popular for the same technical and environmental reasons that justified ozone-depleting fire extinguishing agents. There was a need for chemicals with high performance and low toxicity, and ozone-depleting solvents often fit that description.[12] In the early 1900s two chemicals were available to dry-clean leather and delicate fabrics: gasoline and carbon tetrachloride (CCl_4). The drawbacks of gasoline are self-evident. Carbon tetrachloride, by comparison, was effective, non-flammable and low-cost, and could be used to dissolve fats, oils and greases in metal and fabric cleaning. However, it was also toxic when inhaled or absorbed through the skin, and at high temperatures it could react to form highly poisonous phosgene gas.

By the 1960s carbon tetrachloride was prohibited in most developed countries for all consumer products due to its toxic effects on livers and kidneys. Methyl chloroform, a low-toxicity ODS, grew in popularity in the 1970s and 1980s as a dry-cleaning chemical, metal degreaser and adhesive ingredient. Less expensive but more toxic chlorinated solvents were controlled by environmental authorities. During the 1970s and 1980s CFC-113 was also used to clean certain delicate fabrics, leather and garments with decorations that are adversely affected by other solvents. Methyl bromide was occasionally used as a degreasing solvent for wool.

For military applications, ozone-depleting solvents were extremely important in the manufacture and cleaning of electronics, controls, rockets and weapons systems, where any contamination could decrease reliability or durability. Until the 1970s, electronics and precision products were cleaned with alcohol and aqueous and chlorinated solvents. Once CFC-113 was commercialized in the late 1970s, however, it quickly dominated the market. While not the most effective solvent, it is a gentle solvent ideally suited to cleaning products with delicate components; it is compatible with a wide range of metals and plastics; it dries quickly, leaving no solvent residue; it has low toxicity; and it is very easy to use. The military demanded the use of CFC-113 for its products, and suppliers of high-quality civilian electronics soon followed with their own specifications prescribing cleaning with CFC-113. CFC solvents were essentially a 'badge-of-honour' for manufacturers wanting to produce the best products. By 1989 a full 50 per cent of CFC-113 used in the electronics industry resulted directly or indirectly from military specifications.[13]

Box 5.3 Military use of ozone-depleting solvents

Manufacturing

- Printed circuit boards.
- Communications systems.
- Lasers, radars, infrared tracking devices.
- Composite and honeycomb fabrication.

Maintenance

- Aircraft and missiles.
- Ships and submarines.
- Ground combat vehicles.

Air-conditioning and refrigeration

The military uses some ODSs in exactly the same way as the civilian sector. It uses them to air-condition for human comfort and refrigerate for food storage and cold drinks. In addition, however, the military has unique uses like cooling weapons during combat and dehumidifying recirculated air in submarines. CFC-114 turned out to be a particularly useful refrigerant for the US Navy. This substance was chosen for navy vessels because it provided quiet operation to escape sonar detection, reduced equipment volume, and offered compatibility with existing submarine atmosphere control equipment – all important features for warship systems.

MILITARY MOTIVATION FOR LEADERSHIP

ODSs were widely used in military applications. When ODSs were proposed for regulation, the armed forces in developed and developing countries quickly discovered that virtually every weapons and support system in their arsenal used ODSs – in refrigeration, for fire protection, as solvents or to perform some other vital function. Since many weapons and support systems relied on ODSs and could not be manufactured or repaired or function effectively without them, the use of these chemicals was directly linked to military readiness. Accordingly, armed forces had to attach a high priority to ensuring that their use of ODSs was properly managed and that the transition to alternatives under the Montreal Protocol would be smooth.

By the time ozone depletion became an issue, ODSs were thoroughly embedded in military and civilian technologies. Phasing ODSs out would be a daunting challenge. Military organizations had some of the most technically demanding uses for ODSs, including fire and explosion protection, weapons systems electronics, and rocket and aircraft manufacture. These challenges were further complicated by harsh operating conditions, difficulty of repair of field equipment and the catastrophic consequences of technical failure. Moreover, the military aircraft, ships or armoured vehicles cannot be removed for repair or for evacuation of soldiers in combat situations, requiring fire extinguishing agents, refrigerants and solvents that are non-toxic and non-flammable.

Nevertheless, military organizations were motivated to take the lead on ozone protection. There were several reasons for this. First and foremost, ministries of defence in many countries, and particularly the US Department of Defense (US DoD), were some of the largest consumers of ODSs and recognized the strong global scientific consensus that stratospheric ozone depletion was caused by emissions of ODSs. At the time, military and aerospace organizations knew as much about the science of ozone depletion as almost any other sector because military scientists had became involved in the issue during the early 1970s, when it was hypothesized that supersonic transports (SSTs) would deplete stratospheric ozone. Scientists from the US National Aeronautics and Space Administration (NASA) had investigated whether rocket exhaust would be a contributing factor. As early as 1975 the North Atlantic Treaty

---

BOX 5.4 THE CHALLENGE OF FINDING REPLACEMENTS

Bill Holder, US Air Force

A successful halon substitute for the US Air Force must be able to operate under temperature extremes from -65°F to 250°F, under airflows as high as 18,000 cubic feet per minute and 200 knots, and still be able to extinguish a fire in a fraction of a second. Furthermore, the chemical must be effective in cluttered spaces where volume can vary from a single foot to several hundred cubic feet, and should not excessively increase the weight aboard an aircraft.[14]

BOX 5.5 PERSPECTIVE ON THE NEED FOR EARLY ACTION TO PHASE OUT HALONS AND ODSs

Gilbert F. Decker and Robert L. Walker, Assistant Secretaries, US Army[15]

Halons and other ODSs are used extensively in industrial processes, in ground and air combat weapon systems, and in facilities of virtually every type. The need to plan now for the absence of these chemicals is obvious. Failure to do so will result in a catastrophic impact on Army readiness.

Organization (NATO) was briefed on ozone depletion issues by the EPA Administrator and was periodically updated.

Second, financial, mission support and logistical concerns played a very significant role in motivating military organizations to find alternatives to ODSs. If ODSs were to become unavailable in the future, military and aerospace missions – including those to space – would be at risk. Because there was a strong likelihood there would be an international agreement to control theses substances, the US DoD began assessing its uses of ODSs and looking for alternatives. It soon became apparent that military facilities and their contractors were the sources of some of the highest ODS emissions in the world. The US DoD also quickly went public, accepting its responsibility to be part of the solution: 'the DoD response was to acknowledge its contribution to the problem and commit to finding solutions'.[16]

Finally, the US DoD and other military organizations viewed ozone depletion as an environmental security concern because the decline in agricultural productivity and adverse human health affects would increase poverty and human migration and disrupt geopolitical status, and thus had the potential to increase armed conflict.

EARLY MILITARY LEADERSHIP

In the early 1970s it was hypothesized that nitric oxide (NO) and nitrogen dioxide (NO_2) from high-altitude aircraft would react in a catalytic cycle to destroy ozone, and that added water from the exhausts would both deplete ozone and increase global warming. It was estimated that military high-altitude flights alone were unimportant to the stratospheric ozone layer, but that civilian supersonic flights could damage the climate and ozone layer. That controversy was solved when the US Government abandoned financing of its supersonic transport programme, and most global airlines cancelled orders for Concorde.[17] In the late 1970s, when it was hypothesized that CFCs depleted the ozone layer, the military was not yet dependent on CFC solvents for electronic and aerospace applications, and halons were not yet implicated. That era resolved regulatory concern when the US and several other governments banned CFCs

---

BOX 5.6 DECISIVE ACTION BY THE US DoD

Gary D. Vest, Former US Principal Assistant Deputy Under Secretary of Defense

The US DoD took decisive action to protect the ozone layer even before the Montreal Protocol was ratified or entered into force [...] the DoD encouraged the commercialization of technology to replace ODSs and concentrated its own research and development on eliminating the use of ODSs from unique military applications.

---

from most aerosol products, and environmental organizations moved on to other issues.

However, when concerns resurfaced in the mid-1980s and the Vienna Convention was signed in 1985, most modern US military organizations were heavily dependent on ODSs and aware that future regulations would surely impact their operations. In 1986, mindful of this history and these concerns, Stephen O. Andersen of the US EPA and Gary Vest from the US Department of Defense agreed to cooperate on protecting the stratospheric ozone layer. Andersen pledged that the EPA would not jeopardize military operations or the safety of military personnel, and Vest pledged to aggressively pursue every technology to reduce or eliminate the use and emissions of ODSs.[18]

Military leadership included policy directives, financing, staffing and creation of new special-purpose offices such as the US Navy CFC and Halon Clearinghouse. Phaseout efforts included research and development, changing military specifications to allow the use of ODS alternatives, and developing strategies to comply with the Montreal Protocol.

This leadership first became conspicuous at the September 1987 meeting in Montreal when the Protocol was signed. US Air Force experts staffed a technical display showing how to reduce halon emissions through use of alternative testing and training methods, by increasing the reliability of systems to improve fire protection levels and reduce accidental discharge, and by switching to chemical and not-in-kind substitutes for some applications. Up until that point, halons were considered the most irreplaceable ODSs; in fact, industry experts had been reluctant to include them in the Protocol. As a result of military and industry leadership, however, halon ultimately became the first ODS to be phased out by developed countries (in January 1994).[19]

Civilian and military collaboration in the early phaseout of halon

The halon phaseout is a good example of military and industry leadership to protect the ozone layer. Since the new halons had been introduced in the 1960s, their use had become ubiquitous. They were popular because of their effectiveness in extinguishing fires and suppressing explosions without significantly endangering people from chemical exposure. They were also aggressively marketed both for applications where they had clear economic and safety advantages and for applications where there was no fire protection advantage but

---

Box 5.7 Dedication to fire protection and ozone protection

E. Thomas Morehouse, Jr., Institute for Defense Analyses

People dedicate themselves to the fire protection profession because they believe passionately in protecting lives and property. Once the threat halons posed to mankind was understood, the community worked to find solutions with that same passion they bring to fighting fires.

---

where convenience and the perception of sophistication justified the high price and high profits for halon chemical and equipment suppliers. Military organizations used significant quantities of halons to protect their personnel, infrastructure, aircraft, ships and tanks from fire and explosion. Civilian organizations used halon to protect a wide variety of equipment, including manufacturing facilities, oil and gas production facilities, ships, aircraft, and electronics equipment. Halon was strongly preferred to protect property in occupied spaces. Before it was recognized that halon was destroying the ozone layer, it was widely used and emitted during the testing of firefighting equipment and training firefighting personnel.

Military organizations had already begun efforts in the mid-1980s to identify halon alternatives, reduce halon use and find even better agents than halons. This was mostly done for economic reasons, because halon costs and usage were increasing. Various military organizations contributed to technical and educational innovations through the use of simulators, computer modelling and surrogate agents.

In 1986 the US National Fire Protection Association (NFPA) was planning to require mandatory discharge testing of new halon-1301 fire protection systems because of concerns that halon systems would be improperly designed or installed. The US EPA was concerned about those emissions, as well as the possibility that owners of existing systems, local fire authorities or insurance companies might also require such testing. The EPA contacted Gary Taylor, NFPA Chair of the Halogenated Fire Extinguishing Systems Committee. It was agreed that the NFPA would act as a gathering point and clearinghouse for information with the immediate goal of reducing halon emissions. Dozens of meetings ensued. By 1987 the Committee reversed its recommendation for full-discharge testing of halon-1301 systems and instead began engineering and demonstration programmes to reduce halon use and emissions.

The EPA also contacted the US Air Force in 1986 and explained its concerns about halon use. The Air Force, under the leadership of Gary Vest and Captain E. Thomas Morehouse, worked quickly with EPA and university experts to develop and implement changes in their use policy, including:

- restricting use to 'mission-critical essential applications';
- implementing new techniques for testing systems and prohibiting discharge testing;

- modifying training techniques to limit discharge;
- ultimately replacing non-critical uses and banking available halon for essential military uses; and 5) working with the Aerospace Industries Association and Boeing Commercial Airplane Group to organize a worldwide working group of military and civilian aviation corporations and regulatory authorities to find alternatives.[20]

Once the scientific consensus was accepted and the Montreal Protocol signed, industry and other military organizations stepped up their efforts. Important collaborators included the NFPA, insurance underwriters, industry trade associations, Underwriters Laboratories and the enterprises most dependent on halon, including AT&T, GTE, British Petroleum Exploration, NASA, Nortel and the US military services. Fire protection standards and codes were changed to discourage or prohibit the use of halons for testing and training; substitute tests such as the door fan test were developed to test fire protection systems without releasing halons; and a greater focus was placed on improving fire detection equipment in order to decrease the number of unnecessary halon discharges.

In May 1989 the Environment Protection Authority of the State of Victoria, Australia, organized the first international halon/ozone conference, called 'The Future of Halons'. Australian citizens were demanding action on the ozone layer, and labour unions announced plans to refuse to install or repair halon systems. During the conference, the fire protection industry abandoned its strong opposition to controls on halon and issued a consensus conference agreement that halon consumption and emissions needed to be reduced to protect the stratospheric ozone layer. The Victoria EPA published the detailed conference recommendations covering halon use, product labelling, firefighting training, recovery and recycling, decommissioning, storage, disposal, and phase-out dates. These recommendations were the basis of the Australian approach of defining essential halon use through a panel of experts.

The search for halon alternatives quickly became a global effort among military organizations and the private sector, including Navy researchers in Norway, the UK and the US. The Halon Alternatives Research Corporation was organized by the US Department of Defense, the US EPA, the National Institute of Standards and Testing and the fire protection industry as a publicly funded research programme to develop substitutes for halon. Its first project was to develop and screen a large list of chemicals as potential substitutes. The project was funded by a combination of the US military, the EPA and halon users, and was performed at the National Institute of Standards and Testing and the University of New Mexico's Engineering Research Institute. Once it became clear that industry was developing halon alternatives, the Halon Alternatives Research Corporation transformed itself into an international information clearinghouse, a facilitating organization focused on removing barriers to the development and approval of environmentally acceptable alternatives and a coordinator of halon bank planning and management programmes.

MILITARY TECHNICAL LEADERSHIP

After the ratification of the Montreal Protocol, military organizations through-out the world got down to the hard work of researching alternatives, eliminating requirements that compelled the use of ODSs and stewarding stocks of ODSs to meet requirements that could not be immediately phased out. For the US DoD, this 20-year effort was successful, above all, because of the central role that key offices in the Office of the Secretary of Defense and the Military Service Headquarters played in overseeing and coordinating all ODS elimination efforts. An example of this central leadership is the Ozone Depleting Substances Services Steering Committee, chaired by representatives of the US Department of the Navy, which has met regularly in Washington, DC, for more than 16 years, performing key supervisory and coordination roles and sustaining senior leadership emphasis on ODS elimination and responsible ODS management.

The US DoD in particular adopted an aggressive policy concerning technology efforts aimed at seeking alternatives to ODSs. A 13 February 1989 directive required DoD Components to conduct R&D to identify or develop alternative processes, chemicals or techniques for functions currently being met by CFCs and halons. The Director of Defense Research and Engineering was assigned to coordinate these R&D programmes among the military departments and defence agencies.

The Director of Defense Research and Engineering established the Halon Alternatives Research and Development Steering Group (Halon Steering Group) in September 1991 to formulate and oversee the execution of an integrated DoD near-term technology strategy and technology development plan to identify suitable alternatives for all ODSs. The Halon Steering Group formulated the DoD Technology Strategy for Alternatives to Ozone-depleting Substances for Weapon Systems Use, approved on 31 August 1992, to guide the near-term investigation and performance testing of commercially available chemicals: CFCs used in weapon systems applications for refrigeration and environmental control and for general and precision cleaning, as well as the halons used for firefighting. The Halon Steering Group then developed an execution plan, the DoD Technology Development Plan for Alternatives to Ozone-depleting Substances for Weapon Systems Use (Technology Development Plan), which was approved on 28 June 1993 and updated annually. Executed by the Army, Navy and Air Force, the Strategy and the Technology Development Plan together formed the overall DoD near-term response to the impact of the Montreal Protocol on weapon systems.

Research under the Technology Development Plan was completed in 1997 and resulted in several important technology alternatives to ODS which are discussed later in this chapter. However, many of the alternatives identified in the Technology Development Plan had limitations in terms of their implementation for existing weapons systems. Recognition of these limitations led to the formulation of a new research programme to develop improved options. In 1997 the Halon Steering Group initiated the Next Generation Fire Suppression

---

**BOX 5.8 US MILITARY TRANSFER OF HALON RECOVERY
AND RECYCLING TECHNOLOGY**

Bella Maranion, US EPA

In 1992 the US EPA partnered with the United States Marine Corps (USMC) and the United States Navy (USN) to plan and conduct regional halon-1211 training workshops. Halon-1211 was widely used in fire extinguishers, with significant emissions losses each year from unnecessary testing and improper equipment servicing and maintenance practices. The goal of the workshops was to reduce emissions through the use of recovery/recharge equipment and improved servicing practices. One workshop was held in Okinawa, Japan, for Asia–Pacific countries, and a second workshop was held in South Carolina for Latin and Caribbean countries. In total, 100 participants attended these workshops, which were designed to provide information on the role of halons in ozone destruction, strategies to reduce halon emissions, and training on the operation of halon-1211 recovery/recharge machines and improved fire extinguisher maintenance. These workshops initiated halon phaseout activities in 15 countries in these regions.

Our partners in the USMC and the USN brought their energy, commitment and passion to this project, which was just what we needed. Students from everywhere bonded with our Marine Corps sergeant instructor and studied hard to master the techniques. Perhaps we could make even bigger strides in global environmental protection if we could recruit for such passion.

In 1993 China asked for US EPA assistance to address its annual use of approximately 4000 metric tons of halon-1211. In cooperation with China's Ministry of Public Security, the United Nations Development Programme Global Halon Project and the USN, the EPA China Halon Partnership conducted three halon-1211 recovery/recharge training workshops in 1993 in the major cities of Beijing, Hangzhou and Guanzhou. The workshops provided local firefighting groups with the skills and advanced equipment needed to prevent unnecessary halon emissions.

The workshops focused on teaching improved maintenance procedures for fire extinguishers and provided halon recovery/recharge equipment originally developed for the USN. Over 130 fire officials and fire extinguisher maintenance technicians attended the three-day training workshops. A demonstration network of 23 halon recovery/recharge machines was provided to operate in the Beijing area and in the provinces of Zhejiang and Guangdong. It was estimated that these machines would reduce halon-1211 emissions by 41 metric tons annually. With full implementation of halon-1211 recycling in China, it was estimated that emissions could be reduced by 600 metric tons annually. As a complement to the work in China, several senior Chinese fire officials were trained in fighting challenging fires without halon at schools operated by American fire protection equipment manufacturers.

Technology Program to develop and demonstrate retrofitable, economically feasible, environmentally acceptable and user-safe alternatives to halon-1301.

Halon recovery and recycling equipment globally developed and introduced

Technology to recycle halon-1211 was initially developed by the US Air Force at its Engineering and Services Center at Tyndall Air Force Base in Florida. The

US Naval Air Warfare Center Aircraft Division in Lakehurst, New Jersey, took the concept from a laboratory test to a commercial prototype and then to the commercial market through partnership with the private sector. Members of the US Marine Corps were among the key leaders who not only worked with private industry to ensure commercialization of the equipment, but also travelled to many developing countries to demonstrate its use, often as part of a military-to-military exchange programme.

Walter Kidde Aerospace, Aer Lingus, NASA and others developed and commercialized halon-1301 recycling equipment that could separate halon from the nitrogen used for cylinder pressurization without emitting halon to the atmosphere. The US Army, in cooperation with petroleum companies and other critical industry users, fast-tracked the development of a new standard for recycled halon-1301. At the same time, they encouraged the American Society for Testing and Materials to publish a standard of purity for recycled halon-1301 and rapidly cited this standard for military operations. The Halon Alternatives Research Corporation completed a study that showed that halon banking within the US was a viable concept and detailed the blueprint for a halon banking clearinghouse scheme that was subsequently adopted by several countries, including Australia, Denmark, Italy, The Netherlands and the UK. These efforts cleared the way for global halon banking as a viable technique for managing existing halons, thereby ensuring a supply of halon for critical uses without the need for new production. The US Defense Logistics Agency established the world's most extensive, secure and sophisticated physical reserve of ODSs to supply critical military uses. Ron Sibley, Director of the Defense Logistics Agency ODS Reserve, serves on the Halons Technical Options Committee (HTOC) of the Montreal Protocol and continuously updates globally distributed information on successful halon banking.

One significant critical halon use is in passenger aircraft. In 1992 the military, the Federal Aviation Administration (FAA), the International Civil Aviation Organization, NASA and the private sector began a programme to enable aircraft to obtain airworthiness certification without the use of halon. The Air Force Wright Laboratory Aircraft Halon Replacement Team used instrumented aircraft engines, nacelle mock-ups and other test equipment to develop new agents and delivery systems under the Joint Military Service Aircraft Halon Replacement Programme. In 1993, at the request of the US Air Force, the Aerospace Industries Association and Halon Alternatives Research Corporation co-sponsored a two-day conference that led to the formation of the International Aircraft Systems Fire Protection Working Group under the leadership of the FAA. This group has produced minimum performance standards for replacing halon in cargo compartments, engines, passenger compartments and lavatories.

Development of alternatives to halon

Extensive research and development led to the identification of multiple halon alternatives. Three primary alternatives identified for general purpose halon-1211 fire extinguishers were a powder rated for type A (combustible materials),

B (flammable liquids) and C (electrical) fires, carbon dioxide and halocarbons (HFCs and HCFCs). For total flooding fire suppression systems, halocarbons initially became the preferred alternative, but, over time, water mist and inert gases turned out to be viable alternatives as well.[21] Even so, military and public sector organizations had a long way to go: none of the alternatives identified were drop-in replacements, and the military still had some highly specialized applications for which these alternatives would not work. Furthermore, fire protection authorities and insurance underwriting organizations had to be assured of the performance and safety of new alternatives in each system or application.

The US Air Force Air Base Fire Protection and Crash Rescue Systems Branch was the lead US DoD agency to fund significant research in the development of halon-1211 replacements. Its development of a halon recycling system, innovative firefighter training methods without halon-1211, and the removal and replacement of all halon-1211 with dry chemical extinguishers eliminated over 70 per cent of the halon-1211 used by the US Air Force. The Wright Laboratory Aircraft Halon Replacement Team, with funding and cooperation from the US Army, the US Navy and the FAA, developed test procedures and full-scale test apparatus for joint use by military and civilian aircraft designers. In addition, the Naval Air Systems Command conducted tests on an innovative new fire extinguishing technology based on an offshoot of automobile air bag inflation devices. These modified air bag inflation devices, known as 'solid propellant inert gas generators', proved so successful that they were immediately implemented in the newest aircraft designs such as the F/A-18E/F Super Hornet and the V-22 Osprey. The Norwegian Navy developed and demonstrated halon-free fire protection for ship engine rooms on combat vessels, and the US Navy developed and implemented several alternatives to halons, including HFC flooding systems, water-mist systems, and combination HFC and water spray systems. This technology was rapidly accepted for civilian applications because combat vessels face more demanding fire threats. Thus military technology, support and aggressiveness in implementing alternatives played an integral role in the successful phaseout of halons.

The identification of halon alternatives and the establishment of halon use standards gave industry the confidence it needed to accomplish an accelerated halon phaseout. Industry's development of hydrofluorocarbon and inert-gas alternatives and the standards promulgated by the National Fire Protection Association were essential. So were the actions of key international military players. Military organizations in India, Norway and the UK were also instrumental in enabling the halon phaseout. The Indian Defence Institute of Fire Research built public awareness and brought together fire protection organizations to establish standards for dry chemical firefighting foams and portable extinguishers for use by industry as replacements for halon. The Royal Norwegian Navy Materiel Command approved the use of foam to replace water halon in combat ships. The UK also established one of the first national commercial halon banks and prohibited many halon applications.[22]

BOX 5.9 HALON USAGE AND MANAGEMENT IN INDIA'S AVIATION SECTOR

Air Marshal A. K. Singh, PVSM AVSM VM VSM (Retd)

India's first major step towards implementation of the Montreal Protocol was to stop production of halons. Halons are clean, non-toxic, fire-extinguishing agents. These gases contain chlorine or bromine which, when released in atomic form, can destroy ozone molecules in great numbers. The ozone layer is thus directly affected with the release of this gas in the atmosphere.

All over the world, parties to Montreal Protocol began a phaseout of halon production with effect from 1 January 1994. The fire protection community has established global information networks and coordinated halon banks. A period of halon recycling thus commenced.

Halon phaseout in India has been a combined effort of manufacturers, defence users, legislatures, research establishments, insurance and fire authorities, due to which the production and consumption of this ozone-depleting substance has been reduced from 750Mt per annum in 1991 to almost zero in 2004. Today no halon is being produced in the country.

A facility has been set up in Delhi to pool all the existing halon for recycling and redistribution. This authority strictly monitors that the halon is being used for essential purposes only. Use of halon is restricted to critical areas such as protection of atomic power installations, oil platforms and military equipment. The halon for this essential use is imported from the countries which are recycling it since manufacture of halon has been stopped internationally.

The Indian Air Force mostly uses halon-1211 in aircraft-mounted fire extinguishers and in crash fire tenders. At high temperatures, halons decompose to release halogen atoms that combine readily with active hydrogen atoms, quenching the flame even when adequate fuel, oxygen and heat is available to sustain the fire. Halons are considered the best because they suppress the radicals that propagate the fire and are able to 'snuff out' the fire at much lower concentrations when compared with other fire suppressants using the more traditional methods of cooling, oxygen deprivation or fuel dilution. For aviation requirements this translates into reliability, as well as efficiency at a much lower weight, both of which considerations are crucial to military aviation. For example, halon-1301 is used at concentrations no higher than 7 per cent by volume in air, to suppress fires. By comparison, carbon dioxide fire suppression systems need over 34 per cent concentration by volume, which adds to the weight penalty. For any alternative to be a viable replacement for halon in the aviation sector it would have to be both lighter and less toxic.

Halons are very effective in fighting class A (organic solids), class B (flammable liquids and gases) and class C (electrical) fires, but not metal fires (class D). Small volumes are sufficient for big fires. Halon-1301 is injected into the fuel tanks of the F-16 fighters to prevent the fuel vapours in the fuel tanks from becoming explosive when the aircraft enters combat zones. However, due to environmental concerns research is being conducted to use CF3I as an alternative.

The Indian Air Force's inventory of aircraft has been procured from either the developed countries in the West or from Russia. Some of these aircraft are also being produced in India under a licensing agreement. However, the technology and material used in the fire protection system continues to be based on use of halon.

Broadly speaking, India currently does not require very large quantities of halons.

However, with the passage of time the requirement will increase significantly. In the next twenty years the predicted quantities are as given below:

	Halon 1211	Halon 1301	Halon 2402
Existing Systems	50 Tons	50 Tons	50 Tons
Future (20 Years) Requirement	100 Tons	175 Tons	100 Tons

The significant factor in the projected increase is that halon-1211 is the gas used in crash firefighting vehicles, the number of which will increase greatly as more large aircraft are brought into use. In the past one or two such vehicles were considered adequate at an airfield, but today, with vehicles such as the Airbus A380, an airport would need a minimum of 10 to 12 vehicles. We can see that 75 per cent of the total requirement of halon-1211 is for crash fire tenders. Therefore, it is imperative that an equally effective non-toxic alternative is found at the earliest opportunity.

There are three options for limiting halon use:

- **Conserve the existing supplies.** Existing supplies of the three types of halons that are in use in the military for weapons systems, communication systems, detecting systems and storehouse must be conserved by ensuring that there are no leaks either in the storage or in the operational system. Proper servicing is carried out by trained personnel, who are able to minimize spillages without compromising efficiency.
- **Recycle.** By recycling, no fresh halon is created and all the old halon available in the storage and as well as in use is recycled to original specifications and fed back into the operating appliances after these appliances have been refurbished to avoid spillage.
- **Search for alternatives.** In the cases of non-critical areas where alternatives to halons can be used, the users must be taught to replace halon after retrofitting the appliances with alternative firefighting substances which are technically suitable to meet their requirements.

India's strategy for limiting halon use has two main strands:

- The Bureau of Indian Standards, a national standards developing agency funded by UNEP, is in the process of preparing Indian standards on halon alternatives; and
- a national halon banking and management system is in place for recovery, recycling and quality assurance; this is being carried out at Centre for Fire Environment and Explosive Safety (CFEES).

The Government of India issued notification to all user agencies by a gazette dated 19 July 2000 mandating phaseout by the year 2010. Since then, HFCs and HCFCs have been considered as replacements for halons, but it is evident that they are also not environmentally friendly. They are relatively safe as far as the ozone layer is considered but their global warming potential means that a really safe alternative still needs to be developed. Our quest must continue.

We need to take a holistic view, because phasing out of halons may not provide us with all the answers. The existing alternatives to halons may prove to be harmful to the environment, albeit at a lower degree, but the cost of using alternatives may become prohibitive, especially for the developing nations. Therefore, we need to plan a 'road map' for the future, to learn to manage within our existing stock levels. There is a dire need for collaboration between private manufacturers in the aviation industry and other aviation specialists, to seek lasting solutions.

Fire protection community promotes halon alternatives

Military and industrial leadership culminated in the early phaseout of halon production in developed countries. In fact, it was the HTOC itself that proposed the 1994 phaseout of halon production, two years before that of any other ODS. There was remarkably little industry opposition from manufacturers or users to the accelerated halon phaseout, but many diplomats and national environmental authorities were reluctant to take such drastic action, and some were confused by the very concept of military and industry environmental leadership. Participation by fire protection experts from developed and developing countries at crucial Meetings of the Parties ensured acceptance of the early phaseout date.

The fire protection community, with financial assistance from the Multilateral Fund for developing countries and from the Global Environment Facility for countries with economies in transition (primarily the former Soviet states), will soon complete the global phaseout of halon production. However, the production phaseout of halon is currently possible only through halon banking for the remaining critical uses where no alternatives are available. The military recognized that while the vast majority of halon uses could probably be transitioned to alternatives over time, some uses would probably have to remain until the equipment that depends on them is retired from use. This includes most military aircraft and some armoured vehicle and shipboard applications. The military, through its sponsorship of one of the HTOC co-chairs and several members, worked with the HTOC to introduce the concept of halon banking as a mechanism for enabling an early phaseout. It was intended as a means for managing the small but critical uses for which technically and economically feasible alternatives were not considered likely to be developed in the foreseeable future. As a result, today critical halon uses are supplied from existing sources while the search for halon-1301 substitutes and alternatives continues. As predicted, small quantities are still used in critical applications in both the military and civil sectors. Critical civilian uses include commercial aircraft, oil and gas production facilities, pipeline pumping stations, and medical and communication facilities. Critical military uses remain in armoured vehicles, ships and aircraft.

MILITARY CONTRIBUTION TO THE PHASEOUT OF OZONE-DEPLETING ELECTRONICS AND AEROSPACE SOLVENTS

Just as military and industry leadership was indispensable in the halon phaseout, so too was it essential in the phaseout of solvents that deplete the ozone layer. After the Montreal Protocol was signed, the magnitude of military influence on the global use of ozone-depleting solvents became evident. A survey of industrial solvent uses in high-technology applications such as electronics discovered not only that the US military itself used significant quantities of ozone-depleting solvents, but also that military specifications and standards prescribing ozone-depleting solvents had been adopted as industry standards around the world. A team of soldering experts estimated that up to half of the CFC-113 used globally could be replaced with readily available alternatives if military standards were changed. Military organizations responsible for prescribing solvent use joined the effort to help define standards for cleanliness and materials compatibility which alternatives would have to meet. A unique partnership was formed between the military, private industry and the US EPA to identify and verify the acceptability of non-ODS solvents for military uses. Private companies, including AT&T, the Digital Equipment Corporation, Ford, IBM, Motorola, Nortel and Texas Instruments, assigned their brightest engineers and environmental performance managers to work with the partnership. The Defense Logistics Agency, the Defense Electronics Supply Center, the Air Force Guidance and Metrology Center and the Naval Weapons Center Electronics Manufacturing Facility were leaders in changing the military specifications and standards on electronic components and products to eliminate ODS requirements.

Some ozone-depleting solvent uses proved to be particularly difficult to replace, such as cleaning oxygen life-support systems on board aircraft and submarines and in diving applications. These systems consisted of long runs of thin tubes, assorted valves and complex geometry. Any contamination posed possible flammability problems because of the oxygen-enriched atmosphere. The US Naval Sea Systems Command and its private sector partners developed and commercialized a non-flammable, ozone-safe cleaner that can be recycled and disposed of easily. Military experts also made breakthroughs in non-ODS cleaning of sophisticated electrical, optical and precision components. In conjunction with its private sector partners, the US Department of Defense reported the phaseout progress to Congress and identified barriers to prompt action. This self-imposed reporting mechanism motivated military managers to prioritize solving most ODS phaseout problems and provided leadership support for overcoming barriers.

NASA, military and industry collaboration in aerospace solvent uses

After the US Clean Air Act was amended to include a provision for ozone layer protection in November 1990, military and civilian experts formed a team to coordinate the ODS reduction efforts related to the Space Shuttle, Titan, Delta and Atlas programmes. These programmes support global scientific, geographic and weather-related space missions. They are also critical for defence purposes. By August 1995 the use of ODSs in these programmes had been reduced to 1 per cent of 1989 levels, and the Parties to the Protocol granted an essential use exemption for remaining minor uses. This is significant because not one of the 191 Parties to the Protocol objected to the exemption. It was testament to the trust the Parties placed in the efforts made to eliminate ODS use and in those making the case that the essential use applications were necessary. The Titan IV Programme ODS Reduction Team, with the Department of Defense, the US EPA, the Industry Cooperative for Ozone Layer Protection and NASA, subsequently published a handbook, *Eliminating Use of Ozone-Depleting Substances in Solid Rocket Manufacturing* (1996). It is now used worldwide as a guide to reducing the use of ODSs in rocket motors.

Essential use exemption for methyl chloroform used in aerospace applications

US military and civilian space programmes faced daunting challenges when production of CFCs and methyl chloroform was halted in developed countries on 1 January 1996. CFC-113 solvent produced before the phaseout could be stockpiled for future use, but an essential use exemption was needed for methyl chloroform (1,1,1-trichloroethane) because it is chemically unstable and any technical deficiency could jeopardize lives and cargo, including scientific equipment necessary for monitoring 'Spaceship Earth' and its stratospheric ozone layer.

ODSs had been routinely used globally for the manufacture of space launch vehicle solid rocket motors because of their excellent cleaning properties, low toxicity, chemical stability and non-flammability. Non-flammability is of critical importance to the safety of operations involving highly energetic propellant materials.

In the US, large solid rocket motors are used to launch into space communication, navigational and scientific satellites and the Space Shuttle orbiters and their payloads and human crews. The solid rocket manufacturing industry is unique in that there is no method to test the performance of an individual rocket prior to use. The only way a solid rocket motor can be tested is a static firing or an actual launch. Accordingly, success can be assured only through rigorous manufacturing that includes detailed material specifications and continuous quality control.

At the Sixth and Seventh Meetings of the Parties (Decision VI, Nairobi, 1994, and Decision VII/28, Vienna, 1995), Parties granted an initial essential use exemption (EUE) to the US for the use of methyl chloroform for aerospace applications, including the manufacture and assembly of solid rocket motors used on the Space Shuttle and Titan. At the Tenth Meeting of the Parties (Decision X/6, Cairo, 1998), Parties agreed that the remaining quantity of methyl chloroform, authorized for the US at previous meetings, be made avail-

---

BOX 5.10 NASA LEADERSHIP TO ELIMINATE
OZONE-DEPLETING SOLVENTS

Stephen Newman, NASA

NASA is very proud of our leadership in eliminating ODS solvents from its solid rocket motors and in the transition from CFC-11 foam to HCFC-141b foam on the oxygen tank of its liquid rocket motor. But such changes are never easy. On 1 February 2003 the Space Shuttle Columbia was lost during its re-entry into Earth's atmosphere, causing the death of the seven astronauts onboard. An extensive accident investigation determined that the failure was in one of the only remaining applications where CFC foam was used – on the struts that attach the oxygen tank to the Shuttle. This tragic accident reminds us of how hard it is to phase out substances and how careful we must be in balancing environmental and safety risks.

able for use in manufacturing solid rocket motors until such time as the allowance is depleted, or until such time as safe alternatives are implemented for remaining essential uses.

Until January 2005 Thiokol/NASA used stockpiled methyl chloroform for uses not requiring the highest levels of purity and purchased small batches of freshly manufactured methyl chloroform for the most critical applications, where absolute purity and chemical stability are essential. However, the US Clean Air Act banned the manufacture of methyl chloroform even for essential uses after 1 January 2005. To accommodate this situation, in late 1994 Thiokol/NASA purchased newly manufactured methyl chloroform granted under the EUE authorization and placed it in a sophisticated leak-proof, refrigerated storage system designed to maintain chemical purity. NASA plans to draw from the refrigerated inventory until suitable alternatives and substitutes are identified and will then destroy any methyl chloroform manufactured under the terms of the EUE that is unneeded or unusable.

APPLICATION OF AEROSPACE TECHNOLOGY TO SOLVE ODS ELIMINATION CHALLENGES – A TEN-YEAR QUEST

When NASA first became aware of the requirement to eliminate ODSs from the processes their prime contractors were using to build and maintain their launch systems, it was with a bit of irony that the data used to sound the ODS warning came from satellites launched on their own space systems (such as the Space Shuttle). To NASA's credit, it mobilized a multidisciplinary task force and

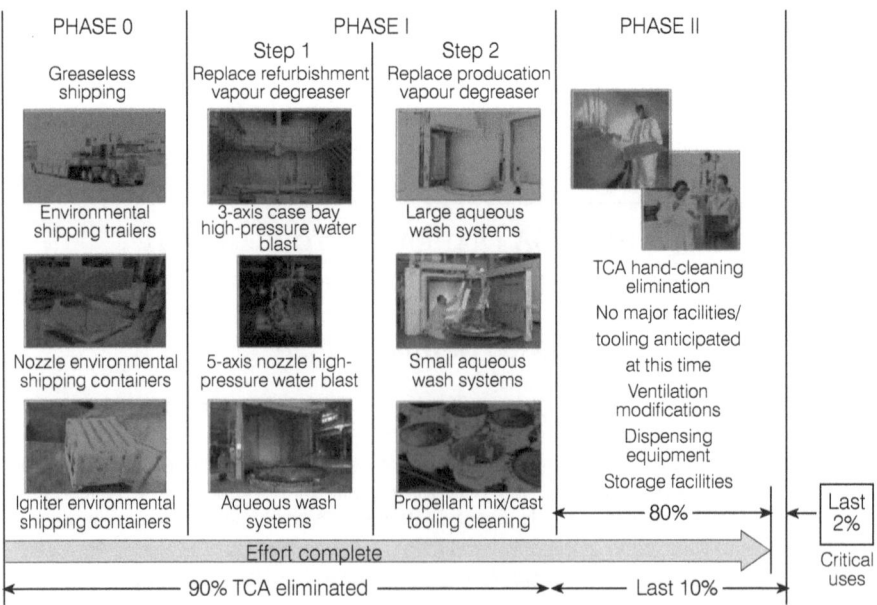

Figure 5.2 *NASA's reusable solid rocket motor ODS elimination programme overview*

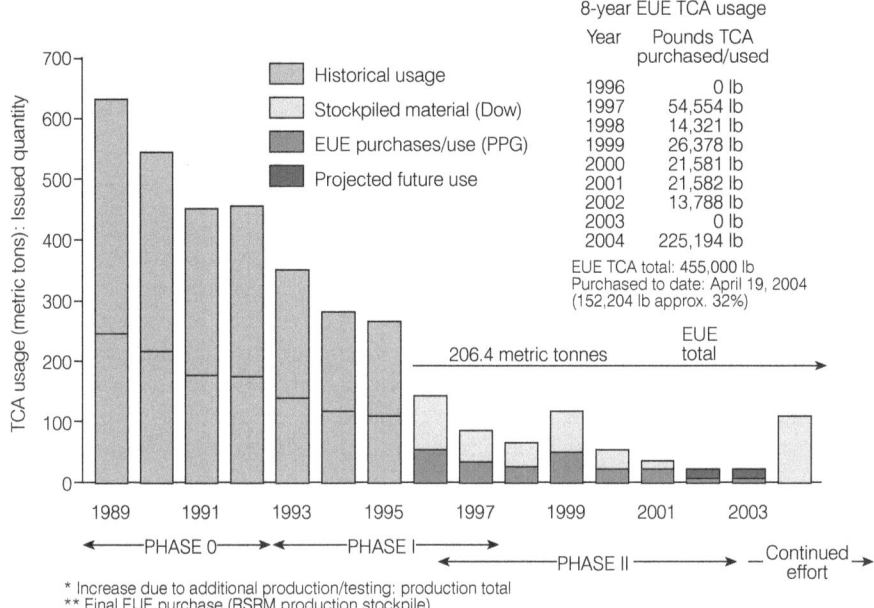

Figure 5.3 *Total solid rocket motor TCA usage/reduction history*

provided funding and programme priority to immediately embark on the complex task of eliminating all ODSs from its operations and identify suitable alternatives that would still get the job done without compromising the safety or reliability of the human-rated space access systems. This task was daunting in scope and complexity, not to mention cost. To tackle it, the project was broken down into three phases, with the second of these phases further broken down into two steps. By doing this, NASA was able to take advantage of immediate reductions, the 'low-hanging fruit', and get the reduction ball rolling within the first year. Figure 5.2 shows the main attributes of each step and phase. Figure 5.3 shows the impressive reductions made from a starting point in 1989 of 635 metric tonnes (1.3 million pounds) and how the use of TCA (methyl chloroform) was reduced to a low of 16 metric tonnes (30,000 pounds).

Along the way, this project yielded several surprises, some good and some bad. One thing was evident up front, and NASA knows this all too well: you must test what you are going to use, but the test results don't always point in the right direction. After testing over 100 different TCA replacement solvents and cleaners, the candidates that worked really well on the hardware had other negative attributes (such as toxicity or shelf-life) that rendered them unsuitable for long-term programme use. In the end, the programme realized that there was really no other 'universal solvent' similar to TCA that could provide a direct 'one size fits all' replacement and that a number of solvents and cleaners had to be employed. This was a major paradigm shift for the NASA programmes and the prime contractors. Eventually, though, the benefits of a family of discrete solvents and cleaners became evident and more robust bond strengths and enhanced processing became the norm. Furthermore, the technology that was invented to implement the new materials reaped other benefits through applications to other projects.

Chemical fingerprinting provides crucial data to develop ODS replacements

NASA and its prime contractor for the solid rocket motor programme in particular, ATK/Thiokol, investigated the use of advanced analytical laboratory techniques and equipment in order to thoroughly characterize materials used to manufacture the solid rocket motor (the Chemical Fingerprinting Program), while at the same time both organizations embarked on their ten-year journey to eliminate ODSs from their processing of rocket motors and associated payloads, and launch systems in general. The use of these techniques allowed early identification of what substances would be compatible with existing processes and materials and, more important, what substances would not agree with current manufacturing protocol. As NASA's ODS Elimination Programs matured, chemical fingerprinting became invaluable to identifying and evaluating the long-term degradation storage potential of existing ODS stockpiles (methyl chloroform (TCA) in particular).

TCA itself is not stable and will readily degrade. To combat this, it is typically blended with small amounts of various stabilizers, the most important of which is butylene oxide – an acid scavenger. Another additive in the TCA formulation used by ATK/Thiokol is 2-butanol, which is primarily important if the TCA is used for vapour degreasing (as part of ODS elimination, vapour degreasing is no longer in use at ATK/Thiokol).

This testing ultimately led the NASA–ATK/Thiokol team to conclude that the TCA could be stored long-term without the 2-butanol additive and still be effective as a precision application solvent for critical applications. ATK/Thiokol is still monitoring the chemical stability of this TCA stockpile and under specific storage conditions (glass-lined tanks with nitrogen purge and at lower temperatures) it appears to be very stable. Prior to the application of chemical fingerprinting, this conclusion would not have been as obvious, and the programme risk of storing this material long-term would neither have been proposed nor accepted.

NASA and its contractors are very proud of what was accomplished in a relatively short amount of time (as far as large aerospace programmes such as the Space Shuttle are concerned) and is continuing the development and implementation of alternative ODS processes and materials. As new programmes, such as the new Lunar and Martian Exploration Initiatives, are initiated, the groundwork already established in these previous ODS programmes will be fully exploited.

A significant amount of research was conducted worldwide to find alternatives for critical space-related uses of ODS solvents; this has produced important scientific understanding that enabled alternatives to be identified and implemented in other solvent use areas, leading to:

- advances in the overall knowledge of materials and processes that apply to the functions performed by solvents;
- unique solutions to cleaning exotic materials, and with complex soiling, to the most demanding standards;
- improved bonding and adhesives technologies, resulting in more robust systems with higher reliability;
- better ability to deal with a wide variety of chemical and compound programme issues based on related ODS elimination experience;
- increasingly stringent worker safety issues, employee preferences and ergonomic considerations;
- win–win benefits such as increased hardware life due to less aggressive refurbishment processes and elimination of repetitive grit blasting operations; and
- inventing refrigerated inventory systems to store unstable solvents.

Factors that might prevent a solvent from being eliminated from military, space and critical civilian uses include:

- **No technically feasible alternative.** Available replacements cannot meet technically rigorous performance requirements or pose unacceptable health and safety hazards in specific applications.
- **Environmental issues.** Candidate alternatives meet technical performance requirements but pose unacceptable environmental risks.
- **Undocumented performance.** Defence and aerospace systems involving human life and mission-critical operations require a high degree of reliability that can sometimes be validated only by actual use over a prolonged period of time.
- **Timeline and unforeseen circumstances.** In some cases, aerospace and military applications will end before an alternative to ODSs can be fully assessed, so the financial and environmental cost of finding an alternative for the period of use that remains is not justified. In other cases, unforeseen aerospace and defence situations may require emergency use of ODSs in applications where alternatives have not yet been approved.

BOX 5.11 ELIMINATING ODSS IN SOLID ROCKET MOTORS

Jeffery J. Norton, Colonel, US Air Force and Program Director, Launch Programs

Our team has taken enormous strides in reducing and eliminating the use of Class I ODSs. Our insights from having worked with the four leading manufacturers of solid rocket motors, combined with the EPA, industry and international experts, have allowed us to eliminate more than 99 per cent of the original ODS baseline of 1330 metric tons of ODS equivalent. Since this effort has worldwide implications, we developed a Solid Rocket Motor Handbook to share our elimination successes with all countries involved in space flight.

Box 5.12 Eliminating ODSs from the processes used to manufacture Space Shuttle solid rocket motors

Ross Bowman, Space Operations, Thiokol Corporation (company statement, 1993)

Eliminating ODSs from the processes used to manufacture Space Shuttle solid rocket motors was a great challenge to the NASA–Thiokol team. With outstanding help from the US EPA, industry leaders and the Montreal Protocol Solvents Technical Options Committee, a plan was prepared and implemented. The excellent results were possible due to great teamwork on the part of many government and industry leaders who are dedicated to protecting the stratospheric ozone layer. We appreciate their unselfish sharing of information and technology.

- **Complex compliance requirements.** Some complex and potentially confusing provisions of the Montreal Protocol and its implementation by Parties can unnecessarily increase compliance costs or can compromise other environmental performance criteria.

Defence contractors also took the lead in eliminating ODS solvents from their production lines. This often required them to take risks in order to convince their military customers to accept the changes they proposed. General Dynamics and Lockheed Martin rapidly eliminated more than 90 per cent of their ODS solvent use by painstakingly selecting alternatives that met all their performance requirements. They developed and implemented the non-ODS technologies for cleaning of gaseous oxygen and hydraulic tubing for aircraft and space launch systems. Later they eliminated all ODS use. Technologies developed by military contractors have been implemented at major civilian and military manufacturing facilities around the world.

MILITARY LEADERSHIP IN THE PHASEOUT OF OZONE-DEPLETING REFRIGERANTS

The military also played a significant role in the phaseout of ozone-depleting chemicals as refrigerants. As with other ODSs, the military often faced greater challenges than civilian refrigerant users due to the need for equipment to perform and maintain reliability in the harshest of wartime conditions aboard mobile weapons platforms such as ships, aircraft and armoured vehicles. In addition to the normal facility air-conditioning and refrigeration systems, cooling is required on ships and armoured personnel carriers to keep critical weapons control and communications systems functioning.

The US Navy faced a particularly difficult challenge with many of its systems, which were specifically designed to use CFC-114. This substance was chosen for navy vessels because it provided quiet operation to escape sonar detection, reduced equipment volume and offered compatibility with existing

---

Box 5.13 The CFC phaseout leads to
superior technology

Richard Helmick, Former Director of Climate Control Systems,
US Naval Sea Systems Command

The Montreal Protocol CFC refrigerant phaseout provided the opportunity for the Navy to develop and transition a superior technology that will benefit the Fleet and the global environment far into the future.

---

submarine atmosphere control equipment (all important features for warship systems). Since CFC-114 was not used extensively by the private sector for refrigeration, there was little incentive for commercial companies to develop alternative refrigerants to replace it. The primary use of CFC-114 by the private sector was for foam-blowing applications, an emissive use. This meant there were no supplies of CFC-114 in private sector equipment that could be recovered as the equipment reached the end of its life, as was the case for halons and refrigerants. Once the foam industry converted to other alternatives, the market for CFC-114 would disappear and its availability to support existing shipboard refrigeration systems would become a major concern.

Military organizations worked diligently to identify alternatives. After several alternatives proved unsuccessful, HFC-236fa was eventually formulated as a viable alternative. A fleet-wide programme was subsequently undertaken to convert nearly 1100 existing air-conditioning systems and refrigeration units on US Navy ships. While HFC-236fa was still a significant global warming gas, the Navy was able to more than offset its direct emission contribution to global warming by also making other improvements during the conversion process that made the air-conditioning systems about 10–15 per cent more energy-efficient. In parallel to the work to identify alternative refrigerants for existing equipment, the Navy worked to develop non-CFC systems for new ships which were also 20–35 per cent more energy-efficient than the CFC systems they replaced and incorporated technologies to reduce refrigerant leakage. In 1995 and 1996 the US Navy also provided support to Taiwan and Spain for conversion of their equipment from CFC-12 to HFC-134a.

VINTAGING

Military equipment is declared to be 'vintaged' if it has remaining useful life but is not valuable enough to be updated and modernized. If the ODS alternatives identified to date could not be retrofitted into existing weapons systems because of space and/or weight constraints, or for some other reason, then that equipment was vintaged. The ability to vintage important military equipment depended on responsible use of ODS stocks by the market at large. Minimizing use and emissions of the remaining ODSs would enable armed forces to vintage their equipment for that much longer. Access to the ODS reserve (or bank) would be necessary to support each vintaged weapons system through to the end of its useful life.

Halting training and avoiding unnecessary uses

Quick and inexpensive measures were implemented so that ODSs were not used in training (especially halons) or for system maintenance and performance checks (for example pressure testing or fire protection system checks).

Implementing recycling/reclamation

In the US, a national ODS reserve (or bank) was established for those military uses for which alternatives did not currently exist. Other military organizations used national banks that supply both civilian and military critical uses. The size of this bank was determined by the number of systems which had to be vintaged or retrofitted, the length of time such systems were expected to remain in service, and the estimated annual quantity of ODSs lost to the reserve as a result of maintenance or use.

Identification of mission-critical uses

Mission-critical uses are those ODS applications which have a direct impact on combat mission capability and for which no alternative has yet been identified, developed or implemented. They include uses that are integral to combat mission systems or that directly affect their operational capability.

Mission-critical uses may include:

For CFCs:
- shipboard chilled-water air-conditioning and/or refrigeration systems; and
- shipboard cargo refrigeration systems.

For halons:
- shipboard room-flooding applications;
- aircraft fire protection and explosion-suppression systems;
- portable extinguishers on board aircraft;
- flight-line fire protection; and
- explosion suppression in crew compartments of armoured combat vehicles.

For HCFCs:
- vintage weapons systems with exotic materials incompatible with solvent alternatives;
- oxygen systems with complex geometry and blind passages; and
- radiological decontamination of humans.

For methyl chloroform:
- surface preparation and solvent carriers in solid rocket manufacturing.

Securing supplies to meet critical needs

Securing mission-critical supplies normally involves establishing or joining an ODS bank. However, one of the most important steps in reducing the quantities of ODSs needed is to minimize current emissions. Armed forces in developed countries found this a useful first step in creating awareness that these substances were a problem and had to be handled with respect. That awareness was quite often responsible for field-initiated reduction efforts. The most dramatic reductions in ODS use during 1987 to 1990 were achieved simply by adopting more careful maintenance practices. ODSs not released into the atmosphere as a result of poorly maintained systems or poor maintenance procedures then became available to service critical military applications.

UNEP Division of Trade, Industry and Economics (DTIE) OzonAction Programme and the Multilateral Fund published a 60-page handbook, 'Maintaining military readiness by managing ozone-depleting substances: Guidelines for armed forces in developing countries'[23], based on first-hand experiences of, and lessons learned by, armed forces in developed countries. It sets forth a strategy for compliance that includes procurement, retrofit, servicing, training and early equipment retirement.

ESSENTIAL USE EXEMPTIONS

One of the requirements for an essential use exemption for an ODS under the Montreal Protocol is that supplies adequate to meet the need for the ODS are not available from stockpiles of recycled material of that ODS. If there is no bank of recycled material of an ODS, an applicant cannot prove that he is unable to get recycled material. If a bank exists, and the applicant is a participant in a functioning banking scheme, he may be able to establish that the bank is unable to supply him with the needed ODS. Thus, participation in an ODS bank is a virtual prerequisite for satisfying the essential use criteria. The terms 'mission-critical use' and 'essential use' are not synonymous. Just because the armed forces declare a use 'mission-critical' does not mean it will qualify as an essential use under the Montreal Protocol. An 'essential use' can only be determined by the Parties.

HALON BANKS

Many developed countries have established military ODS reserves or banks. While they might begin as a small stockpile with warehouses and storage tanks, they tend to develop into agencies with lists of ODS users who no longer require their ODSs and users who still have mission-critical needs. As the armed forces begin to implement their ODS management programmes, recovered ODSs are taken to the bank, where they are kept for mission-critical applications. Reports from the Halons Technical Options Committee (HTOC) and the Technology and Economic Assessment Panel (TEAP) discuss in detail the various models used by countries to establish and operate ODS banks. Details

---

BOX 5.14 THE US DEPARTMENT OF DEFENSE ODS RESERVE

The Defense Logistics Agency manages the DoD ODS Reserve which will support military ODS uses now that production has ceased. In some cases, it is not practical or affordable to retrofit the ODS alternatives presently available into existing weapons systems. Alternatives may have consequences in terms of space loss or additional weight that impact adversely on weapons-system performance and make a retrofit costly or technically impossible. Based on analyses of such issues, each branch of the US armed forces has decided to 'vintage' some of its existing weapons systems. The DoD ODS Reserve will support these mission-critical ODS applications through to the end of their useful lives or until technological and economically feasible alternatives are developed. The DoD ODS Reserve was established by the Defense Authorization Act of 1993 to provide support to the military services and DoD agencies for their mission-critical requirements involving halons, refrigerants and solvents until alternatives can be fully implemented. The Reserve is managed by the Defense Logistics Agency through its Defense Supply Centre in Richmond, Virginia. The Reserve of Class I ODSs has been built up from recycled ODSs held by the armed forces in non-mission-critical applications and from recycled ODSs purchased on the open market. The armed forces and DoD agencies recover material from their non-mission-critical systems or from systems withdrawn from use because they have reached the end of their useful life. Recovered ODSs are turned over to the Reserve for reclaiming, storage and future issue. The Reserve purchases recycled ODSs on the open market only when absolutely necessary; it relies primarily on ODSs recycled from internal sources. It is important to note that the Reserve has safeguards in place to audit the source of the recycled ODSs and so avoid purchasing 'black market' material. The programme began its operations in January 1994 with halons and then expanded into refrigerants in January 1995 and finally into solvents in January 1996. Halons constitute the largest ongoing military ODS requirement.

---

of websites with additional information on establishing military ODS banks can be found at the end of this chapter.

THE ROLE OF GOVERNMENT AGENCIES AS CUSTOMERS AND MARKET LEADERS

Government agencies and military organizations were large consumers of ODS products in electronics, refrigeration, air-conditioning, vehicles and fire protection. As such, they often influenced product design and manufacturing techniques used in commercial products. When government organizations implemented ODS phaseout programmes, they transformed electronics, refrigeration air-conditioning and fire protection markets, making it less expensive and more reliable for other enterprises to convert to ozone-friendly products.

Once markets were transformed away from ODSs, the phaseout became almost automatic. For example, since about half of the CFC-113 solvent used worldwide was directly a result of US military endorsement of its technical performance, once the suppliers were free to adopt alternatives, demand fell quickly. This was partially due to the fact that some companies used CFC-113 to qualify products for military markets and used the same cleaning process on products for civilian markets. Other companies used CFCs because the military had determined they were technically superior. The US Department of Defense transformed markets away from ODSs by working with the EPA to certify alternatives and then prohibit the use of CFCs. This and other industry actions accelerated the commercialization and implementation of ozone-safe solvents.

Many industry participants in the sector-specific partnerships encouraged military and government agencies to adopt alternatives as a strategy for building market share for alternatives and to help convince their own companies to change. The perception was that DoD adoption meant high performance cost-effectiveness. In an intergovernmental meeting on the importance of military organizations in stratospheric ozone protection in 2001, military and civilian leaders observed that alternatives and substitutes often perform as well as, if not better than, ODSs if systems are carefully designed and engineered to optimize performance. Additionally, in many cases alternatives reduced costs and simplified manufacturing.[24] In this way the DoD had a significant indirect effect on the market, and the size and timing of market shifts to alternatives support this view.

THE CHALLENGE FOR GOVERNMENTS

National governments had a role to play in supporting industry in its efforts to develop effective programmes. Governments did this by creating a constructive policy framework to facilitate phaseout at the national level and by acting internationally in cooperation with other Parties to the Protocol. The armed forces are often the largest and most visible agency of government. As such, they had an opportunity to provide leadership to the rest of the nation and to demonstrate the technical and economic viability of phasing out ODSs and adopting specific alternative technologies. In many developed countries, the armed forces proved to have the technical and organizational capacity for implementing alternatives to ODSs more quickly than industry. In addition, because armed forces purchase large quantities of equipment and supplies, procurement policies favouring non-ODS technologies provided a clear and convincing message to industry that it needed to abandon ODS products.

The value of collaboration with industry and government

In many countries, the armed forces are among the most technologically sophisticated and stable organizations. Because the Montreal Protocol affects both the armed forces and industry alike, it offered the military a unique opportunity to act as a catalyst to improve the strength and capacity of both government and the country's established industries. By working together, both the armed forces and industry phased out ODS use faster and at a lower cost than either could have done by working alone.[25]

**Box 5.15 US Navy clearinghouse helps
to coordinate ODS phaseout**

*Pete Mullenhard, Science Applications International
Corporation Contractor for the US Navy Shipboard
Environmental Information Clearinghouse (formerly the CFC
and Halon Information Clearinghouse)*

The US Navy quickly recognized the need to not only coordinate internal ODS phaseout efforts, but to also exchange information with industry and other government agencies. This is why they included a 'clearinghouse' function in their original 1990 ODS Elimination Program Plan.

Some of the organizations established in developed countries to coordinate the efforts of government, industry and the armed forces proved so successful that they continued to operate once their initial short-term goals had been achieved, even expanding their mandate to take on board other environmental issues of mutual concern – issues such as the handling and disposal of hazardous materials, air pollution, pollution prevention and climate change.

Workshops by NATO

International collaboration has also proven a useful channel for sharing experiences and solutions. For example, the Committee on the Challenges of a Modern Society (CCMS) of NATO was set up in 1969 to address environmental issues relevant to armed forces. Under the CCMS umbrella, NATO established a range of technical groups to address specific military ODS uses. These included the use of halons in military aviation, halons and refrigerants on board ships, halon systems in armoured combat vehicles, solvents in the maintenance and production of weapons systems, and methyl bromide as a fumigant on both ships and aircraft.

NATO co-sponsored four workshops in 1991, 1994, 1997 and 2001 to assess military ODS uses, exchange information about alternatives, share policy strategies, and discuss the best ways to implement banking and recycling programmes. Sponsorship was available to facilitate the participation of representatives from developing countries, particularly in 2001, when the Multilateral Fund (MLF) co-sponsored the workshop. In addition, many bilateral and trilateral military groups were formed to work on specific military environmental issues of common concern. These workshops offered participants the opportunity to interact with military officials from other countries who were trying to solve the same problems.

BOX 5.16 OZONE AND CLIMATE PROTECTION CONFERENCES

The First International Conference on the Role of the Military in Protecting the Ozone Layer – sponsored by NATO, the US Air Force and the US EPA; 11–13 September 1991, Williamsburg, Virginia.

The Second International Conference on the Role of the Military in Protecting the Ozone Layer – sponsored by NATO, the US Department of Defense, the US EPA, ICOLP, the Aerospace Industries Association, the American Electronics Association, the Electronic Industries Association and the University of Maryland Center for Global Change; 24–25 January 1994, Brussels.

The Third International Workshop on the Role of the Military in Implementing the Montreal Protocol and the First International Workshop on the Military Role in Climate Protection – sponsored by NATO, The Netherlands Ministry of Defence, the Norway Ministry of Defence, the US Department of Defense, the US EPA, ICOLP and the University of Maryland Center for Global Change; 6–7 November 1997, Herndon, Virginia.

The Fourth International Workshop on the Importance of Military Organizations in Stratospheric Ozone and Climate Protection – sponsored by the Australia Department of Defence, the Center for International Environmental Law, the Climate Institute, the Department of National Defence Canada, Environment Australia, Environment Canada, the Institute for Defense Analyses, the International Cooperative for Environmental Leadership, the UK Ministry of Defence, the UK Department of Environment, Transport and the Regions, UNEP, the UNEP Multilateral Fund of the Montreal Protocol, the US EPA and the US Department of Defense; 6–8 February 2001, Brussels.

USEFUL PHASEOUT STRATEGIES ADOPTED

The phaseout strategies used successfully by armed forces in developed countries include:

- announcing policies restricting purchase of new ODS-using equipment once alternatives are available;
- modifying technical specifications to phase out ODS requirements as soon as alternatives are identified and as an incentive for research to develop alternatives in uses still requiring ODSs;
- coordinating efforts with private industry and other branches of the armed forces;
- conducting training and awareness for maintenance personnel working on systems using ODSs;
- monitoring the ODS bank, reviewing deposits, and restricting or controlling withdrawals;
- using existing technical assistance and cooperative agreements with other organizations and armed forces in other countries to exchange experiences in phasing out ODSs;
- conducting periodic programme reviews to ensure the phaseout is on schedule and keeping high command abreast of any problems;
- funding the phaseout as an integral part of military forward budget planning; and

- using national and international forums on military issues to address ODS phaseout issues.

CONCLUSION

When the Montreal Protocol was signed in 1987, every sophisticated weapons system worldwide was entirely dependent on the use and emissions of ODSs, particularly halon fire extinguishing agents and CFC-113 and methyl chloroform solvents. Military organizations went beyond the call of duty in protecting the ozone layer. They were in Montreal advocating that halon be among the first chemicals regulated. They were founding members of the most effective ozone phaseout partnerships, among the first to declare that procurement would favour ozone-safe products and processes, and among the first to prohibit ODSs in training and servicing. They spearheaded the elimination of solvent standards compelling the use of CFC-113 and they financed the research leading to halon phaseout. A consortium of military organizations, including those of Australia, Canada, The Netherlands, Norway, Sweden, the UK and the US, persuaded NATO to officially write to the United Nations advocating a complete phaseout tempered by essential use exemptions and ODS banking.

Military organizations shared information within and outside the military community through publications, conferences, websites, workshops and training sessions. Today military websites are a prime source of information on environmentally superior technology. Although some older legacy weapons are still dependent on ODSs for service and operation, the US and other nations have eliminated ODS dependence for all next-generation weapons systems and have shared the technology making this possible worldwide.

It is significant that military organizations faced the most daunting challenges in phasing out ODS and it is extraordinary that military organizations and individuals earned the lion's share of the US EPA Stratospheric Ozone Protection Awards. Never before, and never since, have military organizations so quickly embraced the precautionary principle and engaged so aggressively in defending the Earth against threats to prosperity and sustainability.

Technology Transfer to Phase Out ODSs in Foams[1]

INTRODUCTION

When the Montreal Protocol was signed in 1987, the transition to alternatives for ozone-depleting substances (ODSs) began in earnest. New blowing agents had to perform well and meet safety, cost and environmental criteria. From an environmental standpoint, alternatives were evaluated for impacts on global warming, local air quality and human health, in addition to ozone depletion. Alternatives also had to be safe, with any additional risks mitigated, and had to come with a reasonable price tag and create a product that satisfied technical and consumer standards. Balancing these criteria was not always easy.

This chapter documents the extraordinary transformation that has taken place in the foam industry, focusing in particular on the experience of countries with economies in transition (CEITs). Incredibly difficult research and development was carried out over many years, mostly by the major polyurethane chemical suppliers. This hard work resulted in technical innovation which could cope with a wide range of blowing agent boiling points (from -40°C to +50°C) and the use of environmentally advantageous and cost-effective but highly flammable hydrocarbons.

This chapter also analyses alternative technologies in terms of environmental and safety performance, cost and technical suitability and provides concrete examples from technology cooperation projects sponsored by the Multilateral Fund (MLF) and Global Environment Facility (GEF). These investigations reveal factors that led to the selection of certain technologies over others, thus providing an understanding of the current technologies and where the future of foam-blowing technology is headed.

A BRIEF HISTORY OF ODS FOAM

During World War II, the Dow Chemical Company introduced the first foam product manufactured with CFC-12 under the brand name StyrofoamTM. In order to make foam:

a gas (or volatile liquid) is introduced into a liquid plastic. The gas forms bubbles in the plastic; when the plastic hardens, a cellular structure remains – the finished foam. The gas which makes the cells is the blowing agent.[2]

Chloroflourocarbons (CFCs) were used as blowing agents because 'they have suitable boiling points and vapour pressures, low toxicity and low thermal conductivity and because they are non-flammable, non-reactive and cost-effective'.[3] CFC-12 also helped the polymerizing mixture to form closed cells containing CFC gas. In thermal insulating applications, the CFC-12 improved the foam's thermal insulating properties because it was less conductive than most other foam-blowing agents.

Polystyrene foam insulation boardstock using methyl chloride as the blowing agent was invented in Sweden in the early 1940s. It was further developed in the US using CFC-12 in an extrusion process. CFC-12-based foam was extensively used in food service, packaging and insulation.

In the late 1950s flexible foams made from polyurethane quickly became the predominant cushioning material for furniture, bedding, carpet underlay, and automobile seats and dashboards.

In 1959 Imperial Chemical Industries UK (ICI) implemented the first use of CFC-11 foam for insulation of the holds of ships used to transport meat from Australia and New Zealand to the UK. In 1961 the same company implemented the first use of CFC-11 insulating foam to replace glass fibre in refrigerators manufactured at a Danish plant. By 1985 CFC-11-based foams represented a high market share in refrigerators, building insulation and other applications.

By the late 1980s foams were used in almost every aspect of our daily lives – they were in the seats in our vehicles and the soles in our shoes; they provided insulation for our refrigerators, freezers and water heaters, and also for building walls and roofs. Foams were also used for packaging materials, food containers and flotation devices. At the time the Montreal Protocol was signed, foam manufacturing accounted for 25–30 per cent of global CFC use.

TECHNICAL DESCRIPTION OF FOAM DESIGN, MANUFACTURING AND USE

Four kinds of plastics are used to make foam: polyurethane, polystyrene, polyolefin and phenolic. Most foam manufactured with ODSs in countries with economies in transition (CEITs) and developing countries was made with polyurethane. Polyurethane foams can be broadly divided into two types:

1   flexible foams, typically used for mattresses, upholstered furniture and transportation seating; and
2   rigid foams, mainly used in refrigerator, freezer and building insulation.

In between these two classes are various expanded elastomers with intermediate rigidity, used for applications such as shoe soles and car steering wheels. The range of foam products is shown in Table 6.1.

Table 6.1 *Uses of ODSs in foams*

Foam type	Process	Application
Polyurethane Flexible	Slabstock	Furniture, mattresses
	Block foam	Furniture, mattresses
	Moulding	Car and bus seats, auto bumpers, flotation/buoyancy, carpet underlay, footwear
	Integral skin	Furniture, automotive, footwear
Polyurethane Rigid	Injection	Domestic and commercial refrigerators and freezers, insulated transport containers, picnic coolers, sandwich panels, flotation/buoyancy, pipe-in-pipe
	Continuous lay-down	Boardstock/laminate and sandwich panels
	Discontinuous lay-down	Block foam – panels, pipe section
	Spray/pour-in-place	Roofs, walls and tanks
Polystyrene Rigid	Continuous lay-down	Boardstock and sandwich panels
	Injection	Flotation/buoyancy
	Sheet	Food and food service packaging,
	Moulding	miscellaneous packaging
Polyolefin Rigid	Injection	Flotation/buoyancy
	Boardstock	Cushion, packaging, pipe insulation
	Moulding	Cushion, packaging, auto bumpers
Polyolefin Flexible	Sheet	Life jackets
	Moulding	Carpet underlay, furniture
Phenolic	Continuous lay-down	Boardstock/laminate, pipe section

Polystyrene rigid foams, which were also made with ODSs, are made using a variety of processes, including continuous lay-down, injection, sheet and moulding. Polystyrene foam products include flotation and buoyancy, packaging (particularly for food) and sandwich panels.

EARLY VOLUNTARY PHASEOUT OF POLYSTYRENE RIGID FOAM

Polystyrene rigid foam that used CFC as a blowing agent was voluntarily phased out in the US by the US Environmental Protection Agency's (EPA's) first voluntary partnership. The cost-saving alternative technology developed under the partnership was shared worldwide by the Foodservice and Packaging Institute and was stimulated by pledges by McDonald's and other global food enterprises to halt CFC use worldwide.

FACTORS THAT INFLUENCED CHOICE OF TECHNOLOGIES

Replacement technology was first developed and commercialized in the 'developed' countries and was then adopted by 'developing' countries. In developed countries, the drivers of foam technology choice have included:

- The Montreal Protocol, which determined the direction of technology transfer in its initial version and various amendments because measures were always earlier and/or more stringent in developed than developing countries;
- national regulations on CFC replacement; and
- societal reaction to environmentally superior chemicals like hydrocarbons that have increased safety risk.

Box 6.1 The US EPA Energy STAR™ and High GWP Partnerships grew out of Montreal Protocol cooperation

Sally Rand, US EPA

In the mid-1990s, my job at the EPA was to implement the Montreal Protocol under the US Clean Air Act, but my passion was the voluntary partnerships with industry that phased out most ODSs well ahead of schedule. Enterprises and countries went 'beyond compliance' in halting ODS use and in choosing alternatives with the highest environmental performance. In 1996 I replaced founding Technology and Economic Assessment Panel (TEAP) member Jean Lupinacci as Co-Chair of the Foams Technical Options Committee (TOC).

Jean Lupinacci and Stephen O. Andersen organized the first ever voluntary partnership at the EPA – an agreement to phase out CFC foodservice packaging within one year, to switch from hydrochloroflourocarbons (HCFCs) as soon technically feasible and to share technology worldwide. It was the first voluntary national sector phaseout in the world. The early success of voluntary partnerships for ozone protection is well known. What is less well known is that the people and strategies from the Montreal Protocol made a smooth transition to climate. The energy conservation and climate partnerships grew out of TEAP, the EPA, and the Montreal Protocol industry non-governmental organizations (NGOs).

Energy STAR™, which promotes energy efficiency across a wide spectrum of consumer products; the Semiconductor Partnership, which reduces perfluorocarbons (PFCs) emissions; and the Aluminum and Magnesium Industry Partnerships, which target sulphur hexafluoride (SF_6), were launched by EPA managers who had successfully used industry cooperation, voluntary agreements and corporate pledges to help protect the ozone layer.

Energy STAR™ was started by John Hoffman, who with Eileen Claussen and Stephen Seidel managed EPA's stratospheric ozone protection programmes from before the Vienna Convention through the early days of the Montreal Protocol. The Energy STAR™ brand is now a leading symbol for energy efficiency around the world, rewarding enterprises with environmentally superior products and increased profits and market share; rewarding customers with less cost of ownership; and rewarding government with the appreciation of its taxpayers.

See Boxes 6.2–6.4 and the section in this chapter entitled 'Sustainability, suitability and environmental properties of alternative technologies', which elaborates the safety risks of alternatives.)

BOX 6.2 THE EU GREEN LEADERSHIP

Mike Jeffs, ISOPA Foams TOC

After about 1990 the European Union (EU) viewed itself as a 'green' leader and set phaseout dates for both CFCs and HCFCs ahead of those required by the Protocol. This resulted in many foam replacement technology developments being pioneered in the EU and then being transferred elsewhere.

NGOs, such as Greenpeace, also had a strong influence in Europe. European industry and legislators had a tendency to look to the future and introduce low GWP (global warming potential) alternatives, such as hydrocarbons (pentanes), at an early stage rather than use HCFCs for an interim period. This also has the advantage of fewer technology changes over time and allows enterprises to get on with their main business.

BOX 6.3 TECHNOLOGY CHANGE IN THE JAPANESE FOAM INDUSTRY

Akira Okawa, JICOP

Achilles Corporation, Bridgestone, Kaneka Corporation, Asahi Kasei Life and Living, and Asahi Kasei Construction Materials worked independently and with coordination from the Japan Industrial Conference for Ozone Layer Protection (JICOP) and their foam industry associations to produce ODS-free foams that had suitable technical properties such as energy efficiency. Each enterprise ultimately rejected HCFC and HFC blowing agents and chose water, carbon dioxide or hydrocarbons. Each enterprise overcame the problems of adhesion, thermal conductivity and rigidity by developing 'micro-bubble' cell structure. Closed-cell products blown with hydrocarbons were able to resolve flammability concerns by adding flame retardant to the foam matrix or by using surface covering.

Achilles Corporation had been very ingenious in making ODS-free foam that achieved the same energy efficiency as foam made with CFCs or HCFC. So many other enterprises took one step from CFC-11 to HCFC-141b and a second step to HFC – only to find themselves facing the Kyoto Protocol. Achilles is making thermal insulating boards using pentane as a blowing agent and achieving the same level of insulating properties as CFCs by micronizing the cell structure of the foam and mitigating product flammability by improving the mixing of the urethane and other ingredients. Achilles Corporation is making ODS- and HFC-free spray urethane foam using super-critical CO_2. The spray foam achieves the same level of dimensional stability as CFC foaming by increasing the strength of foam bubbles by isotropizing with super-critical CO_2 and achieves desired adhesive properties at temperatures as low as 0°C by decreasing the density by using a smaller quantity of water. This technology was diffused nationwide with the financial support of the Ministry of Economy, Trade and Industry (METI).

In CEITs and developing countries, the drivers of foam technology choice included not only the factors listed above (the Montreal Protocol, national regulations and societal reaction to alternatives), but also:

- the time of choice;
- the influence of the Ozone Operations Resource Group;
- policies of financing institutions;
- laws passed in important export markets; and
- market forces and competition among foam system suppliers.

The time of choice: Impacts on CEIT and developing country technology selection

The time when an alternative was selected often influenced technology choice. Early on, alternative technologies were not all well developed, with many underperforming ODS technologies.

The Ozone Operations Resource Group's influence on developing country technology selection

The Ozone Operations Resource Group (OORG) was a special organization in the World Bank which evaluated new technologies and judged whether they were ready to be deployed in developing countries.

The OORG was set up in April 1992 by the Bank's Montreal Protocol Unit and composed of industry experts from all sectors. Over the years, the OORG team assessed technologies, evaluated specific enterprise and country phaseout projects, and developed implementation guidelines.[4]

The OORG was often decisive in guiding phaseout investment. For example, it conducted a detailed technical and economic analysis of pentane technology for foams in refrigerators in October 1993, at a time when the technology had been in commercial use in Germany for only about six months. The OORG concluded that the technology was indeed ready to be transferred to developing country enterprises via the MLF.

> *The early OORG expert judgment that hydrocarbon foam is environmentally and economically superior has been proven to be absolutely correct. At the time, the decision even surprised the green NGOs!* (Mike Jeffs, ISOPA Foams TOC)

The OORG was also instrumental in publishing technical information that had never previously been assembled. In January 2000, for example, it published detailed information on foams density that contrasted the technical performance of alternative technology against the CFC baseline for all relevant types of rigid polyurethane foams.

> *The World Bank's OORG made a huge difference to technology transfer, and it has been a privilege for me to be a member throughout its life. Some expert working groups had 150 years of foam experience amongst 5 or 6 members, all concentrating on bringing the best technology to developing countries.* (Mike Jeffs, ISOPA Foams TOC)

BOX 6.4 TECHNOLOGY CHANGE IN THE DOMESTIC REFRIGERATOR AND FREEZER INDUSTRY

Mike Jeffs, ISOPA Foams TOC

CFC replacement in rigid polyurethane foams for refrigerators and freezers

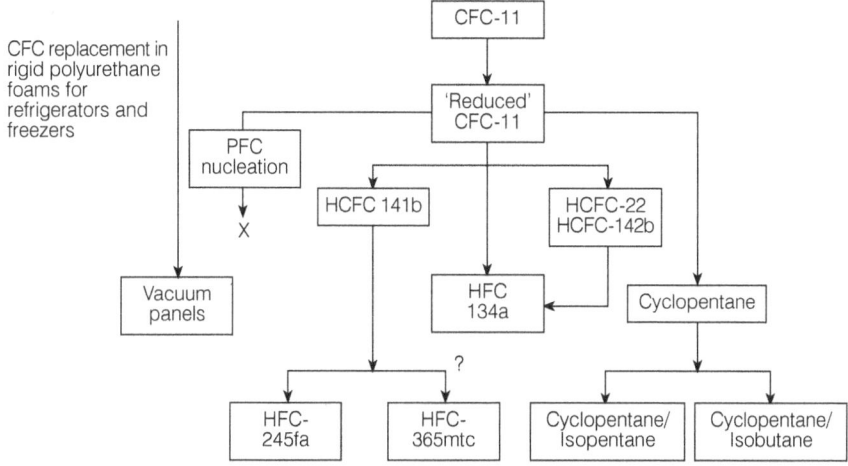

Source: Mike Jeffs

Figure 6.1 *Foam technology innovation*

The first point illustrated by refrigerator foam replacement is that the universal blowing agent CFC-11 was replaced by a range of alternatives. Second, many refrigerator producers in developed countries have undergone several technology changes. Such changes are expensive to execute and not without risk. The risk is that durable products with an expected (by the purchaser) lifetime of 15–20 years are placed on the market with only a year or two of product lifetime testing. As an example, 'reduced CFC-11' foams were introduced in the EU in 1989 to follow the first 50 per cent cut in CFCs of the original Montreal Protocol. Fortunately, these models actually perform well and are still in use.

Another factor is that these domestic appliances use significant amounts of electricity, with economic and climate consequences. The US was the first country to introduce appliance energy-efficiency requirements in 1997 and was followed by the EU. The thermal insulation performance of the replacement blowing agent initially favoured HCFC-141b as this was closest to CFC-11 in overall performance.

Pressures against the use of fluorocarbons in the EU led to the use of cyclopentane as early as March 1993 by Liebherr of Germany and in later, in 1993, by Bosch-Siemens. The use spread rapidly across the region and then to developing countries. The technology has been further refined (cyclo-isopentane blends), and this technology is now the global industry standard, except in North America. In the US, CFC-11 was replaced by fluorocarbon alternative HCFC-141b, which in turn has been replaced by HFC-245fa. The simple truth is that refrigerator manufacturers have been able to follow increasingly stringent energy-consumption requirements, and refrigerators in 2006 consumed less than half the energy, per unit of cooled space, than those made before the Montreal Protocol was introduced, irrespective of the blowing agent used.

Impact of policies of financing institutions on technology selection in CEITs and developing countries

Policies of the institutions that helped finance conversion projects influenced technology choice in CEITs and developing countries. Close investigation reveals that CEITs were much more likely to select 'final' technologies and that the transitional technology HCFC-141b was rarely used. This was due in part to the Ozone Layer Depletion section of the GEF Operational Strategy, adopted in 1995, specifying that 'the GEF will fund the conversion to the technology with the least impact on global warming' so long as it is 'technically feasible, environmentally sound and economically acceptable'. This would have discouraged the use of substances with high global warming potentials, such as HCFC-141b, which is a potent greenhouse gas.

Policies of the MLF also had a profound impact on technology selection in developing countries; these policies are addressed in the analysis of the costs of Fund projects and counterpart contributions below.

Operational strategy of the GEF on ozone layer depletion

According to the GEF operating strategy, there are two potential ways in which the phaseout of ODSs might add to the risk of climate change. The first is the use of substitutes that have a high global warming potential. The second is the introduction of less energy-efficient technologies that do not use ODSs – if energy is being supplied from fossil fuels, decreasing energy efficiency will increase emissions of greenhouse gases. Therefore the GEF funds conversion to the technology with the least impact on global warming that is technically feasible, environmentally sound and economically acceptable.

MLF principles that guided foam project financing

Three principles guided foam project financing by the MLF:

1 **Cost-effectiveness thresholds.**[5] The Executive Committee of the MLF set upper limits on the cost per kilogram of ODSs phased out ('cost-effectiveness thresholds'); see Table 6.2. Enterprises were paid actual project costs up to the threshold, beyond which they were required to pay any additional costs ('counterpart costs').
2 **No extra payment to meet the safety costs for selection of hydrocarbon technology for foams.** The Executive Committee decided that

Table 6.2 MLF cost-effectiveness thresholds[7]

	Large enterprises (US$/kg ODP)	Small and medium-sized enterprises and low-consuming countries (US$/kg ODP)
Miscellaneous foam	9.53	14.30
Flexible polyurethane	6.23	9.34
Integral skin	16.86	25.29
Polystyrene/polyethylene	8.22	12.33
Rigid polyurethane	7.83	11.74

because projects using hydrocarbon technology were already below the cost-effectiveness threshold for rigid polyurethane foam projects, there was no need to account for the additional safety costs.[6]

3   **Prefer to avoid HCFCs but no objection if the enterprise and the government make an informed choice to use them.**

Summary of MLF decisions

At the 12th Executive Committee meeting it was decided that HCFCs should be approved for use only in applications where more environment-friendly and viable alternative technologies were not available.[8] Also at the 12th meeting the MLF required the implementing agencies to ensure that adequate information on all alternative technologies was provided to enterprises converting from CFCs.[9]

At the 15th meeting it was decided that, whenever possible, HCFCs should not be used;[10] the MLF required implementing agencies, where HCFC projects were proposed, to fully justify the choice and to include an estimate of the potential future costs of second-stage conversion.[11]

At the 19th meeting it was decided that the implementing agencies must provide a full explanation of the reasons why conversion to HCFCs was recommended, together with supporting documentation that the criteria laid down by the Executive Committee for transitional substances had been met. The agencies were also obliged to make it clear that the enterprises concerned had agreed to bear the cost of subsequent conversion to non-HCFC substances.[12]

At the 27th meeting the MLF requested implementing agencies to provide letters from their governments verifying that they had reviewed the specific projects proposing to use HCFC, understood national commitments under Article 2F, had nonetheless determined that the projects needed to use HCFCs for an interim period and understood that no funding would be available for the future conversion from HCFCs for these enterprises.[13]

The 38th meeting decided that for projects to phase out CFCs by conversion to HCFC technologies, governments had to officially endorse the choice of technology and agree that it had been clearly explained to them that no further resources could be requested from the MLF for funding any future replacement for the transitional HCFC technology that had been selected.[14]

Executive Committee policy on use of methylene chloride

The Executive Committee also had a policy on the use of methylene chloride that affected developing country technology transfer. Some enterprises chose the high-cost, safer, 'final' technologies, which often required them to pay counterpart costs. Other enterprises that did not have the financial resources chose the lower-cost methylene chloride technology. The Executive Committee required projects using methylene chloride to take the necessary precautions to safeguard workers' health.

The influence of laws passed in important export markets on technology selection in CEITs and developing countries

Enterprises in CEITs and developing countries had to satisfy European product specifications in order to sell products in the EU.[15] Furthermore, CEITs that wanted to join the EU had to bring their laws into conformity with EU regulations that called for faster ODS phaseout schedules.[16] Particularly influential was EC Regulation No 2037/2000, which came into effect on 1 October 2000. This regulation phased out the use of all HCFCs in foam blowing by 1 January 2004 and required future recovery of ODSs from foams where practicable.[17] For these reasons, HCFC-141b was not selected as a technical solution in any country considered eligible for EU membership.

Market forces and competition from foam system suppliers

From about 1975 to 1990, the technology of foams systems had become commoditized, and technology advances had slowed. After the Montreal Protocol was signed, global foam producers and their suppliers realized that there was now an opportunity for substantial competitive advantage through technology advances. These enterprises produced foam systems tailor-made for specific applications, and the basic CFC-replacement technology was ensconced in these systems.

To engage in this new situation, additional research and development (R&D) staff were hired, and innovation was again under way. The timetable was set not only by legislation but also by the market response to lower cost and environmentally superior new technology. For example, the development in the EU of pentane-based technology for rigid foams enabled earlier HCFC replacement to be targeted and achieved in the EU.

This resurgence in the development of technology was critically important to technology transfer because it coincided with industrial development in developing countries. The polyurethane products were used in consumer goods such as mattresses, furniture, refrigerators and automobiles – some of the very products now in demand in developing countries as their economies grew and they started to take advantage of lower wages and increased exports to developed countries.

To take advantage of the new business opportunity, from the early 1990s onwards the polyurethane chemical system manufacturers established assets such as systems houses in many developing countries, staffed initially with European or American staff. The expatriate staff trained local chemists and engineers to become familiar with the technology and to apply it with local enterprises.

A prime example is China, where foreign multinational enterprises have established such strong local capabilities that basic innovation can now originate locally. Local manufacture of diisocyanates chemicals for blowing is currently growing rapidly in China to meet burgeoning demand internally and for exports.

Another interesting example is India, where local systems houses had been established for many years. These systems houses worked with international enterprises to some extent but were also capable of developing technology on their own, in particular for unique local applications such as thermoware.

SUSTAINABILITY, SUITABILITY AND THE ENVIRONMENTAL PROPERTIES OF ALTERNATIVE TECHNOLOGIES

Alternative (in other words non-CFC) foam-blowing agents currently include HCFCs, hydrocarbons, carbon dioxide and HFCs. The major HCFCs used in foam blowing are HCFC-22 and HCFC-141b, which was introduced in 1993. Hydrocarbon foam-blowing agents include butane, cyclopentane, pentane and isopentane, cyclopentane being by far the most popular. Carbon dioxide blowing agents come in two varieties, classified by method of injection. CO_2 can be inserted either in liquid form or by a chemical reaction between water and isocyanate. HFCs are becoming popular blowing agents as well – options include HFC-134a, HFC-152a, HFC-227ea, HFC-245fa and HFC-365mfc. HFC technology was largely unavailable during the initial transition to alternative foam-blowing technology, but now that HCFCs face strict phaseout deadlines, many enterprises that initially transitioned to HCFCs are considering HFC alternatives.

The technical suitability, environmental acceptability, safety and cost-effectiveness of alternatives guided technology selection. Prior to the Montreal Protocol, foam products and processes used CFC-11 as the blowing agent, and the technologies were thus fashioned around the properties of CFC-11. The replacement technical specification included:

- suitable boiling point (near 24°C would be ideal);
- miscible with other chemicals in the formulation and with the capacity to reduce the overall viscosity; and
- low solubility in the polyurethane matrix.

In closed-cell rigid foams, the blowing agent remains as a gas in the foam cells and will increase the thermal insulation efficiency if it is less conductive than air. Consequently, additional requirements are low diffusivity through the foam over the life of the product and low gaseous thermal conductivity.

From an environmental standpoint, the best alternative foam-blowing agent would have no ozone-depleting impact, a very low or no GWP, low thermal conductivity (for thermal insulating applications), and a minimum of other environmental impacts.

HCFCs, while superior to CFCs in both ozone depletion potential (ODP) and GWP, still deplete the ozone layer and will have to be phased out by 2020 by non-Article 5 Parties and by 2040 by Article 5 Parties.

CO_2 has a low GWP but it increases the thermal conductivity of foam. When foam is used for thermal insulation, the higher thermal conductivity can create environmental problems indirectly: less insulation performance means more energy has to be used to keep the desired level of heat or cold in a building (or appliance) and hence more emissions of greenhouse gases.

Box 6.5 The TEAP Foams TOC

Sally Rand, US EPA, Foams TOC Co-Chair, 1993–1998

The TEAP (Technology and Economic Assessment Panel) and its Foams Technical Options Committee were forums for global discussion regarding the environmental acceptability of foam-blowing options. From the start, there was great concern about the toxicity of the foam-blowing agents and about their thermal performance, which would affect the energy efficiency of appliances and buildings where many foam products are used. Environmental authorities made sure that TEAP had the latest toxicity information, and findings from the TOC were quickly reported through UNEP, foam associations and expert networks.

Methylene chloride does not contribute significantly to atmospheric pollution, to the depletion of stratospheric ozone or to global warming. However, it is more toxic than other alternatives. The US Occupational Safety and Health Administration set a very strict worker exposure limit on methylene chloride: no more than 25 parts per million averaged over an eight-hour time period. Similar limits exist in other industrialized countries.

Hydrocarbon foam-blowing agents are volatile organic compounds that can adversely affect air quality. On the other hand, they are relatively non-toxic and do not harm the ozone layer or contribute to global warming.

HFCs are ozone-safe, with zero ODP, and are far superior to CFCs in terms of GWP. However, they are still strong greenhouse gases with higher GWPs than certain other substitutes and alternatives.

From a safety standpoint, the best alternative would bring the fewest additional risks. Under normal manufacturing circumstances, HCFCs and HFCs do not pose fire risks above and beyond those associated with CFCs.[18] Carbon dioxide has toxicity and asphyxiation risks, and some CO_2 foam-manufacturing processes, such as low-index additive technology, increase fire safety risks due to the heat generated inside the foam from the chemical reaction. Hydrocarbons and methylene chloride are more flammable than the CFCs they replace, and methylene chloride is toxic. Though the risks can be managed, many consider methylene chloride to be an interim technology. Enterprises selecting these technologies must take proper safety measures, including improved containment, ventilation, training and the provision of protective equipment for personnel.

From a cost-effectiveness standpoint, the best alternative would maintain profitability even if the cost of production and product price changed. The cost of the alternative is dictated both by the price of conversion (including equipment, safety, training, technology licensing fees, permits and much more) and by the cost of the blowing agent itself. The costs of mitigating flammability or toxicity have a large influence on the ultimate economic feasibility. Operating costs must be factored in as well, although these may be higher or lower for the replacement technology over time. Generally speaking, conversion to HCFCs required the least capital investment but had high foam ingredients costs.

Hydrocarbons, particularly cyclopentane, are available in ample supply at affordable prices, but the associated safety requirements made economic conversion to hydrocarbon technology difficult for small enterprises. Similarly, methylene chloride is commercially available and economical but requires substantial safety investments. Liquid carbon dioxide, HFC foam-blowing technologies and recently developed additive technologies require more investment than other alternatives.[19]

From the technical standpoint, the best alternative would have performance equal to or better than that of the agent replaced for each critical aspect of the foam product, including appearance and thermal, acoustic and structural performance over the product lifetime. Each foam-blowing agent creates specific foam qualities, and this limits the application of certain alternatives. For example, the higher thermal conductivity of carbon dioxide relative to CFC-11 cannot always be fully compensated for by improved cell structure or the application of thicker foam insulation. In practice, HCFCs tended to increase foam density and increase thermal conductivity, although in some cases these effects could be mitigated through blending. For example HCFC-141b, when blended with HCFC-22, gives advantages of reduced density and cost with minimal impact on thermal conductivity and energy consumption. CO_2-blown water-based foams typically increase densities and reduce the flow properties of the foam mixture due to higher viscosity. Typical problems associated with liquid CO_2 foams include achieving high hardness at low density, control of cell structure (pinholes), achievement of optimum block profile and producing foams with solid particles. CO_2 storage can also be an issue; requiring permanently installed storage tanks for slabstock processes rather than returnable O_2 bottles that would require frequent resupply.

TECHNOLOGIES SELECTED IN DEVELOPING COUNTRIES

The analysis covered 943 enterprises in 631 foam projects. Over half of these projects (332) chose HCFC-141b, 100 chose water, 90 chose methylene chloride (MC), 65 chose hydrocarbon (HC), 35 chose liquid carbon dioxide (LCD) and 9 chose HCFC-22. The total ODP tonnes phased out by each alternative technology are presented in Figure 6.2.

Did choice of technology evolve over time?

Even though the first commitment of developing countries to freeze their consumption of certain ODSs started in 1999, many started their projects early – and were far in advance of their commitments – in order take advantage of the MLF support that was available. Findings regarding the choice of foam technologies over time are illustrated in Figure 6.3. HCFC-141b was a preferred choice throughout the entire period. Water and LCD were first selected in 1994, and the number of projects grew after that. HCFC-22 was selected only in India and only for a brief period (1994–1995).

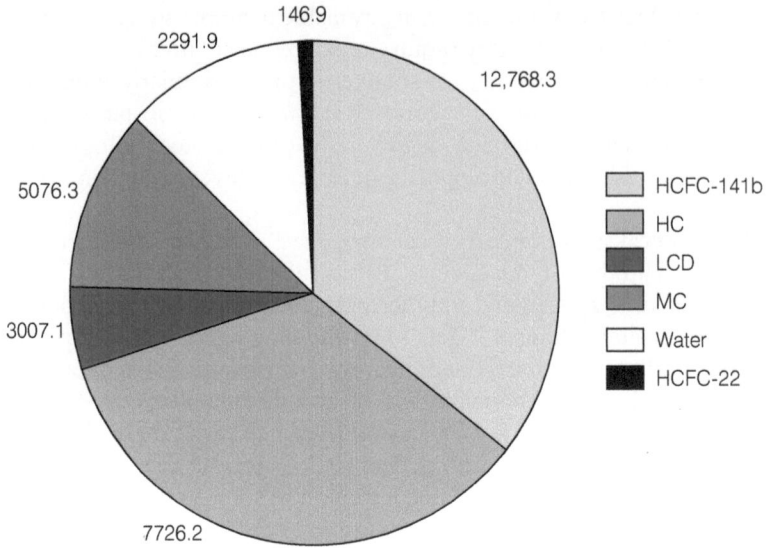

Figure 6.2 *Ozone depletion potential tonnes phased out by alternative foam-blowing technologies*

Did small enterprises choose different technologies? If so, why?

Small enterprises in developing countries phasing out less than 10 ODP tonnes overwhelmingly chose HCFC-141, whereas larger enterprises selected a variety of technologies (see Figure 6.4).

Under rules decided by the Executive Committee of the MLF, firms financed to phase out CFCs with HCFC alternatives will eventually have to phase out the HCFCs at their own expense. Why, then, did any firm choose HCFC technologies? One hypothesis is that HCFCs were chosen because the initial cost of capital investment was lower. The MLF put an upper limit on the cost per kilogramme of ODS phased out. These limits varied for each ODS-using sector and sub-sector. If a project cost more than the limit, the enterprise had to pay the difference.

There was a formula[20] for increasing the cost-effectiveness threshold for HC projects in the domestic refrigeration sector (which often included a foam-blowing element) that had the effect of increasing the maximum grant level by almost 50 per cent.[21] However, this was practical only for larger enterprises because the costs of safety equipment for HC technologies were about the same for any size of factory. Due to MLF funding limits, small enterprises frequently could not afford their portion of the high safety equipment costs. HC safety technologies for foam and refrigeration were also complex, and considerable knowledge and employee discipline was required to maintain safety during operation and maintenance. And small industries often do not have such capabilities. Some large enterprises could easily have opted for HC technology but still insisted on HCFC technology because it was less expensive and easier to implement. Possibly they also perceived a preference in the market for HCFC

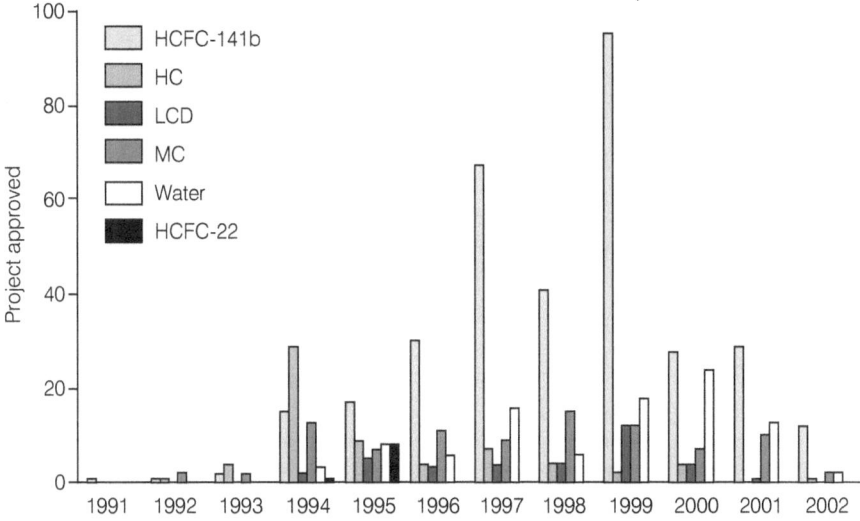

Figure 6.3 *Evolution of foam technology choice in developing countries*

products. In some cases, enterprises also favoured HCFCs because they could claim incremental operating costs (for the HCFCs themselves, as they are more expensive than CFCs), instead of reducing their overall grant due to incremental operating savings arising from the lower costs of HCs.

To answer whether smaller enterprises chose HCFC-141b to avoid payment beyond the MLF contribution, reports were examined from enterprises phasing out 10 ODP tonnes or less with HCFC-141b technology. Information was available for 30 of the 42 projects in this category, representing 97 small and medium-size enterprises. We found that almost all of these enterprises preferred a zero-ODP alternative, but that none of the preferred choices were available at the time they selected HCFC-141b technology. This suggests that costs were not the only or most important factor in technology selection among small enterprises.

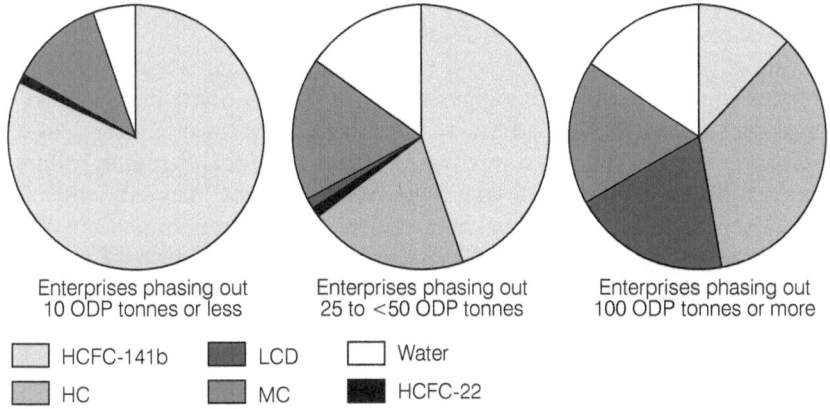

Figure 6.4 *Technology selection and project size*

Developing countries selected a variety of technologies

With the exception of Nigeria, enterprises in all the developing countries selected a variety of technologies, with HCFC-141b, hydrocarbons and water the three most frequent choices. Companies in Nigeria selected a conspicuously different mix of technologies, with 40 (out of the 46) Nigerian foam projects selecting methylene chloride, 5 selecting HCFC-141b and 1 selecting LCD. The reason why Nigeria selected a different mix of technologies is that it has a particularly strong flexible-polyurethane foam industry; hence methylene chloride was the preferred choice as the least-cost technology, even though precautions for worker health and safety were necessary.

India was the only country where enterprises selected HCFC-22 foam; all nine HCFC-22 projects were located there. The majority of these projects (six out of nine) were implemented together to save time, to achieve an economy of scale and to avoid duplication of effort.[22] These Indian enterprises produced a range of rigid polyurethane foam thermoware items, such as containers, water bottles, jugs and casseroles. They chose HCFC-22 because it was a proven foam-blowing technology and manufactured locally, and because support was available from local vendors. The other Indian enterprises that chose HCFC-22 produced rigid polyurethane foam panels. They selected HCFC-22 over other alternatives because it was locally available, had a lower ODP than HCFC-141b, and did not require major modifications or investments.[23] Other reasons cited included operational safety, its mature technology and the support from local system suppliers.[24] The tendency of HCFC-22 to diffuse rather rapidly from the foam was not seen as a major disadvantage in applications which had surface materials which would inhibit diffusions.

TECHNOLOGIES SELECTED IN COUNTRIES WITH ECONOMIES IN TRANSITION

Unlike the developing countries, CEITs were much more likely to select 'final' technologies. Hydrocarbon and carbon dioxide foam-blowing agents were the most popular alternatives selected. Methylene chloride was not selected, and the transitional technology HCFC-141b was rarely used.

Cost, safety, product quality and environmental acceptability were major factors influencing firms' technology choices in CEITs. Some firms settled for technologies they may not otherwise have chosen because alternatives had higher operating and installation costs.[25] Others, like JSC Stroidetal (Russia) provided substantial co-funding to get the technology they wanted, rather than settle for second best.[26] In a few cases, hydrocarbon technology was avoided due to flammability and safety concerns, even though it might have produced better foam or a more permanent solution.[27] Some enterprises also cited difficulty in getting a permit to use hydrocarbon technology due to proximity to local residents.[28] LCD was more popular where insulation was not a concern, and transitional technologies, like HCFCs and HFCs, were generally avoided due to anticipation of future phaseout.

It is interesting to note that several of the CEITs acceded to the EU in May 2004 and had immediately to adhere to the Regulations then applying in the EU, including the need to phase out the use of all ODSs, including HCFCs, from the date of accession. These countries included the Czech Republic, Hungary and Poland.

SPECIFIC EXPERIENCES WITH ALTERNATIVE TECHNOLOGIES

Experiences with hydrocarbon technology in CEITs and developing countries

Hydrocarbon blowing agents are used to produce foam for insulation in domestic and commercial refrigerators and freezers, water heaters, and picnic boxes; insulated trucks and refrigerated transport; boardstock/flexible-faced lamination; sandwich panels; pipe-in-pipe and slabstock/preformed pipe; flexible polyurethane (PU) foam slabstock; integral skin foams; extruded polystyrene board; polyolefin foam; and extruded polystyrene sheet.

Hydrocarbon foams typically have higher density and higher initial thermal conductivity than the ODS foams which they replaced. However, optimization of the foam systems using blends or increased thickness can achieve comparable insulation performance in many applications. Safety requirements make economic conversion to flammable hydrocarbon technology difficult, particularly in the case of small factories. However, many foam production facilities in both developed and developing countries are large enough to make conversion to hydrocarbons profitable.

Hydrocarbons were particularly challenging to small and medium-sized enterprises located in crowded urban areas, which would have had to shift to industrial areas in order to safely adopt hydrocarbon technologies. The use of hydrocarbons in any existing or new facility usually entailed delay in order to get regulatory permissions from the local governments, fire safety authorities and worker safety inspectors. Foam-making firms in rented factory space were reluctant to make investments in safety improvements that would be lost when the lease expired and often did not have necessary funds to purchase a suitable building in a location where hydrocarbons could be safely used. In addition, the grants of the MLF, limited by the cost-effectiveness thresholds, were a factor in the small firms shying away from the more costly hydrocarbon technology and choosing the cheaper HCFC technology.

An evaluation by the MLF found that hydrocarbon technology used for foams in refrigeration enterprises was proven suitable for large-scale enterprises with sufficient manpower and technical resources at their disposal. There were, however, serious concerns in the polyethylene and polystyrene packaging industries, which consisted mostly of small enterprises lacking the skills needed to handle such highly flammable materials.[29]

The risk of firms resuming the use of CFC-11 and CFC-12 after transition to hydrocarbons is very low. Hydrocarbon polyol systems are well developed, and the prices of foam-grade hydrocarbons are low relative to the price of HCFCs or dwindling CFC supplies. Equipment previously using CFCs has been destroyed, new equipment is not easily adaptable to CFCs, and market conditions and local legislative measures are effective.

In the case of one CEIT enterprise, the introduction of hydrocarbon foam-blowing technology had the incidental benefit of improved worker safety.[30] Previously, refrigerator doors at this enterprise had been put together with pneumatic riveting machines. These machines had caused stress fractures in the wrists of over 100 female workers over the course of 15 years. The new foam-blowing technology eliminated the riveted design and also eliminated worker exposure to CFC-11 vapours in the door foaming workshop.

Increased fire risk was a challenge in all projects.[31] Measures taken to cope with this varied from country to country. For example, in Bulgaria and Poland, all sub-projects followed industrial safety guidelines in order to mitigate flammability risks. Also in Bulgaria, an environmental impact assessment of health and safety plans was carried out. In Hungary, where safety requirements are particularly strict, some components (valves and electric motors) had to be made explosion-proof in order to gain approval.[32]

In both CEITs and developing countries, there was general acknowledgement that hydrocarbon foam-blowing technology was a suitable and long-term solution, unlike HCFC-141b. The quality of hydrocarbon foam products was about the same as the quality of CFC foam products. Nevertheless, some developing country enterprises were dissatisfied with hydrocarbon technologies for various reasons: high capital investments for safety, unreliable or unavailable supply of hydrocarbon foam-blowing agents (specifically cyclopentane), and complicated fire-safety regulations that required approval from many different statutory agencies and could often take several months.

In Thailand, for example, the US$8.22/kg threshold set by the Executive Committee of the MLF did not provide sufficient funding to allow the purchase and implementation of all the required safety-related equipment.[33] As a result, one enterprise had to rely on its own capital reserves to supplement the grant, and the project was carried out piecemeal and at a slow pace. Companies in several developing countries reported that the upfront costs of converting to hydrocarbon technology were not fully recovered by the long-term operating cost advantages. The upfront costs and uncertain payback discouraged other developing country enterprises from adopting hydrocarbon technologies.

Most CEIT enterprises found the technology to be suitable to their needs and were able to adapt it when necessary; only one enterprise in a CEIT reported minor technical problems, and these were resolved after consultation with technology suppliers.[34] The most common administrative complaint was that World Bank procurement training was insufficient for the project management unit and for participating enterprises.[35]

Experiences with HCFC-141b foam-blowing technology in CEITs and developing countries

HCFC-141b is commonly used to produce rigid polyurethane foam for insulation in domestic refrigerator and freezers, commercial refrigerated appliances, insulated trucks and refrigerated transport trucks (reefers), boardstock/flexible-faced lamination, sandwich panels in construction, spray polyurethane foam insulation, pipe-in-pipe, slabstock/preformed pipe, phenolic foams and integral skin foams. There is also minor use of HCFC-141b for rigid integral skin foam. It is used because of the low capital investment required by manufacturers, many of which are small enterprises with limited production capacity. The 'near drop-in' nature of this liquid blowing agent is a major advantage.

Many developing country enterprises chose to switch to HCFC-141b foam-blowing technology, even though it is a transitional substance that will ultimately be phased out under the Montreal Protocol. Almost all enterprises selecting HFC-141b reported that lower investment costs, reliable technical performance, fire safety, familiarity and simplicity influenced their choice. Some enterprises were unwilling to make large investments in fire safety improvements because they operated in leased facilities and would be vulnerable to higher rent when the lease was renewed or would lose their investment if the lease were not renewed.

Some conversion projects in CEITs and developing countries encountered a number of problems that caused delays. One enterprise in India experienced major problems due to the late shipment of one piece of equipment from the US. An enterprise in the Czech Republic had problems with the higher temperatures reached during the foaming process and the fact that considerable development was necessary before an acceptable product could be made.

In most countries, the chances of businesses reverting from HCFC-141b to CFC use are quite slim: improved HCFC-141b polyol systems are available, the cost of CFC-11 is increasing relative to HCFC-141b, and the MLF required that old CFC equipment be dismantled and destroyed in every funded project.

Experiences with HCFC-22 technology

HCFC-22 is used in the manufacture of foams used in sandwich panels and in low-density rigid foams for flotation devices. It was selected by only nine enterprises, all in India. Mixing ratios were difficult to control and the cell structure of the polyurethane was initially poor in quality. The process also consumed 13 per cent more foam matrix material than the CFC-based process. In some cases, HCFC-22 was blended with HCFC-141b in order to get better results.

Experiences with LCD foam-blowing technology in CEITs and developing countries

LCD foam-blowing technology is used for flexible PU slabstock, moulded foams, and 'cold cure' and extruded polystyrene board. The basic principle of LCD technology is the blending of liquidized CO_2 with other foam components under pressure prior to the initiation of the chemical reaction. This blend is then released through nozzles, where decompression triggers the release of CO_2, resulting in froth. This froth further expands because of the CO_2 released from a water/isocyanate reaction. All LCD equipment suppliers have developed patented technologies to manage these issues. LCD technology has proven to be commercially viable for a significant variety of foam grades in the 15 to $35kg/m^3$ density range.

An MLF analysis found that, although LCD technology conversion projects generally work well, the LCD technology does not forgive errors with regard to foam pressure, chemical mix and mechanical handling. Another reported problem is that segments of the market in developing countries continue to require very cheap foam products, implying very low foam densities (11–12 kg/m^3), and it is difficult to generate such low densities with LCD technology. Accordingly, enterprises sometimes reverted to the use of methylene chloride, purely for reasons of profit.[36]

Sustainability

All developing country enterprises and all but one CEIT enterprise selecting LCD technology judged it environmentally superior and sustainable, with little possibility of reversion to ODSs. This sustainability is because LCD is less expensive than CFC-11.[37] Furthermore, older CFC equipment was often destroyed after the new technologies were installed. Companies successfully implementing LCD conversion technology reported no new workplace safety, health or environmental risks. In at least one CEIT, safety was improved by switching to CO_2 because CFC fumes were eliminated.[38]

Suitability and problems encountered

LCD technology had many attractive properties: it allows for the possibility of manufacturing foams of low density; it is often locally available (usually produced by air separation plants or by the CO_2 units at soft drinks factories); and it is an environment-friendly chemical material that was often chosen by enterprises seeking a non-ODS alternative when government authorities or the enterprises themselves considered methylene chloride too toxic and acetone too flammable. The technology proved its cost-effectiveness and technical and commercial advantages at many large-scale industrial enterprises in different countries. It has better physical properties than CFC-11 or methylene chloride blowing foams and better compression set and resiliency. In addition, CO_2 eliminates the odour problem associated with hydrocarbon blowing agents, which contain strong stenching agents so that leaks can be easily detected.

Developing and CEIT enterprises reported that it was a challenge to introduce sufficient CO_2 to allow the production of the lowest density foam and to control the froth created by the reaction mixture. Combining this system with the production of foam types that need no auxiliary blowing agent was also difficult. Overall, the technology was complicated and required a substantial learning curve.

The suitability of the CO_2 technology varied according to use. In the CEITs, enterprises who used it for non-insulating foams found that the technology suited their needs. Those who used it for insulating foam, on the other hand, often found they had to re-design their products, and at least one Hungarian enterprise was forced to continue using HCFC-141b until a better alternative could be found.[39]

Experience with water-derived CO$_2$ in CEIT and developing countries

CO$_2$ derived from a water/isocyanate mixture is used to blow integral skin and non-insulating rigid foams. Almost all non-insulating rigid foams have moved to

---

BOX 6.6 QUOTES FROM THE PROJECT COMPLETION REPORTS OF THE GEF AND MLF ILLUSTRATING THE REACTIONS TO LCD FOAM TECHNOLOGY

The CO$_2$-blown adhesive foam makes an excellent light building panel; no negative impacts of CO$_2$-blown foam were mentioned. Conversion to CO$_2$ did, however, lengthen the curing time required for each panel. As a result, the enterprise had to build a new extension to the factory where the panels could cure before being moved to storage. GEF funds paid for part of this, and they also partially financed a gantry crane used to move the boards [...] the project served as a catalyst for in-house technical developments that have generated interest worldwide – technicians from several European countries, Japan, Russia and Saudi Arabia visited Trimo to learn more about its innovation. (Slovenia)

Metisol planned to use CO$_2$; however, trials revealed an unacceptable degree of dimensional instability in the foamed panels. (Hungary)

The expense and effort involved in the conversion is significantly greater than conversion to methylene chloride. But LCD is cost-effective (CO$_2$ is low-priced and less is needed) and environmentally acceptable (CO$_2$ is obtained from natural sources, no ODP and low GWP). (Argentina).

The product quality of the foam produced on the Maxfoam is much improved from the boxfoam, as it is a fully automated, continuous process. The LCD conversion allows production of the full range of products. (Brazil)

It takes time for local users to get used to new LCD foam [...] The rate of unqualified (defective) products was higher, more raw materials were consumed than planned amounts. (China)

Companies with low technical ability should not try to adopt this kind of technology. (Turkey)

The use of LCD as a blowing agent has many advantages which encouraged the counterpart to select it. (Iran)

Despite there being a very experienced and reliable technology partner (Cannon), there is a possibility that the chosen technology option will not work in a satisfactory manner. Alternative technologies are HCFC-141b or [water-derived] CO$_2$. In addition, Cannon can offer a glass fibre-filled polyurethane. (Russia)

water-based systems with the exceptions of low-density rigid foams for flotation devices and floral foam. CO_2 (water)-based systems have superior environmental performance (CO_2 ODP=0; GWP=1), low health and safety hazard, almost unlimited CO_2 commercial availability, and low or no capital outlays. Disadvantages – which can be overcome by equipment, chemical and process modifications – include the potential of increased densities and reduced flow properties of the foam mixture due to higher viscosity.

Sustainability

The incentive to revert to CFC use was higher for enterprises that switched to water-based CO_2-blown foam technology than for enterprises that selected other alternatives. CO_2 foam technology is less sustainable because of increased operating costs, more expensive polyol and additives, and initially higher scrap generation. In some locations, local regulations and compliance monitoring helped prevent enterprises from reconverting to ODSs.

Suitability and problems encountered

A number of developing country enterprises encountered problems with the water-blown foaming technology. Two enterprises reported that consumption of raw material increased by almost 20 per cent; several reported higher product rejection rates due to air voids; and some reported that water-based spray foam was lower in quality than baseline foam. One enterprise in Brazil found that the quality of the end product was not suitable using only water-based formulations and had to convert to methylene chloride-blown formulations in order to elimi-

---

BOX 6.7 QUOTES FROM THE PROJECT COMPLETION REPORTS OF THE MLF ILLUSTRATING THE REACTIONS TO WATER FOAM TECHNOLOGY

The consumption of raw material [...] increased by almost 20 per cent [compared to CFC foam]. The enterprise faced problems of higher rejection rates due to air voids found in foam products. (India)

The enterprise found that the quality of the end product was not suitable using only water-based formulations [and therefore] converted to methylene chloride-blown formulations. (Brazil)

The production of this type of panel is based on a US patent, and the plant was operating under a licence. The level of product knowledge was relatively low because everything was received under the licence agreement. The change involved the development of in-depth knowledge of the production process and resulted in several improvements by the enterprise that were not directly related to the project but were directly based on the fact that such an in-depth analysis of the production process was necessary to come to an acceptable replacement option. The final product, although higher in cost per kilo, was cost competitive because it turned out to be possible to reduce the density. This project is an example of how technology transfer, although focused on ODS replacement, can benefit an enterprise in its entire process. (Mexico)

nate the use of CFC-11. A similar situation occurred in another enterprise in Brazil, where production trials and practical experience determined that a small amount (up to 20 per cent) of methylene chloride was required to provide the proper cell structure and to enhance adhesion to the vacuum-formed vinyl seating surface.

Experiences with HFCs in CEITs and developing countries

HFC-134a is used to create foam sandwich panels for construction, integral skin foams, extruded polystyrene sheet, extruded polystyrene board, phenolic foam, flexible integral skin and rigid polyurethane foam, and for domestic refrigerator and freezer insulation. HFC-152a is used in extruded polystyrene sheet and extruded polystyrene board, and HFC/LCD blends are used in extruded polystyrene board.

No enterprise in either developing countries or CEITs selected HFCs alone as a foam-blowing agent. One Czech enterprise (BHL) chose a blend of CO_2 and HFC-134a on the basis of fewer environmental risks and non-flammability. BHL solved technical problems, which caused three foaming beds to warp, with little outside help. The GEF paid for a comprehensive foamed panel testing laboratory at BHL, which was made available to other foam users for test and development work and proved invaluable in solving technical problems caused by the foaming agents. BHL successfully transferred its HFC-134a and CO_2 technology and techniques to other Czech manufacturers that adopted the foam panel process. The HFC-134a/CO_2 blend produced foam that had about an 8 per cent higher thermal conductivity than the original CFC-11 blown foam, meaning that the panels had to be made thicker and cost more.

Although the HFC-134a/CO_2 technology was satisfactory, the enterprise predicted that other zero ODP options (for example HFC-245a) would soon be available that would probably offer better final product without introducing flammability concerns. It also predicted that the laboratory supplied under the HFC-134a/CO_2 project would probably be well used in the future as new foaming technology is developed.

Despite the difficulties with the HFC-134a/CO_2 technology, it is highly unlikely that the enterprise will revert to CFCs due to its desire to maintain access to markets that discriminate against ODSs. These markets have allowed the enterprise to increase its market share from about 10 per cent to 30 per cent, making it the largest foamed panel maker in the Czech Republic.

Experiences with methylene chloride in CEITs and developing countries

Methylene chloride is used for boxfoam, moulded foams, and 'hot cure' and non-insulating rigid foams, such as packaging foams. It has a combination of desirable foam-blowing properties, such as a low boiling point, relative inertness and virtual non-flammability. It does not contribute significantly to atmospheric pollution through formation of tropospheric ozone, depletion of stratospheric ozone or global warming. Methylene chloride is a widely used industrial chemical and its health effects have been studied extensively both in animals and through epidemiological studies. It is, however, considered 'possibly carcinogenic to humans' (Group 2B) by the International Agency for Research on Cancer (IARC). The volatility of methylene chloride can result in high concentrations in the production area, requiring careful handling to avoid overexposure.

Local and regional exposure and emission regulations vary and affect the use of this auxiliary blowing agent. In several industrialized countries, there are strict limits on worker exposure to methylene chloride. The measures to mitigate human health risks include high-speed factory ventilation and work-station fans, and enclosure of the production line itself. In addition, methylene chloride detectors are installed in strategic places within the production hall and methylene chloride storage rooms.

Results of an evaluation by the MLF [40]

Most developed countries have already halted the use of methylene chloride due to its toxicity, and many developing countries are expected to eventually follow. Because of the cost advantage of using methylene chloride to produce low-density foam products, however, enterprises in developing countries are unlikely to switch to less toxic foam-blowing alternatives. Therefore national legislation, local monitoring and enforcement will be needed to guarantee that safety devices are put in place in these enterprises.

Sustainability

All the countries felt that the methylene chloride technology is economically sustainable because no enterprise will want to shift from low-cost production with methylene chloride to higher-cost production with CFCs. However, many enterprises felt that regulatory authorities would sooner or later require a transition to a technology safer for workers.

Suitability and problems encountered

No CEIT enterprise selected methylene chloride as a foam-blowing agent. It was selected in developing countries because it was allowed by regulatory authorities, commercially available and economical. Methylene chloride results in higher exothermic conditions that require a rapid cooling system, increased/improved process ventilation and fire safety equipment. Methylene chloride was used in many operations, albeit with significant problems related to an inferior cell structure that adversely affects foam quality and risks of fire and toxic exposure. These problems led to a temporary loss of sales in Indonesia. One enterprise in Pakistan that began using methylene chloride was unable to abandon CFCs for foam products where the methylene chloride process had pinhole defects.

Experiences with additive technologies in CEITs and developing countries

Several additives have been developed to modify the chemistry of the flexible

PU foam production process. These additives are predominantly for softening and do not allow very low densities. A special variant of additive technology is the so-called 'low index/additive (LIA) technology', in which the use of certain additives is combined with a lower toluene diisocyanate (TDI) index. The application of additive technologies is limited by the relatively high price. LIA technology is used in flexible PU foam. It is also used in slabstock when the use of methylene chloride is subject to regulatory limitations or poses process problems. It is not used to blow foam; rather, additives were developed to modify the chemistry of the flexible PU foam production process.

A handful of developing and CEIT enterprises chose LIA technologies. Accurate formulation is needed to control the heat generated inside the foam from the reaction between water and toluene diisocyanate, and additional fire safety equipment is necessary. Projects did not report major difficulties and were expected to be sustainable. One project reported low risk of returning to ODSs because the enterprise did not have the technical capabilities to change the formulations back to CFCs.[41]

Other technical options not selected in technology transfer projects funded by the MLF and GEF

Acetone is sometimes used in flexible PU foam and slabstock, with precautions for its flammability. Only about 60 per cent as much acetone as CFC-11 is needed for an equivalent product. Capital outlays and licence fees put the costs close or equal to those of methylene chloride. Organic blowing agents such as ethanol are used in extruded polystyrene board. Several technologies have surfaced that could be classified as 'mechanical' replacement technologies for the use of CFCs in flexible and rigid polyurethane foam, predominantly slabstock. The 'mechanical' technologies allow the integration of the curing area in the emission control, or allow elimination of the use of auxiliary blowing agents altogether. Nitrogen can also be used in extruded polystyrene sheet. These technologies contributed insignificantly to transition away from CFCs in CEITs and developing countries.

CONCLUSION

For buildings insulated with and products manufactured today with CFCs, HCFCs and HFCs, the same or greater energy efficiency can typically be achieved without the use of these ODS and greenhouse gases. Energy efficiency is maintained or improved by using slightly thicker insulation to compensate for the inferior thermal properties of the chemicals in the foam cells, by improving the fundamental design of the products dependent on foam, by not-in-kind substitutes for products and processes, and by regulatory and voluntary programmes that compel the recovery and destruction of foam at the end of useful product life. Alternative foam technology was first developed and commercialized in the 'developed' countries and was then adopted by 'developing' countries.

In developed countries, the drivers of foam technology choice have included:

- the Montreal Protocol, which determined the direction of technology transfer in its initial version and in its various amendments because measures were always earlier and/or more stringent in developed than developing countries;
- national regulations on CFC replacement; and
- societal reaction to environmentally superior chemicals like hydrocarbons, which have increased safety risk.

In CEITs and developing countries, the drivers of foam technology choice, as well as the factors listed above, included:

- the time of choice;
- the influence of the Ozone Operations Resource Group;
- policies of financing institutions;
- laws passed in important export markets; and
- market forces and competition among foam system suppliers.

Markets in developed countries influenced the technology choices in CEITs and developing countries and hastened the adoption of ozone-friendly technologies.

Much can be learned from the different outcomes in CEITs and developing countries. The GEF strategy stressed the importance of choosing alternatives that have both low GWP and higher energy efficiency. Enterprises in CEITs, therefore, chose final (non-ozone-depleting) technologies instead of HCFCs. The MLF left the choice of technologies to enterprises in developing countries but put limits on project funding. The smaller and financially weaker enterprises often chose transitional technologies – HCFCs and methylene chloride – because more sustainable alternatives, like water, were not available at the time technologies were chosen or because they could not afford to put in their own money to top up the grant from the MLF. This shows the need for the agencies promoting technology transfer to ensure that alternative technologies must be friendly to worker safety as well to the environment.

Technology Transfer to Phase Out ODSs in Refrigeration and Air-Conditioning[1]

DIFFICULT DECISIONS: THE HISTORY OF REFRIGERANT CHOICE

Dawn of civilization to 1910

The first controlled refrigeration of perishable foods was accomplished in caves, cellars, wells or artesian 'springhouses'. Simple air-conditioning was first achieved with flowing cold water or by evaporating water on porous surfaces. Natural ice was originally stored locally for warm-weather use, and by the early 1800s globally marketed to temperate regions. Carbon tetrachloride and methyl bromide were the first ozone-depleting substances (ODSs) used as refrigerants, and even these only in obscure instances.[2] In the middle of the 19th century, mechanical ice-making equipment – primarily using sulphur dioxide, ammonia and hydrocarbon refrigerants – became competitive at locations far from unpolluted natural ice sources.[3]

In the early 1900s iceboxes competed successfully with electric refrigerators using flammable and toxic refrigerants. Iceboxes were highly reliable, but poor temperature control could lead to food spoilage, ice was sometimes contaminated by water pollution and ice could not keep food frozen. Refrigerators were more convenient, but carried new risks of fire, explosion, toxic exposure, and food contamination and spoilage from refrigerant leaks. Refrigerant leaks, harmful to human health, were common from vibration, machine failure, defrosting accidents and servicing procedures.[4]

1910–1930: Natural refrigerants replace ice

When mechanical refrigeration became available in the 1920s, entrepreneurs rapidly commercialized systems with a wide variety of refrigerants, including carbon dioxide, water, ammonia, isobutane, sulphur dioxide and methyl chloride. Carbon dioxide, water, ammonia and isobutane have been 'rediscovered' in recent years due to their low global warming potentials (GWPs) and have been designated by proponents as 'natural refrigerants'. Mechanical

Table 7.1 *Flammable and toxic refrigerants in use before CFCs*

Refrigerant	Flammability	Toxicity	Comments
Ammonia	Slightly	High but noxious odour promotes safety	Predominant in industrial and heavy commercial refrigeration; competitive in air-conditioning; some light commercial and household refrigerators (particularly models using the absorption cycle)
Carbon dioxide	Extinguishes fires	Low in small concentration; toxic and inhibits human response at higher concentration	Competitive in air-conditioning; high pressure requires sturdy construction; preferred by German vehicle manufacturers to replace HFC-134a; to be phased out under EC f-gas Directive
Dimethyl ether	High	Moderate	One of the first refrigerants used
Ethyl chloride	None	High	Some light commercial and household refrigerators
Methyl chloride	None	High	Competitive in air-conditioning
Methylene chloride	None	High	Distant second preference for light commercial and household refrigerators, competitive in industrial and commercial refrigeration
Isobutane	High	Low	Some light commercial and household refrigerators
Sulphur dioxide	None	High, but strong odour warns of leaks	Preferred for household refrigerators, but even small leaks spoil food; competitive in commercial refrigeration and air-conditioning

refrigeration was vastly superior to ice, which was increasingly contaminated and did not always ensure safe temperatures for food refrigeration. However, leaks of the most common refrigerants of the 1920s – sulphur dioxide and ammonia – typically required rapid evacuation of homes and buildings. People who came into contact with those substances suffered from vomiting, burning eyes and painful breathing. Accidents with sulphur dioxide and ammonia rarely resulted in death, but accidents with methyl chloride refrigerant were frequently fatal.[5]

1928: Discovery of the CFCs that replaced natural refrigerants[6]

In late 1928 executives from General Motors (GM) and its refrigerator manufacturing division Frigidaire assigned Thomas Midgley of the GM Research Laboratory the task of inventing a non-toxic, non-flammable and non-corrosive refrigerant. Midgley determined that elements with boiling points appropriate for refrigeration were clustered on the Langmuir periodic table, which is arranged according to the number of vacancies in the outer shell of electrons.

Working with Albert Henne and Robert McNary, Midgley ruled out unstable and inert elements, leaving carbon, nitrogen, oxygen, sulphur, hydrogen, and the halogens fluorine, chlorine and bromine. Others had dismissed fluorine because chemical substances containing it are often toxic and/or corrosive.

Midgley and Henne, however, were familiar with Frederic Swarts' theory that the toxicity of fluorine could be negated if strongly bonded with chemicals that had complementary valences. Within two or three days of receiving their research assignment, they had identified chlorofluorocarbons (CFCs) as prime candidates and synthesized dichloromonofluoromethane (CFC-21) from carbon tetrafluoride. Within months, it was confirmed that chlorofluorocarbons would be non-flammable, non-explosive, non-corrosive, very low in toxicity and odourless, and that their vapour pressures and latent heat of vaporization made them very suitable for refrigeration applications. Within a year, GM patented the family of CFCs and perfected the manufacturing process for the first commercial substances, CFC-11 and CFC-12. On 27 August 1930 GM and DuPont formed a joint stock company – the Kinetic Chemical Company – to manufacture and market CFCs.

1930–1985: CFCs and HCFCs replace natural refrigerants

In 1930 CFCs were considered perfect in every way because stratospheric ozone depletion was neither understood nor anticipated. CFCs are chemically stable, non-corrosive, non-flammable, odourless, colourless, energy-efficient and inexpensive, and have very low toxicity.

CFCs and hydrochlorofluorocarbons (HCFCs) rapidly replaced natural refrigerants in many applications. CFC-12 soon became the dominant refrigerant in most small appliance applications.[7] In the 1950s HCFC-22 was applied in commercial refrigeration, particularly as an ingredient in R-502, and CFC-11 was applied in large centrifugal chillers. Natural refrigerants continued to be used in certain applications, but CFCs and HCFCs captured much of the global refrigeration and air-conditioning markets. Ammonia continued to be used in cold storage, ice making and ice rinks, and hydrocarbons continued to be used in industrial refrigeration, particularly at oil and chemical works.[8] CFC and HCFC refrigerants captured the rest of the market, and sales increased with expanding global population, wealth and consumerism. Because they were relatively inexpensive, near-absolute containment was not a design priority for equipment manufacturers or users.[9]

The first motor vehicle air-conditioning using an ODS refrigerant was on the 1939 Packard. Global application of air-conditioning for motor vehicles grew from the 1950s to the 1980s; by the time the Montreal Protocol was signed, more than 90 per cent of North American and up to 50 per cent of new Asian and European vehicles had air-conditioning.

Because neither stratospheric ozone depletion nor global warming were prominent global issues until the early 1970s, refrigeration and air-conditioning equipment manufacturers and their customers came to think of CFCs and HCFCs as 'wonder gases'. Refrigerants were typically vented from appliances and car air-conditioners at service to avoid any risk of damage to equipment from refrigerants contaminated with air, acids, water, oil or metal filings. Reliability was the only incentive manufacturers had to encourage refrigerant containment.[10]

Then, in 1974, Mario Molina and F. Sherwood Rowland – building on the work of many other scientists, including Paul Crutzen, who shared their 1995 Nobel Prize – warned that CFCs deplete the stratospheric ozone layer that protects life on Earth from harmful ultraviolet radiation.[11] In the next two decades, the Molina–Rowland hypothesis was scientifically verified; the 1987 Montreal Protocol was signed and ratified and entered into force; and CFC, HCFC and other ODS production was scheduled to be halted in developed countries and eventually in developing countries.[12]

1985–1995: HFCs replace CFCs and HCFCs

When the Montreal Protocol was signed in 1987, hydrofluorocarbons (HFCs) were considered to be the most promising proven substitutes to replace CFCs and HCFCs in a cost-effective manner. As they were from the same family of halocarbons, HFCs required minimum changes in equipment and manufacturing facilities. Stratospheric ozone depletion threatened life on Earth with skin cancer, cataracts, suppression of the immune system, destruction of agricultural and natural ecosystems, and other unimaginable consequences. ODS replacements had to be found immediately. Hydrocarbons (natural refrigerants) were quickly proposed to replace CFCs, but the typical leak rates and service venting practices would have been unsafe. Additionally, no one knew for sure how quickly technology could be implemented to mitigate flammability.

Other natural refrigerants were proposed, but they required further study and technological innovation. Meanwhile, the fluorocarbon chemical industry and its refrigerant customers moved rapidly to market existing HCFC-22 and HCFC-142b to replace CFCs, to commercialize HFC-134a to replace CFC-12, and to commercialize HCFC-123 to replace CFC-11. HFC-134a and HCFC-123 had been identified decades earlier and patented in the 1970s.[13] New chemicals – including HCFC-225, HFC-143a and HFC-124 – were invented to replace ODSs in applications other than refrigeration.[14] By the time Gustav Lorentzen and colleagues filed for their first modern patent for carbon dioxide systems in 1989 (granted in 1993), this technology was too late as further research and development work was required to launch it in practical applications; thus, it could not capture any of the market for the CFC phaseout.[15]

Industry support for phaseout

Major industry support hastened the transition from CFC to HFC refrigerants. In 1988 the mobile air-conditioning (MAC) sector was the first to agree to recover and recycle refrigerants; in 1990 it was also the first to announce plans to replace CFC-12 with HFC-134a.[16] The early MAC commitment to HFC-134a gave chemical manufacturers the confidence to invest in full-scale production, even before toxicity testing and government approval was completed.[17] HFC-134a was quickly embraced by other refrigeration and air-conditioning

Box 7.1 Coca-Cola's impact in India

Radhey Agarwal, Co-Chair of the Refrigeration and Air Conditioning Technical Committee of the Montreal Protocol Technology and Economics Assessment Panel

In the early 1990s there was a steep growth in the refrigerated and frozen food sectors because of increasing demand by increasingly wealthy food customers. Domestic and international food service companies also contributed to a growing demand for refrigerated appliances as they found success in promoting their products in semi-urban and rural areas of the country. Demand for electric bottle coolers to chill beverages was growing at particularly rapid rates in India because of penetration of multinational companies like Coca-Cola and Pepsi, who introduced a scheme to provide the cooling cabinets to their vendors at a very low cost, paid back through beverage sales.

The bottle coolers were manufactured by a large number of small and medium-sized enterprises (SMEs) and even tiny and unorganized enterprises. The enterprises were scattered all over the country, and many were manufacturing a wide range of commercial refrigeration equipment, such as chest freezers and coolers and upright display cabinets serving the dairy, ice cream, beverage, bakery and frozen food industries. The main clientele of these enterprises was the beverage industry. The appliances manufactured by these enterprises were based on CFC-12 as refrigerant and CFC-11 as foaming agent.

Coca-Cola made a global corporate decision to stop procuring ODS-based appliances in 1996 and insisted on moving to non-ODS technology. This early decision of Coca-Cola moved the appliance manufacturers in both developed and developing countries to convert their production to non-ODS technology at the earliest possible time to remain in business. Otherwise, it would have been a challenge to motivate these enterprises to change over so early. More than 20 SMEs in India converted their production facilities to HFC-134a as refrigerant and HCFC-141b as foam-blowing agent. This resulted in the phaseout of about 300 ozone depletion potential (ODP) tons of ODSs which could have been continued for several more years. This intervention of Coca-Cola had a very positive influence in phasing out CFCs in this sector in developing countries.

Coca-Cola also insisted on the quality of the cabinets, including their energy efficiency. Manufacturers had to send their appliances in the initial stage for testing outside India. Later on, the appliance manufacturing enterprises developed their own laboratories for this purpose. This improved the energy- efficiency level in all segments of commercial refrigeration, which also contributed to the reduction of greenhouse gases (GHGs) released into the atmosphere.

applications because it was similar to CFC-12, non-flammable, non-toxic, proven compatible with specific lubricants, competitively priced and widely available. Coca-Cola, the world's largest customer for refrigerator cases and vending machines, also made an early worldwide commitment to HFC-134a, which encouraged its suppliers in both developed and developing countries to take ozone layer protection seriously and to move quickly with the CFC-12 phaseout.[18]

BOX 7.2 THE FIRST JAPANESE MANUFACTURER OF HFC-134A AUTOMOTIVE AIR-CONDITIONERS

Akira Okawa, Japan Industrial Conference for Ozone Layer Protection (JICOP)

In 1991 Denso Corporation was the first company to produce mobile air-conditioners using HFC-134a, a promising substitute refrigerant, for mass-produced cars in Japan. By the end of 1994 Denso and all other Japanese suppliers realized the total conversion to substitute refrigerant in the global market (5,500,000 units/year). The strong environmental leadership by Denso helped expand its global market share to 22 per cent in 1998 by supplying the global market with 8 million mobile air-conditioners.

Denso also made strong technical and leadership contributions to reducing CFC emissions from old systems using CFC-12.

Denso was the first company in Japan to commercialize equipment that recovers and reclaims old refrigerant by eliminating contamination such as refrigeration oil and water and reusing them for recharging air-conditioners at service outlets.

About 13,000 refrigerant recovery facilities were installed at dealers affiliated with car makers, in service shops and so on, contributing to efficient use of CFC-12 and prevention of its emission into the air. Denso was also among the first companies in the world to offer a complete kit of replacement parts necessary to retrofit CFC-12 air-conditioning systems to HFC-134a.

As a member of the Japan Auto Parts Industries Association, Denso helped establish the national Refrigerant Recovery and Destruction System, which has been implemented throughout Japan by the Japan Automobile Manufacturers Association, and the Refrigerant CFCs Reclamation Center of the Japan Refrigeration and Air Conditioning Association, which reclaims CFCs recovered from the market and returns them to the market. Denso has also commercialized a laboratory instrument that can measure the circulation rate of refrigeration oil continuously and without emitting refrigerant gas into the air during measurement. This instrument has been adopted by domestic and foreign air-conditioner manufacturers and contributes to the development of air-conditioners.

Jointly with a manufacturer, Denso developed a new detergent containing an additive which was able to coat and protect the glass surface of hybrid integrated circuits. Thanks to this, at the end of March 2000 Denso also completed the total phase-out of HCFC-225, a transitional substance which the amended Montreal Protocol aims to phase out totally by 2020.

1995 to the present: Natural refrigerants stage a comeback

Natural refrigerants staged a comeback in the 1990s. In 1992 Greenpeace inspired European governments, industry and consumer support for the use of hydrocarbons in domestic refrigerators.[19] Within one year, a hydrocarbon domestic refrigerator was introduced in Germany, and it rapidly penetrated and expanded across the market. Soon hydrocarbon refrigerators gained market dominance in Europe and penetrated markets in Asia, including Japan. Meanwhile, suppliers of equipment using ammonia as a natural refrigerant recaptured market share from HCFC cold storage and food freezing. They also made limited progress in applying ammonia to commercial refrigeration and air-

---

BOX 7.3 HISTORY OF THE BIRTH OF NON-CFC REFRIGERATORS IN JAPAN

Kiwohide Hata, Matsushita Electric Industrial Company

Why was Europe first to commercialize hydrocarbon refrigerators? In 1993 the ex-East German refrigerator manufacturer FORON developed the first commercially successful hydrocarbon refrigerator on behalf of the environmental protection group Greenpeace. In Japan, however, hydrocarbon refrigerators with flammable isobutane refrigerant were commercialized much later than in Europe, since in 1993 the Japanese refrigerator industry and the government decided that the commercialization of hydrocarbon refrigeration would delay the CFC phaseout, and therefore would have to be postponed.

Japan is very different from Europe in climate, natural features and culture, and the refrigerator technology in Japan had evolved in different ways. European refrigerators use direct-cooling (natural cold-air convection) type refrigerators, while Japanese refrigerators use no-frost technology (forced cold-air circulation) with many electrical parts. European laws that were barriers to the use of flammable refrigerants mostly involved the refrigerator design and assembly. Japanese laws on flammable gases involve the complete product life cycle, including design, production, transport, storage, installation, repair, disposal and recycling. In Europe, Greenpeace and its refrigerator manufacturer were very successful in persuading consumers to accept the new products and persuading safety authorities to allow the new refrigerant. In Japan, examination of the refrigerator life cycle would have been too big a job for one enterprise to undertake without the cooperation of other companies in the same sector, safety certification organizations, administrative organizations and university research centres.

In addition, much plant investment for safer production and more outlay on materials for safer refrigerators would have been needed. Moreover, would consumers have bought such refrigerators, environmentally friendly but more expensive and less safe than existing ones, if they had been commercialized?

In the early 1990s there were many problems with Japanese public liability law, such as the possibility that a court would find that fires had started or had been made more damaging by hydrocarbons released from the refrigerators.

---

conditioning, using secondary loops for safety. At the same time, European researchers, with the support of European vehicle manufacturers, particularly the German automobile manufacturers, pursued carbon dioxide for use in mobile air-conditioners.[20]

Despite their comeback, natural refrigerants still faced stiff competition at the turn of the millennium. Although companies in developed countries had halted the use of CFC refrigerants in new equipment by 2000, there were still plenty of CFC refrigerants around. Refrigerant stockpiles and recycling provided ample CFC supply for service of CFC equipment, as did a conspicuous illegal trade in Europe and North America. Some equipment was retrofitted from CFCs to HCFCs or to HFCs, but new equipment provided energy savings which made retrofit financially and environmentally unattractive by comparison. In Australia and elsewhere, some MACs were retrofitted to use hydrocarbon, despite the opposition of vehicle manufacturers and service associations and the findings of the US Environmental Protection Agency (EPA) that CFC MAC

BOX 7.4 CASE STUDY OF MATSUSHITA SUPPORTED BY GREENPEACE JAPAN

Kiwohide Hata, Matsushita Electric Industrial Company

Since early in 1990 the Matsushita Group has actively tackled environmental problems with the aim of coexistence with the global environment through green products. Matsushita was the first Japanese enterprise to convert from CFC refrigerants to non-ODS substitutes such as HFCs and put non-ODS refrigerators on the market in 1993. In 1996, even before the Kyoto Protocol conference to prevent global warming, Matsushita converted blowing agents for insulating materials from CFCs and other greenhouse gases to climate-safe cyclopentane. Greenpeace was so impressed with these efforts that it started an environmental information exchange with Matsushita.

In March 1998, just three months after the Kyoto Protocol was agreed, Greenpeace sent an open letter to eight Japanese refrigerator manufacturers requesting their plans for selling refrigerators not using HFCs or HCFCs. Since that time, Greenpeace has targeted our customers and demanded that our top management commercialize and sell refrigerators free from CFCs, HCFCs and HFCs. We told them that we were working on new technology but we did not know when it would be ready. This answer was unsatisfactory to Greenpeace and they continued to request so tenaciously that we were extremely embarrassed.

---

systems were unsuitable for retrofit to hydrocarbons because of high leak rates, potentially unreliable aging parts and the absence of systems to mitigate fire risk.[21] HCFC refrigerants were also pervasive in 2000 because Montreal Protocol controls on HCFCs were years away, and most countries had not scheduled aggressive early phaseout.

REFRIGERATION AND AIR-CONDITIONING: TECHNICAL OPTIONS AND CHOICES FOR THE CFC PHASEOUT

Refrigerators and freezers originally used CFC-12[22] as a refrigerant and CFC-11 as a blowing agent for foam insulation (CFC-blown foams are covered in Chapter 6). Motor vehicles used CFC-12 as the refrigerant. Ideally, CFC refrigerants would be replaced with low- or zero-ODP refrigerants that provided the highest life cycle climate performance (LCCP). The LCCP is a measure of the cradle-to-grave greenhouse emissions from all aspects of a product's manufacture, use and disposal. This parameter is particularly important because the direct emissions of refrigerant greenhouse gases are generally far less than those of the greenhouse gas emissions from the generation of power to operate the equipment.[23]

Technical options to phase out CFCs

Alternative refrigerants to replace CFCs were selected based on performance, cost, environmental acceptability and safety risk. Regional and national regula-

Table 7.2 *Refrigerant choices to replace CFCs, 1990–2005*

Application	Refrigerant solution(s)
Domestic refrigeration	HFC-134a and isobutane (HC-600a). Next generation choice will probably be isobutane worldwide.
Commercial refrigeration	HFC-134a in stand-alone units, HCFC-22 and mainly R-404A in supermarket systems, hydrocarbons (HCs) in some self-contained units and in a few indirect systems and, to a small extent, carbon dioxide. Next generation choice will include low-GWP blends now proposed for mobile air-conditioning.
Industrial refrigeration	Ammonia, HCFCs, HFCs and, to some extent, carbon dioxide for low temperatures.
Transport refrigeration	HFCs for the majority of applications.
Stationary air-conditioning equipment	HCFC-22 (in about 90 per cent of equipment), with the remainder using the currently produced HFCs and HFC blends and, to a lesser extent, HCs. Next generation will be HCs in small capacity equipment and HFCs in larger capacity equipment. *Note:* HCFC-22 was primary refrigerant in this sector; it is not a replacement for CFC-12.
Chillers	Primarily HCFC-22 in small systems; HCFC-123 in large systems where energy efficiency is paramount; HFC in small to large systems where energy efficiency is less important or where HCFCs are prohibited; and, much less commonly (or perhaps experimentally), ammonia, carbon dioxide and HCs.
Heat pump water heaters	HCFC-22, HFC-134a, propane (HC-290), R-410A and, to an increasing extent in Japan, carbon dioxide.
Motor vehicle air-conditioning	HFC-134a for virtually all new vehicles as the current global choice. Next generation choice will be either carbon dioxide, HFC-152a or low GWP HFC blends.

tions often led to differences in timing and technology choice between countries. The primary refrigerant solutions for new equipment are summarized by application in Table 7.2.

Table 7.2 indicates that transitional HCFCs have been selected in a variety of applications. One reason for this is that at the first stage of the Montreal Protocol, ozone-safe refrigerants were not available for all refrigerant applications. In its 1991 report, the Refrigeration Technical Options Committee of the Montreal Protocol notes that for some applications 'no suitable alternatives or substitutes have been identified except HCFCs'. HCFCs were promoted because they 'were generally accepted as the solution for a rapid CFC phaseout'. Without a quick transition to HCFCs, the Refrigeration Technical Options Committee worried that there would be 'prolonged CFC use awaiting more "certain" solutions'.[24]

For the long term, the Refrigeration Technical Options Committee reports that there are five refrigerant options:

Table 7.3 *Environmental impacts and atmospheric lifetimes of selected refrigerants*

Refrigerant	Chemical formula	Ozone depletion potential	100 years global warming potential	Atmospheric lifetime (years)
Ammonia	NH_3	0	<1	–
Carbon Dioxide	CO_2	0	1	>50
CFC-11	$CFCl_3$	1.000	4600	45
CFC-12	CF_2Cl_2	0.820	10,600	100
HCFC-22	CHF_2Cl	0.034	1700	11.9
HCFC-123	$C_2HF_3Cl_2$	0.012	120	1.4
HFC-32	CH_2F_2	0	550	5
HFC-125	CHF_2CF_3	0	3400	29
HFC-134a	CH_2FCF_3	0	1300	13.8
HFC-143a	CH_3CF_3	0	4300	52
HFC-152a	C_2H4F_2	0	120	1.4
R-404A	HFC-125/143a/134a	0	3800	–
R-407C	HFC-32/125/134a	0	1700	–
R-410A	HFC-32/125	0	2000	–
HC-290	$CH_3CH_2CH_3$	0	~20	–
HC-600a	$CH_3CH_2CH_2CH_3$	0	~20	–

Source: Refrigeration Technical Options Committee Report (2002)[25]

1 hydrofluorocarbons (HFCs and HFC-blends with 400 and 500 number designation);
2 ammonia;
3 hydrocarbons and hydrocarbon blends (for example HC-290, HC-600, HC-600a and blends of HC-600a and HC-290);
4 carbon dioxide; and
5 water.

All of these refrigerants have both advantages and disadvantages; none are perfect. For instance, HFCs have relatively high GWPs, ammonia is more toxic than the other options, and ammonia and hydrocarbons are flammable. Appropriate equipment design, maintenance and use can mitigate these concerns, though sometimes at the cost of greater capital investment or lower energy efficiency.

CURRENT STATUS OF EACH REFRIGERATION SECTOR

The following sections summarize the advantages and disadvantages of alternative refrigerant technologies for each specific application.

Domestic refrigeration (refrigerators and freezers)

The main alternatives for domestic refrigeration are HC-600a and HFC-134a. HC-600a is compatible with historically accepted mineral oil lubricants. It has no ODP and low GWP; it is, however, highly flammable, and equipment design,

---

Box 7.5 THE BIRTH OF OZONE- AND CLIMATE-FRIENDLY REFRIGERATORS IN JAPAN

Kiwohide Hata, Matsushita Electric Industrial Company

Following very considerable efforts by many people, other Japanese refrigerator companies decided to give support to Matsushita to establish its autonomous safety standards by October 2001. Confident that the standards would be acceptable for industry and government, Matsushita promised Greenpeace at the end of 1999 that it would commercialize refrigerators without CFCs, HCFCs or HFCs by the end of 2002.

With the leadership of the Japan Electrical Manufacturers Association, all of the Japanese refrigeration industry worked together to solve all the safety problems for the life cycle of refrigerators and in a short time developed the autonomous safety standards for designing refrigerators. This cooperative and prompt action allowed Matsushita to sell Greenpeace-approved refrigerators in February 2002.

Looking back, it is clear that the persistent demands of Greenpeace helped to unify our industry's values and forced us to band together and develop the new safety standards. Matsushita Electric has basically implemented our theme of coexistence with the Earth and environment. However, success in the future will require that companies, NGOs and consumers work in partnership to share roles and to develop and promote environmentally friendly products. For example, retail stores must correctly understand the features of environmentally friendly products and must actively disclose product and environmental information that is easy for consumers to understand. For coexistence with the Earth, the retail stores must explain such environmentally friendly products to consumers in order to encourage them to prefer such products. Consumers must also be enabled to buy them at reasonable prices.

---

manufacture, repair and disposal must therefore properly deal with the issue of flammability. HFC-134a is non-flammable and has no ODP. However, it has a significant GWP (1300) and uses moisture-sensitive polyolester oils. Manufacturing processes must therefore properly maintain low moisture levels. Long-term reliability requires more careful avoidance of contaminants during production or servicing compared to previous CFC-12-based designs.

Domestic refrigerators and freezers designed for HC-600a and HFC-134a can be safe, energy efficient, reliable and economical. In practice, either refrigerant is able to satisfy regulatory or market demands for energy efficiencies. Current technology designs typically use less than one-half the electrical energy required by the units they replace. This reliable high performance is achieved with better designs and controls and with superior materials for heat exchangers.

Commercial refrigeration

Commercial refrigeration equipment consists of three main system types:

1 stand-alone equipment,
2 condensing units, and
3 centralized systems.

Stand-alone equipment includes integrated display cases, ice machines, vending machines, and an array of small equipment installed in food stores, restaurants or other public areas. HFC-134a is currently the dominant refrigerant replacing CFC-12. Other refrigerants, such as HC-600a, HC-290 and carbon dioxide, are poised to gain market share in response to environmental pledges by Coca-Cola, McDonald's, Unilever and others who have committed to use HFC-free equipment.[26] Like HC-600a, HC-290 is flammable; it can be used safely only through appropriate engineering and by following proper safety practices. Carbon dioxide has no ODP and the lowest GWP (GWP=1) of all alternative refrigerants, but it poses engineering challenges due to the high pressure required. Safety precautions are also necessary to avoid exposing workers to excessive concentrations of carbon dioxide.[27]

Condensing units are typically installed in specialized applications. The refrigerant of choice depends on the temperature to be achieved and the ambient temperatures where the equipment will be located. Both HFC-134a and the HFC blend R-404A are the preferred options for achieving medium refrigeration temperatures, R-404A where lower temperatures must be provided. Due to safety concerns, HCs are not a common option in condensing units because the amount of flammable refrigerant necessary to operate the equipment would pose an unacceptable fire risk if it were to leak into the area where the equipment is operating.

Centralized systems are installed in super- and hyper-markets. These consist of a large and efficient compressor and condensing unit serving individual equipment with separate evaporators throughout the store. The choice of refrigerants largely depends on national regulations. CFC-12 is still being used in developing countries, but supermarkets in all countries are tending toward the use of the same refrigerants in new equipment. HCFC-22 is still widely used in the US, with the HFC blend R-404A gaining market share. In Europe, HCFC-22 has been banned in new equipment since 1 January 2001, and R-404A is now the preferred choice there. In Japan, CFC-12 was replaced by HFC-134a and sometimes by R-407C.

Large-size refrigeration (industrial, cold storage and food processing)

Large-size refrigeration falls into three major categories of applications: industrial refrigeration, cold storage and food processing. Industrial refrigeration covers a wide range of cooling and freezing applications, including chemical, pharmaceutical, petrochemical, metallurgical, plastic moulding, sports (snow making and ice rinks), industrial ice making and air liquefaction. Cold storage is used for both raw materials and finished products in food processing factories and distribution networks.

Food processing refrigeration requires a wide range of cooling and freezing applications, including the processing and storage of meat, fish, cheese, beer, eggs, fruits and vegetables. Refrigeration is used to preserve food from the point of harvest, catch or slaughter; through processing, transport, storage and distribution; and up to the point of retail sale.

Energy consumption is the major concern for large-scale systems. Ammonia refrigerant is used in approximately 75–85 per cent of the current installations, followed by HCFC-22, HFCs and HFC blends.

Transport refrigeration

Transport refrigeration includes refrigeration on aircraft, ships, trucks and railcars. Refrigeration is provided by loading products into refrigerated vehicles or placing separately refrigerated boxes onto the transport vehicle. Before the Montreal Protocol, most transport refrigeration systems used CFCs. Many have been retrofitted or scrapped, but remaining uses on old refrigerated containers and trucks are still significant. In ships, most existing systems use HCFC-22, although R-407C and R-404A/R-507A are options already used in Europe for new systems.

Most existing refrigeration systems on merchant ships use HCFC-22. R-404A/R-507A mixtures dominate new systems, particularly on fishing fleets and naval vessels, which form a significant part of this sector.

Most of the estimated 550,000 refrigerated containers, other than those on ships, use HFC-134a or R-404A, with only a small number of the oldest containers still using CFC-12. The new refrigerated road vehicles use HFC-134a, R-404A or HCFC-22 (to a lesser extent), and some units with R-410A are available.

Railcar passenger air-conditioning is moving from HCFC-22 with relatively high leakage rates to HFC-134a or R-407C, which require extra measures to contain the higher refrigerant pressures. Generally, HFCs are the preferred refrigerant for new systems.

Air-conditioning and heat pumps (refrigerant-to-air)

Globally, air-cooled air-conditioners (including heat pumps) comprise the vast majority of the air-conditioning market. Air-cooled air-conditioners fall into four categories: window-mounted, non-ducted split residential and commercial, ducted split residential, and ducted commercial. Nearly all air-cooled air-conditioners manufactured prior to 2000 used HCFC-22 as a working fluid.

Since 2000 there has been a significant shift away from the use of window-mounted air-conditioners to non-ducted split residential air-conditioners as the entry-level air-conditioning product in developing countries, particularly in Asia. The primary non-ODS refrigerants are R-407C, R-410A and, to a lesser extent, HC-290. A significant shift to non-ODS alternatives has been observed in Europe and Japan, with a much smaller shift – of approximately 5 per cent – in US markets.

The primary retrofit refrigerant is the zeotropic blend R-407C. Hydrocarbon refrigerants are viewed as unlikely retrofit options because of high cost and the complexity of assuring safety.

Chillers and heat pump water heaters

Chillers, also known as water chillers, cool water or other heat transfer fluids

Box 7.6 Using life-cycle climate performance
to assess and minimize climate impacts
of ODS substitutes and alternatives

Donald Kaniaru, Rajendra Shende, Scott Stone and Durwood Zaelke

The Montreal Protocol and its Parties have recognized the need to consider fully the environmental impacts of their strategies, especially those of ODS substitutes on climate, which are often the most significant. Article 2F(7) of the Montreal Protocol sets out the control measures for HCFCs and states that in addition to minimizing ozone depletion, the decision to use HCFCs should meet other environmental standards:

> *Controlled substances in Group I of Annex C [HCFCs] are selected for use in a manner that minimizes ozone depletion, in addition to meeting other environmental, safety and economic considerations.*[28]

This approach was supported by Decision V/8 (Fifth Meeting of the Parties, Bangkok, 1993), which requested the Parties to consider ODS substitutes in the light of Article 2F and their 'environmental aspects'. This was expanded in Decision VI/13 (Sixth Meeting of the Parties, Nairobi, 1994), stating that the assessment panels:

> *should consider how available alternatives compare with hydrochloro-fluorocarbons with respect to such factors as energy efficiency, total global warming impact, potential flammability and toxicity.*[29]

Subsequently, a group of 41 Parties issued a Declaration at the Tenth Meeting of the Parties (Cairo, 1998) reiterating their support for the consideration of climate impacts, noting the 'scientific indications that global warming could delay the recovery of the ozone layer' and that 'environmentally sound alternative substances and technologies are commercially available for virtually all HCFC applications'. The Declaration urged:

> *all Parties to the Montreal Protocol to consider all ODS replacement technologies, taking into account their global-warming potential, so that the use of alternatives with a high contribution to global warming should be discouraged where other, more environmentally friendly, safe and technically and economically feasible alternatives or technologies are available.*[30]

The consideration of environmental impacts is part of a general obligation under principles and concepts of international environmental law. Specifically, the Environmental Impact Assessment (EIA) principle places a general duty on states to consider the cumulative environmental impacts of proposed actions where there are possible transboundary or global impacts. The EIA principle is related to the concept of Integrated Pollution Prevention and Control (IPPC), which was developed to respond to the fact that environmental regulations targeting a single problem can simply shift pollution from one medium to another rather than eliminate it. Broadly, it requires a holistic assessment of environmental impacts when developing regulations, particularly for the use of chemicals, and has been incorporated into numerous multilateral environmental agreements

(MEAs) and other international instruments, including the European Commission's 1996 IPPC Directive.

IPPC requires a life cycle analysis (LCA) of environmental impacts to measure the 'cradle-to-grave' impacts of a product, chemical or technology. This kind of LCA was codified by the International Standards Organization (ISO) 14040 Series. It was described in the Special Report by the Intergovernmental Panel on Climate Change (IPCC) and the Technology and Economic Assessment Panel (TEAP) as involving an 'inventory of relevant inputs and outputs of the system itself and of the systems that are involved in those inputs and outputs (Life Cycle Inventory Analysis)'.[31]

The replacement of ODSs with substitutes and other alternatives, including not-in-kind alternatives, will produce climate benefits to the extent the changes result in higher energy efficiency or otherwise reduce climate emissions.

This is supported by Agenda 21, which calls on Parties to '[r]eplace CFCs and other ozone-depleting substances, consistent with the Montreal Protocol, recognizing that a replacement's suitability should be evaluated holistically and not simply based on its contribution to solving one atmospheric or environmental problem'.[32] This is further supported by the exclusion of gases regulated by the Montreal Protocol from the UN Framework Convention on Climate Change and the Kyoto Protocol.

UNEP, in conjunction with the US EPA, Japan's Ministry of Economy, Trade and Industry, and the Alliance for Responsible Atmospheric Policy, has developed its own version of the substitution principle, known as 'responsible use', which recommends the use of technologies so long as the undesirable effects are minimized and the technology achieves higher environmental performance than its alternatives. Responsible use principles would permit the use of ODS substitutes 'only in applications where they provide safety, energy efficiency, environmental or economic advantage' and where 'undesirable effects are minimized and the technology achieves higher environmental performance than its alternatives'.[33]

that are pumped to the point-of-use for air-conditioning and process cooling. CFCs are being replaced largely by HCFCs and HFCs as the older machines are phased out in developed countries. There is some growth in machinery using not-in-kind non-halogenated refrigerants such as ammonia and hydrocarbons in small machines and CO_2 in some heat pump water heaters. HCFC-123 chillers have demonstrated significantly higher energy efficiency and LCCP than HFC-134a chillers but are currently scheduled for phaseout under the Montreal Protocol.[34]

Vehicle air-conditioning

Vehicles (cars, trucks and buses) built before 1994 used CFC-12 as the refrigerant in passenger compartment air-conditioning. HFC-134a has now replaced CFC-12 as the globally accepted mobile air-conditioning (MAC) refrigerant, and the industry is busy expanding global production to meet the increasing demand for air-conditioned vehicles. HFC-134a is a potent greenhouse gas controlled by the Kyoto Protocol. Due to concerns about relatively high direct refrigerant emissions of HFC-134a from MAC systems and significant indirect emissions of carbon dioxide from fuel used to power MAC, vehicle makers, suppliers, governments and NGOs have formed the Mobile Air Conditioning Climate Protection Partnership to reduce refrigerant leakage, improve energy efficiency and search for replacement refrigerants. The leading replacement refrigerants are carbon dioxide and HFC-152a, both of which have been demonstrated in prototype cars. Recently, Honeywell, DuPont and INEOS have announced new refrigerants with very low GWP and little or no ODP.

On-site recycling of refrigerant at service outlets has proven to be quite effective for HFC-134a systems; 90 per cent or more of the original charge can be recycled and reused during service.

WHAT TECHNOLOGIES DID CEITS AND DEVELOPING COUNTRY ENTERPRISES SELECT, AND WHY?

Over 400 technology transfer projects in countries with economies in transition (CEITs) and developing countries funded by the Global Environment Facility (GEF) and the Multilateral Fund (MLF) were examined for this analysis. The investigation sought to discover what technologies these countries' enterprises selected, and why. It also examines environmental risks of alternative technologies, problems encountered and the enterprises' level of satisfaction with the alternatives.

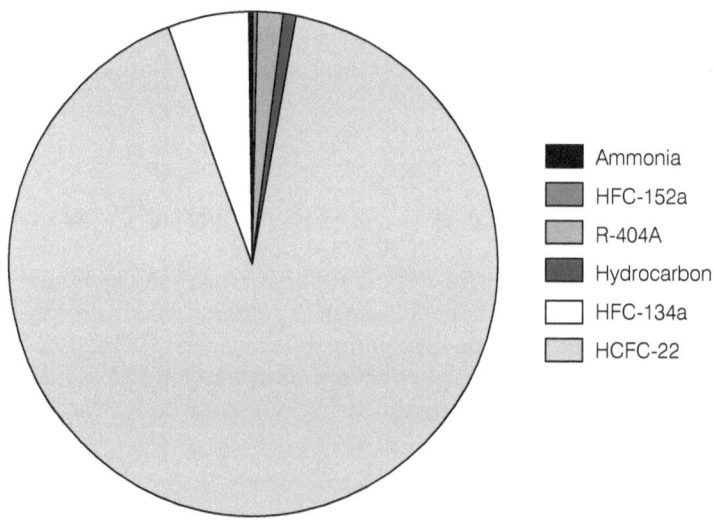

Figure 7.1 *Refrigerant choice in MLF projects*

HFC-134a

Most CEIT and over 90 per cent of developing country enterprises (303 out of 330 projects) chose HFC-134a refrigeration technologies. Developing countries reported that HFC-134a was chosen to replace CFC-12 as a refrigerant since it was a mature and established technology. In addition, hermetic compressors optimized for HFC-134a were commercially available. It was cost-effective under licensing agreements, had zero ozone depletion potential and had low toxicity.

Of the developing country HFC-134a projects, 244 included a foam-blowing element that also eliminated the use of a CFC foam-blowing agent. All but 13 of these projects selected HCFC-141b for the blowing agent. The others chose hydrocarbon (usually cyclopentane) blowing agents.

Several firms reported that they had considered hydrocarbon technology, but ultimately chose HFC-134a because hydrocarbon technology was still in its infancy. Others decided to use HFC-134a instead of hydrocarbon technology due to safety considerations and the high costs of additional safety requirements to manage the flammability of hydrocarbon refrigerants. Several enterprises considered using HFC blends, but ultimately selected HFC-134a because the use of blends would have led to quality and performance shortcomings.

Enterprises generally found the HFC-134a technology suitable to their needs, and few problems and numerous benefits have been reported. One CEIT report notes that the transition to HFC-134a project led to the revival of many enterprises in the commercial refrigeration sector, with co-benefits of increasing employment in areas suffering from high unemployment rates. A CEIT enterprise located in the Czech Republic formed a joint venture with Thermo King US that allowed local engineers to become the lead designers of new HFC transport refrigeration systems marketed worldwide. They also designed retrofit technologies that became popular throughout the Czech Republic and in much of the former USSR.[35]

Many of the enterprises converted to HFC-134a even though CFCs were available and plentiful. (CFC production in developing countries is due for total phaseout only in 2010.) Implementing agencies of the Montreal Protocol were concerned about whether the conversion would be sustainable or whether the enterprises would revert back to CFCs. Happily, few, if any, of the enterprises were expected to revert to CFCs, because of regulations, desire to maintain domestic and export markets that forbid CFCs, and irreversible abandonment of the old, ODS-based manufacturing technology (see Box 7.7). In one case, however, project implementers reported that there was an economic incentive for the enterprise to return to using CFCs as refrigerants because of the higher price of HFC-134a. The lower price of CFC-12 in Argentina gave service technicians an incentive to use CFC-12 in equipment designed for HFC-134a, despite likely long-term damage to the equipment. Compliance and monitoring was necessary to prevent the enterprise from converting back to ODSs and to ensure technicians serviced equipment with the proper refrigerant.

A few CEIT enterprises ran into financial trouble after converting to HFC-134a. Many CEIT companies had attempted to convert to HFC-134a independently and were financially stressed as a result. Although the GEF offered to fund conversion projects in CEIT enterprises, enterprises were not able to recover all of their expenses because GEF rules do not allow retroactive financing. One example of a firm that ran into trouble is Oruva in Lithuania, which converted manufacturing facilities to mass-produce HFC-134a and HC-600a compressors. It had already invested a significant amount of money before GEF funds were made available to complete the conversion. Installation of

---

BOX 7.7 REASONS WHY ENTERPRISES WERE UNLIKELY TO REVERT TO ODSS AFTER SWITCHING TO HFC-134A: EXPLANATIONS OFFERED IN MLF PROJECT COMPLETION REPORTS

Awareness has now been raised, and consumers know that they have to use products made of new technology due to the legislation to ban the use of ODSs in this use. (India, Iran)

The enterprise, a manufacturer of commercial deep freezers, beverage bottle coolers and ice-cream cabinets, mostly supplies its products to multinational organizations like Coca-Cola for the domestic market. When Coca-Cola decided to halt the purchase of CFC-refrigerated cases, equipment suppliers in CEIT, developing and developed countries had no choice but to change to an ozone-safe refrigerant. In addition, new refrigeration components supplied are compatible with HFC-134a/R-404A refrigerants, and hence it will not be possible to revert back to the old, CFC-based system. (Pakistan)

Market conditions dictate which type of refrigerant is in demand – HFC-134a for most applications, or HCFC-22 for low temperature applications. (Argentina, Iran, China)

It is more costly to return to the CFC-consuming technology than to continue working with the HFC-134a alternative. (Colombia, India)

Due to the hygroscopic nature of the oil and its sensitivity to chlorine, it is dangerous to use any CFCs on the new lines. For this reason, partial backward conversion to CFCs (parallel production of compressors with CFC-113 and aqueous solvents) is difficult. (China)

Today's market is asking for products that do not deplete the ozone layer, and demand for the new line of production is therefore ensured. (Colombia)

---

equipment was completed by December 2000, but the enterprise went bankrupt in January 2001.[36] Other companies were similarly affected by general financial depression in CEITs. Another example is the case of EDA in Poland, which lost its former Council for Mutual Economic Assistance market and went bankrupt due to the 1998 Russian crisis, also involving other former Soviet countries, and its deflationary impact. In 2000 Poland's restructuring agency established a new enterprise, EKOPON, to take over EDA's operations in the hope that the new enterprise would gradually gain back its share of the domestic refrigeration market.[37]

Hydrocarbons

Several CEIT and developing country enterprises attempted to convert to hydrocarbon refrigerant technology, with mixed results. Enterprises in Ukraine and Russia attempted to convert to hydrocarbon refrigerants in domestic refrigerators. The Ukrainian project (Nord) was carried out successfully. The

GEF project completion report called it a 'highly satisfactory result' and a model for technology transfer initiatives because 'the choice of hydrocarbon refrigerant ensures that the enterprise can continue to market the technology, even if HFCs are regulated or phased out'.[38]

The Russian project was unsuccessful. The Russian enterprise (KRP) had originally planned to use HFC-134a, but considered hydrocarbons at the GEF's urging because the hydrocarbon has a lower GWP and was preferred in Western European markets. As the project was beginning, the World Bank became concerned about the enterprise's financial viability. The enterprise 'had once been the largest producer of domestic refrigerators and compressors in Russia (750,000 units/year) but had effectively stopped production in 1996 due to market problems and possible restructuring'.[39] As a consequence, the World Bank decided to delay the project until the enterprise could demonstrate its financial viability. In the end, the conversion to HC was not completed and the enterprise converted to HFC instead.[40]

Two developing country enterprises chose hydrocarbon refrigerant technology in combination with other technologies. One of these enterprises was located in China; the other was in India. The project in China was implemented at a 30 per cent foreign-owned enterprise and converted a compressor factory to produce hydrocarbon refrigerant compressors. The Indian enterprise was 40 per cent foreign-owned and converted domestic refrigerators to use a blend of R-600A and R-290 with cyclopentane as a foam-blowing agent for the insulation. Their experience with hydrocarbon refrigerants has been satisfactory. Extensive assistance was required to create safety systems and train employees in all projects, however.

For enterprises that utilized GEF assistance, the GEF Operational Strategy, adopted in 1995, had an impact on the selection of hydrocarbons. The strategy discourages the use of alternatives with higher GWPs when other options are technically and economically feasible. Refrigeration sector proposals submitted to the GEF were re-evaluated after this policy was adopted to ensure consistency, and some projects, such as the one at the JSC Krasnoyarsk Refrigerator Plant in Russia, that had originally planned to use HFC-134a, decided instead to use hydrocarbon refrigerants. Most CEIT enterprises, however, chose to convert to HFC-134a instead of the lower-GWP hydrocarbon refrigerants because of cost-effectiveness thresholds, because of customer preference for non-flammable refrigerants[41] and because HFC-134a was a proven technology highly promoted by its suppliers.

HCFC-22 for compressors

Less than 20 developing country enterprises chose HCFC-22; most of these were commercial refrigeration compressor manufacturers located in China. Although there were few HCFC-22 projects overall, they phased out a large quantity of ODSs – over 3500 ODP tonnes. The history of China's decision to select HCFC-22 can be traced back to the meeting of the MLF Executive Committee held in July 1995. At this meeting, China presented its strategy for phasing out CFCs in the commercial refrigeration sector. It was agreed that the industry should be converted to HCFC-22 as a transitional measure, with the understanding that non-ODS refrigerants would ultimately be used in this sector. The MLF agreed to fund 24 conversion projects, and China agreed that the remaining 49 production lines would either be closed or converted.[42]

Most Chinese enterprises were satisfied with the new compressor manufacturing equipment supplied by the MLF, even though the most advanced HCFC-22 compressor designs were not made available to all Chinese enterprises. An example appears in the completion report of one affected enterprise:

> *The most advanced HCFC-22 compressor designs at that time were not introduced into the Zhejiang Commercial Machinery Factory, either because it was too expensive to purchase the best technology with MLF approved funds for technology transfer, or because foreign enterprises were reluctant to transfer their latest technology to the enterprise. As a result, ZCMF designed its own HCFC-22 compressor and started producing them with the equipment provided by the MLF in 2003. The newly designed compressors perform better than the old CFC-12 compressors.[43]*

Zhejiang and other Chinese enterprises reported that the risk of returning to CFC-12-based compressor production is very low because:

> *1) all dedicated baseline equipment for CFC-12-based compressor production has been disposed of; 2) market preference is for HCFC-22-based products; 3) the price of CFC-12 is higher than that of HCFC-22, and the price of CFC-12 will be further increased as a result of China's CFC production sector plan; and 4), the Chinese Government will ban the sale of CFC-12 refrigeration equipment and compressors (except for compressors for maintenance of existing refrigeration equipment) by 1 July 2005.[44]*

HCFC-22 was also selected for a compressor conversion project in India because refrigerant HCFC-22 is manufactured in India, overall design of the compressor using HCFC-22 is simple, and it was the most economical way to convert from existing CFC-12 compressors.

No CEIT enterprise selected HCFC for refrigeration of any kind. Laws passed in the EU were a major reason for this. The majority of CEIT firms' main markets were in the EU, and they therefore had to ensure their products met EU specifications. Furthermore, many CEITs aspired to join the EU and therefore had to bring their laws into conformity with EU regulations that called for faster ODS phaseout schedules.[45] Particularly influential was EC regulation No 2037/2000 (see Chapter 4). This regulation bans the supply of all CFCs as of 1 October 2000, with use of these refrigerants for the maintenance of existing equipment prohibited from 1 January 2001. As explained by one project, 'markets in developed countries effectively disappeared with bans on ODS-containing consumer goods'.[46] This motivated many CEIT refrigeration equipment suppliers to convert early (even before GEF assistance was available), and spurred innovation for retrofit technology.

HCFC-123 for chillers

One enterprise located in India tried to license HCFC-123 technology but was unsuccessful. Although the best HCFC-123 chillers produced in North America produce substantial energy savings compared to the best HFC-134a chillers, with payback of the higher cost of equipment resulting from the savings in energy consumption, units tested in India failed to achieve comparable energy efficiency. Chillers that used HCFC-123 were manufactured in India, but because the manufacturers did not license the patented technology necessary for near-zero refrigerant emissions and high energy efficiency, their chillers would be expected to have higher power consumption compared to CFC-11 units.[47] Costs were driven up further because HCFC-123 was not manufactured in India and was therefore more expensive. HCFC-123 could not be freely imported into India either, but required a special licence which involved considerable procedural challenges.

HFC-152a

Only one developing country enterprise, located in China, attempted to use HFC-152a as a refrigerant alternative. Because there were no international standards to follow for HFC-152a, the enterprise followed European standards for the more flammable refrigerant isobutane. The safety measures for the HFC-152a refrigerator charging process and storage cabinets were therefore quite strict. The enterprise provided counterpart funding to meet these strict safety measures and satisfy the requirements of local firefighting authorities. All the enterprise's fire safety measures – including the ventilation system, monitoring and alarm system, and anti-explosion and firefighting system of the converted production line – passed the fire inspection conducted by local authorities. The enterprise also organized several safety training workshops for technical staff, operators, mechanics and maintenance workers.

The enterprise was ultimately unable to make the HFC-152a refrigerators profitable. It was difficult to convince locally owned and managed servicing centres to invest in HFC-152a charging and reserves, as well as safety facilities. As a result, the enterprise was unable to obtain enough purchase orders for HFC-152a refrigerators. The enterprise therefore selected HFC-134a as the alternative refrigerant, and purchased and installed HFC-134a charging machines at its own cost. One line converted under the MLF project can now produce both HFC-152a and HFC-134a refrigerators, according to market demand.

HFC blends (R-404A)

The HFC blend R-404A was selected in five commercial refrigeration projects in Mexico, Brazil, Bulgaria and Indonesia. R-404A was selected for a variety of reasons. In Brazil, the enterprise chose to use R-404A instead of HFC-134a because of a switch in production programme to more low temperature applications, for which R-404A is better suited. In Mexico, the factors prompting a supermarket enterprise to adopt R-404A were high CFC prices and regulatory pressure. CFC prices in Mexico had increased from US$11.4/kg for CFC-12 in 1995 to US$16.2/kg in 1998, and from US$26.3/kg for R-502 in 1995 to US$33.3/kg in 1998. Import bans and production quotas led to scarcity and unreliability of CFC refrigerants in Mexico. When combined with awareness that authorities were planning to ban imports of CFC-based equipment, this created a strong incentive for conversion. Supermarkets were motivated to choose sustainable technology because they could not allow refrigeration systems to stop due to a lack of refrigerant or limited equipment supply.

Overall, enterprises were satisfied with the R-404A technology. One enterprise reported minor incompatibilities, concerning the degreasing agent, within the manufacturing process, but these issues were resolved. The enterprise received technical assistance to identify an alternative degreasing agent.[48]

The risk of enterprises reverting to ODSs is very low. Project completion reports rated sustainability as 'likely' or, in the case of Bulgaria, 'highly likely'.[49] Indonesia reported that the risk of the enterprise returning to ODS use is not significant due to regulations and the fact that its main customers are aware of and demand CFC-free technology. This demand constrains the enterprise to sustain CFC-free technology in its production, although, according to the enterprise, CFCs are still cheaper than alternatives.

Ammonia

Only one enterprise chose ammonia. The ammonia project was located at the Yantai Refrigerating Machinery Works in China.[50] The project was unique in that the beneficiary was not a producer of CFC-based equipment. Rather, the enterprise was selected to receive funding to produce a new line of products, small ammonia compressors (under 72kW). These compressors would replace existing CFC-12 compressors of the same size being produced by seven other enterprises, amounting to the production of 3200 sets of small-sized ammonia refrigerating equipment annually. All told, the project indirectly phased out 240 tonnes of CFC-12 each year (initial CFC charge plus servicing requirements).

Complications arose when York Company Enterprise, headquartered in the US, acquired the GRAM company of Denmark. GRAM had signed the technology transfer agreement with Yantai for US$272,000, and when York acquired GRAM, it cancelled the agreement with Yantai. This forced Yantai to search for another partner. The project was significantly delayed, until a technological transfer agreement was finally signed with the CKD company from the Czech Republic.

CONCLUSION

CFCs were invented to be refrigerants and provided the highest social benefits in those uses. They are non-toxic, non-flammable, odourless and energy efficient. However, CFC refrigerants were among the first ODSs to be successfully phased out in developed, CEIT and developing countries because the companies that marketed CFCs developed HCFCs that were virtually drop-in replacements at easily affordable additional cost. Technology and service procedures necessary to safely and efficiently use natural refrigerants were unavailable in 1990, when HFC-134a was embraced for automobile air-conditioning and domestic and commercial refrigeration, and low-GWP replacements were held back by chemical manufacturers until the EC declared a ban on HFC-134a in air-conditioning systems in new cars from 2011. Natural refrigerants faced, and continue to face, fierce resistance in North America but are strongly preferred in Asia, Europe and the rest of the world. New advances in technology using carbon dioxide as a refrigerant may soon overcome its inferior energy efficiency and reliability problems.

While the focus on the single environmental objective of protecting the ozone layer undoubtedly contributed to the phaseout in refrigeration and air-conditioning, more could have been done by environmental authorities, the GEF and MLF, and consumers to promote lower-GWP refrigerants, to discourage the use of high-GWP HFCs and HCFCs, and to encourage energy efficiency. Until recently, the Montreal Protocol logically stayed focused on its successful ODS phaseout, but had not fully embraced the challenge of minimizing the greenhouse gas emissions. During this time, the EC accelerated their phaseout of HCFC mostly with the use of HFC greenhouse gases controlled by the Kyoto Protocol while the use of HCFCs with high GWPs grew more than expected in some of the fast-growing developing countries. Now HCFCs are a concern to both ozone layer recovery and to climate protection, and Montreal Protocol Parties are moving forward with proposals to accelerate the HCFC phaseout in a manner that avoids high-GWP HFCs, creates incentives to minimize substance emissions and allows HCFC use in applications like building air-conditioners (where they provide significant advantage in energy efficiency and can be contained with near-zero emissions). The phaseout of HCFC-22 has the quadruple benefits of:

1   eliminating its ozone depletion effects;
2   eliminating its climate change effects;
3   avoiding the climate change effects of emissions of HFC-23, which is a very high-GWP, inadvertent by-product of HCFC-22 production; and
4   avoiding the joint ozone depletion and climate change effects of the emissions of carbon tetrachloride, which is a feedstock in the production of HCFC-22.

8

Technology Transfer to Phase Out ODSs in Aerosol Products[1]

BACKGROUND AND INTRODUCTION

When Mario Molina and F. Sherwood Rowland published their theory of ozone depletion in the June 1974 issue of *Nature*, aerosol products accounted for about half of all CFCs consumed worldwide. A decade earlier, aerosol products had accounted for an even higher percentage of CFC use; their percentage of total CFC use had decreased because global CFC solvent and foam uses were rapidly increasing, because some aerosol enterprises were switching from CFCs to hydrocarbon propellants, and because other enterprises were innovating with formulations that used a smaller amount of propellant. Nevertheless, in 1974 much of the world's aerosol production was still solidly CFC-based.

Aerosol products achieved early success in the US, where most of the early technology was developed. Major multinational suppliers such as DuPont, Continental Can and Precision Valve helped transfer technology through developed countries, but were much less successful in spreading the technology in developing countries.

CFCs were always perceived as being very expensive. By 1974 the tendency towards reduced propellant formulations for cost reasons was already well established. For example, the original hairspray formulations had up to 80 per cent CFC-11/12 50/50 blend, but by 1974 'modern formulations' used only 30 to 35 per cent CFC-12 plus ethyl alcohol. Some enterprises were also incorporating 10 per cent isobutane (a flammable hydrocarbon) with the CFC-12 to reduce the cost of propellant.

The hydrocarbon propellants were first made available in purified form by the Phillips Petroleum Company in 1954, and were from the start much less expensive than CFCs. Companies that wanted to, or had to, avoid the use of patented CFC aerosol products quickly incorporated hydrocarbons in water-based formulations of shaving creams and the S. C. Johnson Company expanded hydrocarbon propellants into water-based room fresheners and insecticides. But their extreme flammability had essentially limited the hydrocarbons to water-based aerosol products.

Other propellants, such as compressed air, N_2, CO_2 and N_2O were tested, but these suffered from serious technical problems. Vinyl chloride, initially very successful, was hurriedly pulled from the market when studies found it to be a carcinogen. Propellant change was very slow, and only partial, with CFCs continuing to play a major role.

The market for aerosol products changed rapidly after 1974, first through consumer boycotts, then with aggressive marketing of hydrocarbon and not-in-kind alternatives, and finally when the world's largest aerosol producer – the US – decided that the Molina–Rowland theory was credible and therefore prohibited, in 1978, the manufacture or import of most CFC aerosol products. From 1978 on, the US has used hydrocarbons (HCs) in over 85 per cent of all aerosol products produced, although there may be a new shift from HCs to hydrofluorocarbons (HFCs) or not-in-kind products due to volatile organic compound (VOC) laws and concerns. The sudden changeover in the US in 1978 was a clear case of law driving technology, as laboratories throughout the country worked overtime to develop formulations that consumers would accept and that their legal departments would consider safe enough to be sold.

Many of the first HC products to be sold had to accept the words 'extremely flammable' on the front label, because the length the ignited flame extended from the nozzle exceeded the government standard for 'non-flammable' and 'flammable' aerosol products. Later, valve/actuator combinations and formulations were developed that allowed most products to avoid this purchase-discouraging labelling.

When the Montreal Protocol was signed in 1987, Europe and Japan continued to use large quantities of CFCs, although dimethyl ether found an important place among the propellant choices in Japan and several major European countries. Here too, laws and regulations controlled formulators. A very few European manufacturers of upmarket products took the view that they should continue with CFC formulations because they were the best products technically, but the vast majority of enterprises either switched directly to HCs or started replacing a portion of CFC propellants with HCs as soon as good quality HC propellants became locally available. Most enterprises in developed countries rapidly increased HC use to the amounts allowed by their packaging laws.

In 1987 enterprises in some developing countries were also using both HC and CFC propellants, while others were still staunchly using CFCs. India was split between HCs and CFCs, and China and Vietnam were exclusively using CFCs. Africa was mainly using CFCs, as were the Soviet Union and affiliated countries. For cost reasons, Latin America, led by the northern and southern extremes of Mexico and Argentina, had already begun to use some HCs even before the Rowland–Molina theory was published. Conversions proceeded swiftly after the US ban on CFCs, and most of Latin America was already using HCs when the Montreal Protocol was signed.

NOT-IN-KIND ALTERNATIVES TO AEROSOL PRODUCTS

With the important exceptions of certain medical and highly technical uses, there are non-aerosol alternatives to CFC aerosol products. For example, creams and pump sprays are alternatives to aerosol hairspray; sticks and roll-ons are alternatives to aerosol deodorants.

Non-aerosol cosmetic and convenience products, without exception, provide more active ingredients at lower cost, although they have proven less attractive to customers and less profitable to suppliers. Therefore most enterprises in countries with economies in transition (CEITs) and developing countries chose to implement non-CFC aerosol options.

To avoid the strong negative publicity that accompanied the banning of CFCs, marketers developed or reissued not-in-kind substitutes for many aerosol products. Hundreds of millions of CFC-containing aerosol products were eliminated in this fashion. In some cases, the not-in-kind products were themselves later eliminated; in some cases, the aerosol products were later eliminated; and in some cases, the two presentations continue to be sold.

Table 8.1 *Some examples of substitutes for ODS aerosol products*

Aerosol product	Substitute	History
Hairsprays	Pump sprays	Available previously, but with technical problems; these problems were solved and they remain an alternative to aerosol products
Insect repellents	Pump sprays, liquids	Revived after aerosol products were banned
Room deodorants	Liquids, gels	Permanently replaced a sizeable percentage of the aerosol room freshener business
Deodorants and antiperspirants	Roll-ons and sticks	Permanently replaced a significant part of the personal deodorant market and a very sizeable percentage of the aerosol antiperspirant business
Starch	Pump sprays	Technical problems were solved which permitted pump sprays

HC alternatives to aerosol propellants

HCs are the principal substitutes for CFC-12 propellants, and alcohol and water are the principal substitute solvents for CFC-11 used in aerosol products. Suitable mixtures of n-butane, isobutane and propane, with constant pressure, low odour and low olefin levels (unsaturates), are called hydrocarbon aerosol propellants (HAPs).

HAPs have no effect on the Earth's stratospheric ozone layer and negligible effect on climate, with a global warming potential (GWP) of between one and three. Once emitted into the air, they are slowly decomposed into carbon dioxide and water vapour. Since the early 1990s, however, HAPs have been classified in the US as VOCs, which can generate excessive amounts of tropospheric ozone when nitrogen oxide is also present. US federal and state regulations limit VOC emissions, with restrictions particularly stringent in California, where many cities are in 'non-compliance' with air-quality laws. These VOC regulations now greatly affect the US and Canadian aerosol product industries and will challenge them even more strenuously in the future. So far, regulatory authorities in most other developed countries have not yet restricted HAP products, with the notable exception that Switzerland and The Netherlands tax HAP use.

HCs are highly flammable, and care is required during storage, transfer and filling. They are significantly heavier than air, and settle on the floor or in low spots, where they may accumulate. HAPs are refined and 'deodorized' for cosmetic and other consumer product uses, and with virtually no odour in the gaseous form, there is great danger that they may reach explosive concentrations – approximately 2 per cent in air – without being noticed.

HC filling plant and storage facilities must be suitably sited to comply with local planning and other legal requirements. The storage facilities for bulk propellant must be designed and equipped with emergency facilities (for example a liquefied petroleum gas (LPG) detection system, sprinklers and automatic shut-off valves).

For plants located in suitably warm climates, outdoor filling of propellants may be appropriate if sufficient space is available to assure natural ventilation. Attention should still be paid, however, to designing explosion-proof electrical equipment and the elimination of ignition sources.

In several developing countries, there are small and very small 'cottage industry' aerosol fillers that are located in multi-storey buildings or residential neighbourhoods, which makes conversion to HCs inappropriate or prohibitively expensive. The most economic ODS phaseout strategy for these fillers has been to recommend that they assign the filling of their products to local 'contract filling' enterprises that through economies of scale are large enough to afford to locate operations in suitably isolated sites and purchase proper safety equipment.

Non-hydrocarbon alternatives to aerosol propellants

Dimethyl ether (DME) is another alternative aerosol propellant. It is a flammable liquefied propellant with excellent solvency and water compatibility and has found substantial use in European countries as a combination propellant/ solvent replacement. It is also gaining popularity in the US, where it is used to reduce VOC content in aerosol product formulations. DME is itself a VOC, but it allows the use of significant amounts of water, a non-VOC, in some formulations where this would not be possible with HAPs.

Hydrochlorofluorocarbons (HCFCs) are also used to replace CFCs. However, they are controlled under the Montreal Protocol as 'transitional substances' and must be phased out by 2020 in developed countries and by 2040 in Article 5 countries. HCFCs are prohibited for aerosol use in several non-Article 5 countries.

Hydrofluorocarbons (HFCs), including HFC-152a, HFC-134a and HFC-227ea, are used for specialized products in some countries. HFC-152a is slightly flammable and has a medium vapour pressure and a moderate global warming potential (GWP=120). It is not classified in the US as a VOC and makes excellent quality hair mousse, for example. It is also popular as a propellant in the US for aerosol dust blowers, particularly because the price of the preferred non-flammable HFC-134a is prohibitive. HFC-134a and HFC-227ea are non-flammable fluorinated propellants. HFC-134a is replacing CFC-12 in metered-dose inhalers (MDIs). It is also the main non-flammable propellant in certain industrial products. HFC-227ea is used for MDIs as well.

N_2, CO_2 and N_2O are naturally occurring gases that can also be compressed and used for some aerosol products. They are injected into the can in gaseous phase, whereas HAPs and HFC and DME propellants are in a liquid state inside the aerosol can (as liquefied propellants) and go through a phase-change to become gas as the aerosol product is expelled. These naturally occurring substances have slightly increased their share as propellants due to their environmental acceptability and because they are non-flammable and do not require explosion-proof gassing equipment.

From a technical perspective, compressed gases are not very good propellants. Their main limitation is that a wet spray results when the pressure in the aerosol product decreases. This is due to the propellant expansion that occurs as the can is emptied. Accurately controlled compressed gas charge is imperative for performance and safety. Due to high quality-control requirements and innate technical limitations, compressed gas propellants will probably not be widely used in developing countries.

EXAMPLES OF AEROSOL PRODUCTS THAT CONTAINED CFC PROPELLANTS AND ALTERNATIVES FOR EACH APPLICATION

Cleaners and industrial uses

There are a number of industrial technical aerosol products that relied upon the non-flammable and inert characteristics of CFCs. Products in this category include electronic cleaners, dusters, fault detectors, mould release agents (for casting plastic), aircraft pest disinfection, weld anti-spatter, polyurethane foam, spinerette sprays, tyre inflators and aerosol horns.

The majority of these products could be converted using HAPs or DME if flammability was not an insurmountable issue. For some of these products, not-in-kind alternatives that do not use propellant chemicals were available. Where flammability was a concern, HFC-134a was close to being a direct substitute for CFC-12. HFC-152a could also be used in some instances – it is slightly flammable but very difficult to ignite.

Pharmaceuticals

Pharmaceutical aerosol products which used CFCs included MDIs, nasal preparations, local anaesthetics, wound sprays, antibiotics, antiseptics and ancillary products. Most MDIs now use HFC-134a, although some not-in kind substitutes are available for products such as nebulizers and dry powder inhalers. Many topical sprays can use HAPs, DME nitrogen or, in very unusual cases, HFC-134a. Topical pharmaceutical products can be reformulated through the use of these alternative propellants or by using mechanical pump sprays or powders, liquids and creams.

Metered-Dose Inhalers

Asthma and chronic obstructive pulmonary disease (COPD) are the most common chronic diseases of the air passages of the lung and are estimated to affect over 300 million people worldwide. Asthma is a condition of narrowing and inflammation of the airways, with a worldwide prevalence between 2 and 20 per cent. It is common in developed countries and is increasing rapidly in developing countries. COPD is caused primarily by cigarette smoking, but may result from occupational inhalation of certain dusts or environmental air pollution. The prevalence of COPD in developed countries is around 4–17 per cent in adults aged over 40 years; it is increasing in developing countries with rising tobacco consumption.

There are two main categories of inhaled treatment for asthma and COPD: bronchodilators (also called acute relievers) and anti-inflammatory medication (also called controllers or preventers). The preferred method of drug therapy for both illnesses is with medications that are delivered to the airways by means of a hand-held inhaler device, either powered by a propellant (metered-dose inhaler or MDI) or using dry powders as carrying agents (dry powder inhaler or DPI). MDIs contain either CFCs or, more recently, HFCs as propellants. CFC-containing MDIs contain CFC-12 and CFC-11 and sometimes CFC- 114.

While much effort has been focused on developing in-kind replacements for CFC MDIs (HFC-based MDIs), there are other methods of delivering drugs to the lungs. Alternative methods to CFC MDIs for drug delivery other than HFCs include:

- dry powder inhalers (single or multi-dose);
- nebulizers (hand-held or stationary); and
- injected, patch or oral drugs.

The first introduction of an HFC MDI for the short-acting beta-agonist salbutamol was in 1994 in the UK. Today there are over 60 countries where there is at least one salbutamol (short-acting beta-agonist) HFC MDI approved and marketed. In several countries (for example the US), however, there is a large proportion of generic CFC MDIs that are priced significantly lower than the brand-name CFC MDIs and HFC alternatives. Since purchasers, whether individuals, health authorities or insurance enterprises, will continue to favour lower-priced medicines, countries will have to address the means by which CFC-free alternatives gain acceptance. For example, in July 2002 in New Zealand, PHARMAC (the government pharmaceutical purchasing organization) approved for purchase a new CFC MDI Beclomethasone on the basis of relative cost, when a HFC MDI had already been on the market for a number of years.

Technology transfer for ozone-safe MDIs[2]

The main components of all MDIs include the active ingredient, formulation excipients (for example solubilizing agents and suspension agents), lubricants, propellant (a liquefied gas), a metering valve, a canister and an actuator. Historically, the propellants used in MDIs were CFCs (CFC-12 and CFC-11, and sometimes CFC-114), but more recently the ozone-safe HFC-134a and HFC-227ea have been introduced (in the pharmaceutical sub-sector, HFC is referred to as HFA). The first HFA-based MDI for salbutamol was introduced

---

BOX 8.1 PHASEOUT OF CFC-MDIS IN JAPAN

Hideo Mori, CFC Committee of the Federation of Pharmaceutical Manufacturers Association of Japan

Transition from CFC-based MDIs to CFC-free alternative products has been one of the most technically challenging issues for pharmaceutical enterprises in recent years – so difficult, in fact, that since 1996 Parties to the Montreal Protocol have granted essential use exemptions for MDIs after the phaseout of CFC production.

In 1989, 14 Japanese pharmaceutical enterprises which manufacture and/or import MDIs for the treatment of asthma and COPD organized a special committee, the 'CFC Committee', as an affiliated group under the Federation of Pharmaceutical Manufacturers Association of Japan (FPMAJ). The mission of the committee was to discuss ways in which to accomplish the phaseout of CFCs for MDIs, to implement these plans and to oversee the realization of the phaseout in Japan.

The committee and the pharmaceutical enterprises have put considerable effort into the development of CFC-free alternatives and their introduction to the market. In December 1998 the Japanese Government, in cooperation with the CFC Committee, submitted a transition strategy that describes the principles and the programme of the Japanese transition policy. This strategy included a timeline for the phaseout by 2005 and allowed companies manufacturing MDIs flexibility in scheduling the withdrawal of CFC MDIs and the introduction of new CFC-free products. One measure of its extraordinary success is that in 1996 there were 22 brands of CFC MDIs on the market, while today there are a total of 21 brands of CFC-free alternatives. These alternatives are HFC MDIs, based on CFC MDIs, and newly developed DPIs, and they cover the full range of the phased-out CFC MDIs. All enterprises stopped both production and import of CFC MDIs by the end of 2004.

Although the successful development of alternative products by each enterprise is most important, the key success factors for the transition to CFC-free alternatives were the cooperation of the enterprises, the close cooperative relationship between the authorities and industry, and the appropriate transition strategy.

---

in the UK in March 1995 and in mainland Europe in 1997. In 2000 the Chemical, Industrial and Pharmaceutical Laboratories (Cipla) in India launched the first HFC MDI developed by an enterprise based in a developing country.

The CFC-free alternatives of HFC MDIs and DPIs (single dose or multi-dose) are equally efficacious for most patients. HFCs, however, are greenhouse gases, and HFCs for MDIs could come under pressure over coming decades as part of a global climate change response.

Challenges facing developing countries

Under the Montreal Protocol, developing countries are scheduled to phase out CFC production and consumption by 1 January 2010. MDIs are currently manufactured in 16 developing countries,[3] using 1875 metric tonnes of CFCs (equivalent to a production of approximately 75 million MDIs) in 2005. About 68 per cent of this CFC consumption for MDI manufacturing is by local enterprises based in developing countries, and the remainder is by multinational enterprises with local manufacturing plants. Local enterprises based in developing countries face the challenge of converting plants to CFC-free inhaler manufacturing. Multinational enterprises have the choice of either converting factories to manufacture CFC-free inhalers or closing them and importing CFC-free products.

The Technology and Economic Assessment Panel (TEAP) report of May 2006 identifies the following challenges:

- **Countries that rely mainly on imports:** the elements of a transition strategy would include enabling environment and health regulatory authorities to de-register use and/or ban import and sale of CFC-based MDIs; approval of CFC-free alternatives; and patient and physician education programmes. Countries will need to set an end date for transition.
- **Countries that manufacture MDIs:** these countries will need to develop a detailed national transition strategy to phase out CFC MDIs. The strategy would:
  - set a date for cessation of sales of CFC MDIs that takes into account the Montreal Protocol phaseout schedule;
  - involve stakeholders (national departments of health and environment, NGOs, MDI manufacturers, and physician and patient groups) in the education of physicians, other health-care workers and patients. In countries where only a small percentage of patients use MDIs, increasing the use of inhaled medication can be achieved by introducing a single-dose DPI or other low-cost alternatives;
  - ensure adequate supplies of inhaled therapy through phaseout. This will need adequate supplies of bulk pharmaceutical-grade CFCs, which may be affected by the CFC production phaseout schedule from 2007 until the end of 2009 under the Montreal Protocol; and
  - ensure adequate supplies of CFC-free alternatives. Local manufacturing enterprises should avail themselves of technology transfer, which may require funding.

Points to be considered for transition
Reformulation to HFC inhalers has been a formidable technical challenge for the pharmaceutical industry and has already cost approximately US$1–2 billion. All the components of the MDI have required redesign, and the clinical trials and regulatory hurdles have been much more arduous than when the CFC MDI was first introduced 40 years ago. Multi-dose DPIs have also had to meet new stringent standards, especially for precision and repeatability of dosing. The level of technical expertise and financing required to meet these demands is available to very few manufacturers in developing countries, and there will inevitably be manufacturing rationalization in developing countries, as has been the experience of developed countries.

Important factors for successful transition include the following:

- inhaler prices need to be comparable to those of CFC MDIs;
- new World Health Organization (WHO) initiatives to bring affordable MDIs to developing countries (which may even lower prices);
- acceptance by patients and health professionals; and
- timely approval of alternatives.

However, transition to CFC-free inhalers will not occur safely in the absence of a planned and controlled CFC MDI phaseout. Adequate safe supplies of essential inhalers for patients cannot be left to market forces alone.

In some countries, such as India and China, inhaler manufacture is by both nationally owned enterprises and multinational enterprises. One Indian enterprise has developed its own technology for HFC MDIs; for other nationally owned enterprises, however, technologies have to be transferred, and this has not generally occurred so far. These enterprises may need support for regulatory approval, factory/plant design and equipment installation, sourcing of components at acceptable prices, clinical trial and management, trial runs, and validation of quality. Appropriate technical expertise and skills need to be identified, and funds for technology transfer and equipment acquisition are required.

Intellectual property rights
The technology covering the HFC formulations that have been developed is in some cases covered by intellectual property rights owned by multinational pharmaceutical enterprises. The May 2006 report of the TEAP concluded that patents do not appear to provide an insurmountable barrier to transition. However, brand leaders will not give unlimited access to technology as licensing fees might not compensate them sufficiently for lost sales and profits. When the United Nations Development Programme (UNDP) tried to access such technologies for converting manufacturing plants in Cuba (financed by the Multilateral Fund, MLF), it found few interested in licensing the technologies. For example, some imposed conditions that Cuba could not export the new HFC MDIs, while others may have been inhibited by trading conditions in Cuba. The offers received are shown in Table 8.2.

Table 8.2 *Costs of transfer of patented MDI technologies*

	Provider 1 HFA Technology (US$)	Provider 2 HFA Technology (US$)	Provider 3 DPI Technology (US$)
Tech transfer fee	5 million	2 million	1.23 million
Incremental capital cost	1.84 million	1.84 million	3.74 million
Incremental operating cost	5.2 million	4.2 million	12.9 million
Total Project Cost	12.2 million	8.2 million	18.1 million

After negotiations with Provider 2, a project at a cost of US$5.96 million was approved, including a technology transfer fee of about US$1 million. Unfortunately, the selected provider, from the US, could not obtain permission from the US Government to operate in Cuba. After one year's delay, a product development contract (which did not infringe patents) was signed with a Canadian enterprise. As the original patents for HFC products near their expiration date, competition and the desire for cross-licensing may increase.

Developing countries exporting to industrialized countries

Brazil is an interesting case in manufacturing CFC MDIs for national consumption and for exports to both developing and developed countries. It has set a date for phaseout of CFC MDIs for national consumption; however, Brazil's exports of CFC MDIs to other Parties are increasing, and Brazilian industry is understandably reluctant to lose the income. This underscores the importance of synchronized regional and national phaseout plans, including for countries that only import MDIs. This will ensure that manufacturers in Brazil can convert all their production to HFC MDIs without being penalized by losing market share to continuing CFC MDI manufacturing elsewhere.

> *The challenge is to safeguard the early starters so that they do not suffer unreasonable adverse economic effects from going early and losing the market to CFC-based products sold at lower prices. The measures that were available to protect early starters included national regulations to legislate levelling of the playing field and the funding of 'incremental operating costs' by the MLF for a 'transitional period' to offset the increased cost of manufacturing a non-CFC product.* (Tony Hetherington, former Deputy Chief Officer, MLF Secretariat)

Developing countries and essential use exemptions

The developing countries are bound to stop all CFC production and consumption by 2010, but will they ask for an essential use exemption (EUE) to continue to produce CFCs for MDI manufacturing on the grounds of delays in registration or slow patient acceptance? Even if the EUE is available, obtaining pharmaceutical-grade CFCs may be a great problem. As CFC phaseout approaches, only two plants now manufacture such CFCs in developed countries, and plants in developing countries are being closed in accordance with their agreements with the MLF. When a factory makes pharmaceutical-grade CFCs, it also makes less pure grades of CFC. This means that it will become uneconomical to make pharmaceutical-grade CFC: the less pure CFC, which is a by-product, will have to be destroyed, at great cost, adding to the cost of the pharmaceutical-grade CFCs. An EUE may thus not be possible because of lack of availability of pharmaceutical-grade CFCs.

An alternative to an EUE would be to build a final stockpile from the allowable CFC annual production that would allow some additional years for the transition to CFC-free inhaler manufacturing. However, there would be considerable challenges, such as the size of the stockpile, storage, security, ownership, and questions of which drugs and which countries get priority. Thus it may be cheaper and logistically easier for local enterprises based in developing countries to make the transition in time to meet the Montreal Protocol CFC phaseout schedule and for those countries to consider increasing imported CFC-free inhalers.

Pathways of technology transfer

In developed countries, manufacturing laboratory managers studied local rules and regulations, possibly making a visit to another country for conceptual help, but then returning to their laboratories, and developed the formulations that they would use. Some of their assistants left for smaller enterprises, taking with them what they had learned, but they normally stayed within their countries.

In the years following the commercialization of HAPs, country-to-country technology transfer was minimal, and what there was came from multinational enterprises. Right from the start (1947–1948), the CFC suppliers had been the leaders in documenting and teaching CFC aerosol propellant technology. There were no such technology suppliers available when it became necessary to stop using CFCs. In Mexico and elsewhere, enterprises specializing in hydrocarbon products – such as S. C. Johnson – brought the technology for their own use and stimulated local supply of aerosol-grade HCs from local dealers. Eventually, some chemists and chemical engineers left the multinational enterprises where they had acquired experience, and took their knowledge and experience with them.

THE TECHNOLOGY AND ECONOMIC ASSESSMENT PANEL AND ITS TECHNICAL OPTIONS COMMITTEES

The creation of the Technology and Economic Assessment Panel (TEAP) and its Technical Options Committees (TOCs) for each major sector of ODS use was an important first step by UNEP in technology transfer. The Aerosol Products TOC brought together international experts, who catalogued the alternatives and substitutes, identified the barriers, and crafted an action plan for implementing the Montreal Protocol. Many of the TOC experts also served as paid or unpaid consultants who circled the globe, trained other experts, and actually did the work of technology transfer and implementation.

AEROSOL PRODUCTS TECHNOLOGY TRANSFER BY THE MLF

The MLF was set up in 1991, and money began to be available for phaseout actions in the developing countries. In the four UNEP-sponsored Regional Workshops of 1991–1992, government officials from developing countries were introduced to international experts in each sector, which served to delineate future problems.

The policies and rules for the aerosol products sector were simple:

- the MLF would pay for approved incremental capital costs of conversion and would deduct any incremental operating savings (this proved to be a major challenge, however, since in each project the savings in operational costs due to the lower costs of HCs compared to CFCs were significant, and the enterprises would thus have to find additional funds for implementing the projects);
- the MLF would not pay for enterprises to establish new facilities in locations safe for HC filling; and
- the MLF would not pay for inadvertent technical upgrades.

Depending on the approach of the project designer, new technologies would often include upgrade of the entire operation. Accordingly, significant amounts could be deducted from the proposed grants.

> *In many sectors, including aerosol products, the choice of what equipment could be re-used and what needed to be replaced was often contentious. An expert implementing a project and the enterprise might prefer to discard the entire old production line, but this solution was not necessarily the most 'cost effective' in terms of the operation of the MLF.* (Tony Hetherington, former Deputy Chief Officer, MLF Secretariat)

Companies that created products made with or containing ODSs after 1995 were deemed not eligible for MLF funding because they could have set up their new production facilities with non-CFC or non-ODS technologies. However, some MLF projects financed the safety costs of such new factories that started up with HAPs as an alternative to ODSs but did not comply with international safety standards.

Technology transfer in the field

Right from the start, aerosol sector technology transfer work had elements that made it easy and other elements that made things difficult. Because there were normally cost savings from switching from CFCs to HCs, there was some incentive to convert. However, local regulations sometimes complicated matters. In Indonesia, the Health Ministry acted very early, in the 1980s, to prohibit CFCs in cosmetic aerosol products, but was then unable to enforce the ban. Because enterprises manufacturing CFC aerosol products were reluctant to come forward and admit they were breaking the law, it was never possible to work successfully in the Indonesian aerosol sub-sector.

An important early technology transfer effort was the World Bank's Global Aerosols Project. In that project, consultants visited ten countries and then focused in depth on six to determine how aerosol fillers choose their propellants and what challenges had to be overcome to convert from CFCs to alternatives. It turned out that some consultants and implementing agencies were operating under the misconception that HC aerosol propellants were freely available everywhere and that technical information about their safe use was also readily available. The project also learned that the clear profitability of conversion to HCs was often not enough to effect change in the absence of external stimulus from markets and governments. The results of the Global Aerosols Project in 1994 influenced all future aerosol sector work.

Group training, workshops and seminars
There were only three major areas of concern in the aerosol sub-sector:

- availability of properly purified HCs;
- suitable product formulations; and
- safety.

Locating usable HC propellant was a problem in many countries. One project was approved to make pure HAPs for Jordan in its national refinery, but subsequent proposals to finance HAP production in other countries were rejected by the MLF as inconsistent with funding guidelines. Even in circumstances where HAPs were available through centralized production facilities, cost-effective distribution to a large number of very small aerosol enterprises remained a substantial economic challenge. In many cases, MLF project funding was provided to install purification equipment (molecular sieves) at the enterprise level so that locally available stenched or impure liquefied petroleum gases (LPGs) could be utilized. Nonetheless, in several countries, the lack of substitute propellant with suitable quality resulted in degradation in the quality of aerosol products.

Each implementing agency organized local aerosol seminars and workshops, using both foreign and, when available, local qualified experts. However, in many cases, the best local experts worked for enterprises that considered their knowledge proprietary. So most teaching at the events was done by foreign experts, using local translators where necessary.

The programmes at these workshops and seminars were designed to be useful to enterprises of all sizes and product mixes and to deal with areas where group training was possible, such as:

- HC chemistry;
- physical characteristics of HCs;
- formulation advantages and disadvantages; and
- flammability and safety precepts.

These seminars and workshops also had the secondary effect of allowing the consultants to meet the owners of many enterprises that would later become investment projects.

Technical Assistance Programmes

One technology transfer problem was that not all local manufacturers would qualify for MLF investment projects. The deceptively simple answer to this problem was to allow every enterprise to receive individual training under a Technical Assistance Programme (TAP) funded by the MLF. The prototype for this was run in Indonesia and was immensely effective, in part because participating enterprises were given valuable gauges and leak detectors. Subsequent training schemes in Malaysia, Thailand, India, Jordan and Uruguay were run without these incentives, but were still effective.

Investment projects

The most noticeable technology transfer was through MLF-sponsored investment projects. Initially these served as a powerful means to enlist a developing country's cooperation; later they became the principal means of technology transfer in the aerosol sector where it was necessary to do projects one by one, with a 'hands-on' approach by the foreign specialists.

Companies in developing countries overwhelmingly selected HAPs

CFCs were replaced in developing countries by HAPs, and in one case in China by DME. Thus the conversions had to focus on controlling flammability and limiting fire risks. Non-flammable substitutes like carbon dioxide, HFC-134a, nitrogen and compressed air were chosen in only three projects. In North America and Europe, a larger variety of non-HAPs substitutes were tested and adopted. These include highly flammable DME, less-flammable propellants like nitrous oxide and 1,1-difluorethane (HFC-152a), and non-flammable propellants like HFC-134a, carbon dioxide, nitrogen and compressed air. HFCs are greenhouse gases but with GWPs less than the CFC propellants they replace.

Trials by enterprises in developing countries of various non-HAP propellants were, with the few exceptions mentioned above, not satisfactory and abandoned after short periods.

Unlike large enterprises in developed countries that operate their own product formulation research facilities, fillers and marketers in many developing countries lack research capabilities and instead rely on standard formulations. Therefore formula innovations are obtained by copying formulas and packaging specifications used by foreign manufacturers who have their products filled locally. For most developing countries, the leading products are insecticides, followed by personal deodorants, air fresheners, mould releases, hairsprays, spray paints and colognes. The remaining products, less than five per cent, include polishes and shaving creams.

The recommendations of the aerosol consultants always played a key role in the selection of technology and equipment by agencies, subject to certain guidelines laid down by the Executive Committee of the MLF. This modus operandi has generally worked well. Since the consultants live and work in North America (including Mexico) and Europe, where over 90 per cent of the world's aerosol equipment is produced, it is to be expected that they favoured the use of equipment designed and manufactured in these countries. Equipment made in India, China and Japan has been less frequently selected, partly because the MLF was not designed to develop and test equipment, and consultants for aerosol projects were more reluctant to specify unfamiliar equipment than were consultants in refrigeration and solvent sectors.

There are seven international suppliers of aerosol production equipment that have thus far been used in projects. Two are from the US, three from the UK, one from Switzerland and one from Italy. All of these suppliers make equipment ranging from manual to semi-automatic and automatic, with production speeds from about 10 to 400 cans per minute. Most developing country requirements are in the 10–35 cans per minute range.

In most developing countries, due to the bad odour of the LPGs available, molecular sieve columns are required to remove the worst smelling contaminants. Some residual odour of the LPGs is sometimes still present after purification with molecular sieves, however, and this a serious deterrent to aerosol sales, especially in designer and name brand perfumes and colognes. Marketers of such products therefore use finger-pump sprays and thus avoid the use of any propellant. Quality perfumes and colognes are the principal example of where not-in-kind substitutes have thus far eliminated an entire class of aerosol products.

Two small fillers in India purchased thousands of used and emptied aerosol cans from waste dump scavengers. After discarding the unusable ones, they filled them backwards, through the valve, using a complete aerosol formulation compounded in a 1.5m tall cylinder, which hung inverted from the ceiling, near the gasser. The cans were then paper-labelled for the new product. Products such as mould release agents, insecticides and personal deodorants were being filled in this fashion, saving the fairly high cost of the can and valve, but increasing the risk of leakages and the risk that cosmetic ingredients would be contaminated with pesticide and solvents from residues in the recycled cans.

Language differences between equipment suppliers and beneficiary enterprises frequently complicated communication during visits. The problem was compounded by the fact that supplier texts, manuals, layouts, checklists and other important literature was usually in English and could not be understood by many recipients. One reason that Pamasol equipment was preferred in those parts of the Middle East and northern Africa where French is the second language (after Arabic) is that Pamasol employs French- and Arabic-speaking engineers and produces literature trilingually in Arabic, French and English.

Problems encountered by developing countries in HAP technology

The technology for the use of HAPs as a substitute for CFCs is well established in all the developing countries. HAPs are much less expensive than CFC. However, good quality HAP was initially not available in many countries.

Getting clients to accept the much lighter cans that use HAPs is a difficult challenge but has been accomplished. Many projects included the purchase of a complete HC storage system, including a destenching column, hot water test bath for aerosol can leak detection, flammable gas detection and alarm system, and propellant gassing unit with conveyor.

AEROSOL PRODUCTS TECHNOLOGY CHOICES BY CEITS IN GEF PROJECTS

Global Environmental Facility (GEF)-funded aerosol projects in CEITs chose hydrocarbon aerosol propellants, with the exception of one project which chose to use dimethyl ether for perfumes and colognes.[4] HAP would not have been suitable in that application because of the stenching agents applied to the locally available HC. As in the developing countries, conversion projects had to focus on controlling flammability and limiting fire risks. A few enterprises were considering changing to CO_2 or other non-flammable, non-volatile alternatives at the time of project completion.

Conversions helped many enterprises regain business and in some cases facilitated accession to the EU. The civil war in Bosnia–Herzegovina and Croatia that followed the dissolution of Yugoslavia in 1991 had a devastating effect on business. By switching to non-ODS technologies, Slovenian enter-

Box 8.2 Conversion in India

India was a complicated country to convert. At the start of Montreal Protocol activities, half the aerosol product market was CFC-based and half based on very poor grade HCs. Because there was no local HC supplier, each enterprise had to source its material and then purify it.

Many small Indian enterprises worked from multi-storey rented factories that could not be made safe for HC filling. People worked in cramped conditions, often with huge burettes that contained up to 12kg of propellant. This was extremely dangerous, because it could be catastrophic if a large amount of HC was spilled, and to qualify for MLF assistance, these enterprises would have to move. The fact that the MLF did not finance major buildings or land purchase thus eliminated many candidate enterprises.

By the mid-1990s almost 50 aerosol conversion projects had been prepared. Many of these projects were to provide financing to enterprises that had already converted their production lines from CFC to HAP at an earlier date, in some cases many years previously. However, during this period the MLF adopted as a funding policy the requirement that projects could only be considered if the enterprises had used CFCs during the three years preceding formulation of the project. All projects had to be rewritten, and many enterprises were disappointed when their projects were ruled ineligible for funding.

Eventually many projects were written, approved and implemented. Because many enterprises filled CFCs manually on extremely small-scale equipment, and there is no such equipment that is safe for use with HCs, the smallest possible pneumatic equipment was given, which resulted in an installed overcapacity countrywide. Some enterprises came very close to bankruptcy due to overcapacity and excessive competition.

Because of the onerous cost of moving, enterprises requested and required advance opinions from the foreign consultants as to whether their selected lot was eligible, and whether their floor plan would later be acceptable. This avoided future problems, but was costly; many enterprises were not able to obtain the places visited, so the consultant had to visit twice, or, in the case of a few enterprises, three times or more.

India had an inactive industry organization, the Aerosol Promotion Council (APC), which was revived during the Montreal Protocol work and became a valuable ally in technology transfer. The APC published a very acceptable conversion manual in English and Hindi that was funded by the MLF and was of material assistance. Unfortunately, few developing countries had the advantage of such effective industry associations.

prises were able to rebound by gaining access to the European market. For one enterprise, 'the HC aerosol propellant-based Byvacin achieved spectacular results in export markets and virtually restored the enterprise's business position to pre-1989 levels'. According to the GEF, conversions to non-ODS products also helped facilitate Slovenia's ascension to the EU.[5] In Russia, one enterprise expanded production, attracted foreign investment, and developed significant export markets in Western Europe and elsewhere.[6] In Poland, conversion also helped an enterprise gain access to international markets.[7]

The GEF and the World Bank report that:

> *the most successful sub-projects in terms of meeting objectives and offering sustaining economic and social benefits to the country were those where enter-*

BOX 8.3 CASE STUDY: PHASING OUT CFCS IN THE AEROSOL SECTOR IN RUSSIA[8]

In 1996 Russia's aerosol sector was quite large and represented one of the world's better opportunities for immediate, cost-effective CFC emissions reductions. In 1992 the aerosol industry consumed 18,150 metric tons of CFCs, which accounted for nearly half (46 per cent) of Russia's total ODS consumption. Almost all the aerosol CFCs (17,908 metric tons) were used in cosmetics and industrial applications. The small remainder was used in pharmaceutical applications.

Before the GEF offered assistance, one major aerosol producer (Khitan) had already converted to non-ODS propellants using its own resources. The remaining manufacturers continued to use CFCs, and demand for these products remained strong despite declining economic conditions. The GEF financed the elimination of CFC propellants in seven major aerosol product producers. These producers accounted for 85 per cent of the ODSs used in the aerosol sector in 1992. With one exception, all sub-projects utilized HAP as a replacement for CFC. The sole exception was Altaichimprom, where mechanical pumps were proposed. In 1996, because Russian HAP production capacity was limited, GEF initiated projects to add capacity so the country could meet the demand created by the aerosol conversions. HAP was chosen for the conversions because the technology had been developed globally and was readily available.

The Russian project was implemented in three stages, or 'tranches'. One aerosol project was proposed for the first tranche: JSC Arnest. This funded the phaseout of 2456 metric tons of CFC consumption in a 40 million can/year facility. To complete the project, JSC Arnest had to replace its valve manufacturing facility, purchase completely new filling lines to use HAP, purify the HAP, and build facilities in which to store and handle the highly flammable HAP. The enterprise's can-making facilities were not replaced. The GEF provided US$5.7 million for this project, but the enterprise was expected to invest an additional US$10.1 million in order to complete the conversion. The GEF and the World Bank calculated that the enterprise would be able to undertake its portion of the investment out of internally generated funds, but would be unable to complete the overall investment without GEF support. The GEF also believed that the extensive work already done by the enterprise would help it achieve a fast phaseout.

The project at JSC Arnest was completed successfully and achieved phaseout of 3050 ODP metric tons. The enterprise rapidly expanded production, attracting European Bank for Reconstruction and Development (EBRD) and other foreign investment, and developed a significant export market in Western Europe and elsewhere. This project was particularly important because it demonstrated that sustained industrial development was possible in a region (the North Caucasus) with poorer economic prospects. Four projects were funded in the second tranche.

---

prise contributions after implementation were both significant and timely. Conversely, where enterprise contributions were smaller or delayed, poor results were obtained.[9]

The latter was true of the aerosol enterprises switching to HC propellants in Ukraine. Two enterprises were reluctant to make counterpart investments and repeatedly attempted to renegotiate a better rate. As a result, their projects were delayed and their future economic viability remained in question at the time the projects were finally completed.[10]

In the case of Ukraine, 'the project implementation unit's placement at a relatively low level within the overall implementing agency limited its effectiveness in coordinating actual sub-project implementation and in enforcement'.[11] There was also a lot of turnover among the staff, which led to problems and delays in procurement and disbursement.

STERILANTS AND MISCELLANEOUS CFC USES

Several non-aerosol ODS uses are commonly grouped with the aerosol product sector, such as sterilants, laboratory testing, tobacco expansion and minor emissive uses. The Montreal Protocol's TEAP Aerosol Products TOC reports on phaseout activities in these areas as well. For this reason, we consider technical options to replace ODSs for these uses in this chapter.

Sterilants

Sterilization is an important factor in good quality health services. There are a range of sterilization methods, including steam, radiation, ethylene oxide (EO), formaldehyde, chlorine dioxide and ionized gas plasma. Prior to the Montreal Protocol, EO was widely used in a mixture with CFC-12 to inert its flammability and to prevent explosion. EO is toxic, mutagenic, a suspected carcinogen, flammable and explosive. Great efforts have been made to replace EO, particularly in hospitals, where personnel exposure is of great concern. The fact that EO is still used as a sterilant is evidence that in numerous applications the medical benefits of its use outweigh these disadvantages.

HCFC mixtures that replace EO/CFCs are used mostly in the US and in countries that allow venting of HCFCs to the atmosphere. The EO/HCFC blend is actually more effective at sterilization, allowing medical providers the choice of:

- providing the historic level of sterilization with less toxic EO use and emissions;
- sterilizing more effectively or against contagious agents that would not have been killed by the EO/CFC mixture, including new resistant bacteria and viruses; or
- sterilizing more quickly than before, which can reduce the necessary inventory of medical devices and provide faster turnaround of reusable medical devices during emergencies.

Other alternative technologies include use of more devices that can be sterilized by steam, more single-use devices, pure ethylene oxide sterilizers, and other methods that will sterilize or disinfect some of the low temperature devices used in hospitals. Other low temperature processes use:

- vapour phase hydrogen peroxide-plasma;
- steam-formaldehyde (in parts of Europe and South America); and
- liquid phase peracetic acid.

Tobacco expansion (puffing)

Current alternatives here are:

- carbon dioxide, which has been successfully used as an expansion agent for approximately 20 years and is now the most widely used process;
- propane, which is being successfully used by one enterprise in Germany and another facility in the UK;
- nitrogen, which is a high pressure process system being used in France; and
- isopentane, which is being used by one enterprise in the UK.

Minor emissive uses

There are many minor emissive uses of ODSs that are typically resolved in CEITs and developing countries without financing. These include the use of CFC-12 to chill or freeze objects; CFCs and HCFCs to locate leaks in pressurized pipes; CFC-11, CFC-12 and CFC-113 in supersonic wind tunnels; and CFC-12 in dielectric linear accelerators and other devices to allow unaffected microwave transmission while suppressing electrical arcing or in sealed switch gear (common on electric trains).

Many devices are marketed globally which contain small quantities of ODSs. Common devices include expansion bellows used to open skylights and vents or for solar tracking systems, expanding gas in thermostats and thermometers, and electronic controllers and other electronic devices. These devices may be just one component in a complex piece of equipment.

Laboratory and analytical uses

Typical laboratory and analytical uses of CFCs include calibrating equipment; extracting solvents, diluents or carriers for specific chemical analyses; inducing chemical-specific health effects for biochemical research; and acting as a carrier for laboratory chemicals. They are also used for other critical purposes in research and development where substitutes are not readily available or where standards set by national and international agencies require specific use of the controlled substances.

LESSONS LEARNED FROM MLF TECHNOLOGY TRANSFER ACTIVITIES

What worked in the aerosol product phaseout

- The MLF regional workshops initiated activities, the implementing agencies provided technical experts, and the country 'Ozone Officers' provided management and local coordination.
- The Technical Assistance Programmes (TAPs) provided detailed technical information both to enterprises that qualified for investment projects and to those that did not.

- The investment projects obviously worked and were the basis of most conversions.
- Where available, local industry associations made positive contributions. The Aerosol Promotion Council in India and the Mexican Aerosol Institute contributed to technology transfer through seminars and manuals.

What didn't work, or could have worked better, in the aerosol product phaseout

- The MLF professionals did their best to enforce agreements and avoid waste, but might have been more successful if there had also been routine meetings or collaboration with aerosol experts at project sites.
- The expectation that very small family-owned businesses could be preserved without some level of industrial rationalization was unrealistic because they simply could not compete against large-scale production of HC aerosol products.
- The policy of not financing land and buildings was a substantial impediment for enterprises that might have otherwise successfully relocated; non-grant mechanisms such as low-interest loans might have been of use in such circumstances if developing countries had not rejected the inclusion of loans in the modalities of the Fund.
- The MLF policy of funding estimated incremental operating costs while deducting estimated incremental operating savings created challenges in designing projects in this sector, because the transformation of the sector to low-cost HCs resulted in lower prices to consumers and less profits to producers.

CONCLUSION

A decade before the Montreal Protocol was signed, consumer boycotts, aggressive marketing of HC aerosol propellants and not-in-kind alternatives, and product bans did much to spur technology development and transfer to phase out the use of ozone-depleting aerosol products in developed countries. Technology transfer in the aerosol sector was successful in developing countries as well. Enterprises in Mexico and Venezuela phased out CFCs in aerosol products sold in North and South American markets far faster than the phaseout by enterprises in Europe. In fact, the last CFC aerosol products manufactured in Mexico and Venezuela were at the insistence of the European companies that hired contract filling. Spearheaded by workshops and seminars, with the strong support of technical assistance programmes in many cases, and supported by funding from the MLF, CFC phaseout/technology transfer was mainly carried out on an individual 'one-on-one' basis through the investment projects. A few challenges remain, important among them the final phaseout of CFCs in MDIs used to treat asthma and COPD. Effective, non-ozone-depleting technologies have been identified; now is the time for countries to begin planning to be prepared for the final phaseout of all CFC production and consumption in 2010.

Technology Transfer to Phase Out ODSs in Fire Protection[1]

HISTORY OF HALONS AS FIRE EXTINGUISHING AGENTS

Seven halogenated substances, of which six are controlled by the Montreal Protocol, have been used as firefighting agents at one time or the other: carbon tetrachloride (halon-104), methyl bromide (halon-1001), chlorobromomethane (halon-1011), dibromodifluoromethane (halon-1202), bromochlorodifluoromethane (halon-1211), bromotrifluoromethane (halon-1301) and dibromo-tetrafluoroethane (halon-2402).

Carbon tetrachloride was introduced as a fire extinguishing agent around 1900 and by 1910 was widely used in extinguishing gasoline fires from automobiles and other equipment with internal combustion engines. By 1917, however, carbon tetrachloride was suspected to harm human health; in 1919 the first recorded deaths attributed to carbon tetrachloride occurred when two people were overexposed to the fumes of a carbon tetrachloride portable fire extinguisher. Methyl bromide gained firefighting popularity in the late 1920s and was used in both British and German aircraft and ships during World War II. However, it was never popular for use in portable extinguishers because it is even more toxic than carbon tetrachloride. Chlorobromomethane was developed by Germany in World War II to replace methyl bromide, but it is also toxic.

In 1947 the US Army sponsored a contract with the Purdue University Research Foundation to evaluate new fire extinguishing agents based on military requirements. The goals were fire extinguishing effectiveness, non-corrosiveness, low toxicity, low electrical conductivity, low volatility, low freezing point, low cost and commercial availability.[2] Methyl bromide was selected to be the 'control agent' for experiments and assigned a baseline effectiveness of 100 per cent. Halon-1301 was the only material tested that exceeded baseline effectiveness in laboratory fire tests. In addition, halon-1301 had negligible corrosiveness and proved the least toxic of the agents tested. The US Army selected halon-1301 based on these properties and developed automatic total-flooding explosion suppression systems for fire protection in stationary property, and hand-held fire extinguishers for use in the crew compartments of ground combat systems.[3] In addition to halon-1301, halon-1202, halon-1211 and halon-2402 were commercialized after demonstrating both excellent fire extinguishing performance and relatively low toxicity. The US Air Force selected halon-1202 for military aircraft engine protection, and the US Federal Aviation Administration approved halon-1301 for commercial aircraft engine fire protection.

By the 1960s growing concerns with the toxicity of carbon tetrachloride, methyl bromide, and bromochlorodifluoromethane led to their 'official' commercial death as fire extinguishing agents in Europe and the US. Use of halon-2402 continued in what was then the Soviet Union and countries dependent on Soviet weapons systems and commercial vessels.

From the 1960s until the Montreal Protocol initiated the ban on halon production in 1994, halon-1301 was marketed as a total flooding extinguishant for protecting computer and communications equipment, repositories of cultural heritage, shipboard machinery spaces, oil and gas processing facilities, and pipeline pumping stations. Halon-1211 was widely marketed in portable fire extinguishers.

Moving away from dependency on halons

Halon fire extinguishants are no longer necessary in virtually any new installations, with the possible exceptions of engine nacelles and cargo compartments on commercial aircraft and crew compartments of combat vehicles. The very high cost of replacing many existing halon systems with substitutes or other alternative fire protection methods, however, continues to be a major impediment to eliminating continued use of halons.

The movement away from a dependency on newly produced halons can be accomplished in several ways. For existing systems, the practice of halon banking maximizes the use of recycled halons to service and maintain these systems. For new installations, alternative (non-halon) fire protection strategies and designs should be used.

Halon banking

Halon banking exists to promote the use of recycled halons and to minimize the need for new production. In most countries, halon banking is primarily a clearinghouse process, in which halons reclaimed from decommissioned systems are made available for other applications. Such clearinghouse halon banking organizations are not physical banks with warehouses and storage tanks, but are information centres keeping track of halon users that no longer require their halons, and users who still require halon but do not have sufficient stock. These organizations provide the method of matching availability with demand. Such 'banks' trade information on the availability of halons and leave the process of sale and purchase to the individuals concerned. There are also halon banking facilities that do manage physical stocks of halons, with the means to recycle halons for approved essential uses. The largest halon bank is operated by the US Defense Logistics Agency, with smaller physical banking facilities in many other countries.

TYPES OF HALON ALTERNATIVES FOR TOTAL FLOODING AND LOCAL APPLICATION SYSTEMS

Total flooding halon systems function by increasing the halon concentration in a space to a level that prevents fire from burning anywhere in the protected space. This can be contrasted with 'partial flooding' or 'local application' halon systems, which increase the concentration of halon to a level that prevents fire from burning only in a small area in front of the nozzle. Traditional fire protection agents such as dry chemicals, CO_2, water and foams (to protect special hazards) have been promoted as halon replacements.

Halocarbon agents are chemical agents that contain chlorine, fluorine or iodine either individually or in some combination. Classes of agents include hydrochlorofluorocarbons (HCFCs), hydrofluorocarbons (HFCs), perfluorocarbons (PFCs), fluoroidocarbons (FICs) and fluoroketones (FKs). Halocarbon agents share several common characteristics with halons, for example all are electrically non-conductive, all are clean agents, and all are liquefied gases or compressed liquids. They differ widely, however, in ozone-depleting potential, toxicity, volume required to extinguish fires, cost, environmental impact and availability. PFCs, for example, have such high global warming potentials (GWPs) that their use as firefighting agents is restricted in the EU, the US and certain other countries to applications where no other agent is technically feasible due to performance, safety or toxicity requirements. Halocarbon systems need a well-designed system to avoid decomposition, to prevent the production of acid gases and to achieve the concentration necessary to extinguish fires in the protected space. All of these agents can be stored in and discharged from systems that are similar to those used for halon-1301.

Inert gases are designed to reduce the ambient oxygen concentration in a protected space to a level that is breathable but will not support flaming combustion. Oxygen is reduced to between 10 and 14 per cent by utilizing a concentration of non-flammable inert gases at 35–40 per cent by volume. One gas blend uses a small amount of CO_2 to stimulate breathing.[4] These systems use inert gases such as argon and nitrogen or blends thereof. These are applied in total flooding systems. Inert gases are electrically non-conductive, clean agents. However, inert gas systems are not recommended where a rapidly developing fire can be expected, due to technical limitations that usually require long discharge times. Inert gas systems normally require a large tank storage area to supply the volumes of gas necessary to maintain the balance of gases long enough to extinguish the fire in the protected space.

Water mist systems extinguish fires using small amounts of water released as tiny droplets, under low, medium or high pressure. The methods of extinguishing include cooling, oxygen dilution by steam expansion, the wetting of surfaces and turbulence effects. These systems use specially designed nozzles to produce much smaller droplets than are produced by regular standard spray sprinkler systems. The smaller droplets are more effective at extinguishing fires and use less water. There are two types of water mist suppression systems: single and dual fluid. Both systems have been shown to be effective. In addition, when properly installed, they can effectively penetrate where deep-seated fires might develop.

Powdered aerosols are a category of new technology being developed and introduced that use fine solid particulates and aerosols. The different types of powdered aerosol systems include condensed and dispersed aerosol fire extinguishing systems.[5] Condensed aerosols contain finely divided solid particles, generally less than 10 microns in diameter, and gaseous matter generated by a combustion process using a solid aerosol-forming compound. Dispersed aerosols are fine particles of chemicals, generally less than 10 microns in diameter, already resident inside a pressurized agent storage container, suspended in a halocarbon or inert gas. The effectiveness and quality of the different types of powder can differ dramatically and must be closely monitored to match the target hazard and the type of space (occupied or unoccupied) being protected. The advantages of powdered aerosols are a very high effectiveness/weight ratio and low environmental concerns beyond those of the carrier gases.

TYPES OF HALON ALTERNATIVES FOR PORTABLE EXTINGUISHERS

Portable fire extinguishers are typically chemical tanks with either hand-operated or pressurized discharge through a nozzle that can be directed at the fire. They are either hand-held or wheeled. Wheeled units might be mounted on a cart or on a truck or other powered vehicle.

Water extinguishers are effective for use on fires involving ordinary combustible material (Class A). They are not all effective for use on fires involving energized electrical equipment (Class C). Water extinguishers are also not suitable for use on flammable liquid fires (Class B).

Aqueous film forming foam (AFFF), like other types of foam extinguishing products, is acceptable for use on fires involving ordinary combustibles (Class A). This type of extinguisher may not be safe for use on fires involving energized electrical equipment (Class C).

Dry chemical extinguishers are acceptable for use on fires involving flammable liquids and gases (Class B). Multi-purpose dry chemical extinguishers are acceptable for use on fires involving ordinary combustibles (Class A) as well as fires involving flammable liquids and gases. Both types of dry chemical extinguishers are suitable for fires involving energized electrical circuits (Class C).

Carbon dioxide extinguishers are suitable for use on fires involving flammable liquids (Class B) as well as electrical fires (Class C). This type of extinguisher is not effective on fires involving ordinary combustible material (Class A).

Where a clean agent is needed for a portable extinguisher application, most of the commercialized fire extinguishing agents are HCFCs in blends, HFCs or fluoroketones.

HCFCs currently face an eventual regulated production phaseout. Some restrictions are already in place in parts of Europe, and in many cases the EU has accelerated phaseout dates. HCFCs have been marketed in some developing countries as an 'in-kind halon replacement' and have taken over part of the market, especially in Southeast Asia.

HFCs are attractive as replacements for ODSs for three reasons:

1   they are usually volatile, and many have low toxicities;
2   they are not ozone-depleting and have lower atmospheric lifetimes than PFCs; and
3   they have properties similar to those of the halocarbons that have been used in the past.

However, many European countries have announced that they plan to phase out all HFCs within the next ten years because of their GWPs.

The one fluoroketone that has been commercialized is attractive because it is not ozone-depleting and has a low GWP. However, it has a high boiling point and is not as volatile as the HFCs, which may cause problems in some applications.

EVALUATION BY THE MLF OF THE HALON PHASEOUT IN DEVELOPING COUNTRIES

The Multilateral Fund (MLF) examined technology transfer projects in five developing countries (China, India, Brazil, Venezuela and Malaysia) to evaluate their technology transfer experiences. With the exception of China, the countries had not become dependent on halons; the Montreal Protocol was already signed by the time the developing countries began to expand military, aerospace, electronics and petroleum activities where halon would probably have been used.[6] China accounted for about two-thirds of the halon consumption of all developing countries. India produced some halons in the mid-1990s, but its consumption remained fairly low. A few other countries received funding for converting some fire extinguisher manufacturing plants, and during the last few years numerous halon banking and management projects have been approved.

Developing country halon producers

China, India and the Republic of Korea were the only developing countries producing halon. The coordinated phaseout of halon in China is proceeding ahead of the agreed schedule to halt production and consumption by 2010. The World Bank is the implementing agency for this project.

The two Indian halon producers, Navin Fluorine (NFI) and SRF, established their facilities in 1992 and 1995 respectively. Nominal capacities were 500 metric tonnes at SRF and 300 metric tonnes. Production reached a maximum level of 95 tonnes in 1995 at NFI and 109 tonnes in 1998 at SRF, consisting almost entirely of halon-1211. Halon-1301 was not produced in commercial quantities, although its production was technically feasible. The low capacity utilization and early stop of production (NFI in 1996 and SRF in 1998) were due to changes in market demand and supply compared to the original business plan. Demand in India decreased faster than expected because of awareness-raising activities, conversion projects and the ready availability of substitutes, despite their expense. The remaining demand for halons will be covered by low-cost imports from China and from recovered and recycled halons from phased-out installations in the country.

The Republic of Korea continues to produce limited quantities of both halon-1211 and halon-1301.

Fire extinguisher manufacturer conversions in India

The MLF approved fourteen conversion projects in India for fire extinguisher and system manufacturers. The projects converted the production of various sized portable fire extinguishers that previously used halon-1211 to CO_2 and pressurized ABC dry chemical powder-based units. For fixed flooding system applications, the MLF encouraged the utilization of recovered halon-1301 and alternative non-ODS agents such as FM-200 or inert gas systems.[7]

A number of seminars, workshops and major conferences with international speakers prepared the fire equipment market in India for the proposed phaseout of products containing halon substances. Consequently, the fire protection equipment manufacturers very quickly began to embrace the need to produce and market non-ODS alternatives as acceptance for halon-based products by their clients rapidly diminished. There is therefore very little or no risk of a return to ODS use in the future. The regulations in India required that the manufacture of fire extinguishers and fire extinguishing systems using halons be phased out by 1 January 2001, except for essential uses certified by an essential use panel. Servicing can continue until 1 January 2010. The regulations also specify that persons or enterprises that have received technical and financial assistance from the MLF cannot use ODSs beyond an agreed date.

The conversion of halon-1211 fire extinguisher production, and the elimination of the consumption of newly produced halon-1301 at Steelage Industries Limited Minimax Division Chennai (United Nations Development Programme (UNDP) Project No IND/98/G71), required an international technology transfer arrangement that enabled access to a specialized squeeze grip operating valve for the production of fully pressurized dry chemical powder extinguishers, and similar access to a suitable carbon dioxide operating valve. Both of these key valves were designed to suit local Indian conditions and the approval requirements of the fire authorities.

The changeover from halon-1301 to alternative agents for fixed installations was hindered in India by the high cost, which was subject to 67 per cent duty. In India, the alternative HFC agent (FM-200) was three times the price of halon-1301 from China, and permission to continue servicing halon systems was easy to obtain.

Fire extinguisher manufacturer conversions in China

Under the Halon Sector Phaseout Plan in China, a total of 56 conversion and closure contracts were agreed with fire extinguisher and system manufacturers. Although there were some delays in the 2000 programme, 46 were executed on time or ahead of schedule. The volume of halons contracted and funded for phaseout significantly decreased in 2001 and 2002, reflecting that there were fewer enterprises to convert from the original list of 72 fire extinguisher manufacturers and 22 systems manufacturers. The remaining manufacturers have been more difficult to motivate to proceed with the conversion. The average cost is US$0.45 per kg (ozone depletion potential-weighted) of halons phased out, well below the threshold of US$1.48 per kg, set by the Executive Committee of the Multilateral Fund.

Problems experienced in developing countries

Inadequate support from laws and regulations for the use of alternative technologies and agents hampered the rapid conversion of fire extinguisher manufacturing. For example, one MLF project planned to replace halon-1211 with CO_2 fire extinguishers, but found that customers who wanted to change over from halon-1211 were unwilling to trust the firefighting effectiveness of locally produced CO_2 extinguishers with antiquated and cumbersome valves. Funding for equipment to manufacture the modernized squeeze grip valve for CO_2 high pressure cylinders resolved the technical issue, but at considerable project cost. However, enterprises in some countries were unable to proceed because the requisite national test standards had not been approved, and the relevant authorities were therefore unable to certify compliance.

Development and revision of national standards and codes of practice

Lack of standards and codes of practice slowed technology transfer in some developing countries. In the absence of equipment and installation standards and codes for non-ODS fire extinguishers and systems, manufacturers were not sure whether their new products would be accepted by fire authorities and insurance enterprises. This delayed necessary investments.

Each country had the option to adopt international standards, adapt international standards to local conditions or develop new ones from scratch. To avoid duplication of effort and likely choice of inferior standards, the United Nations Environment Programme (UNEP) published a handbook on standards and codes of practice to eliminate dependency on halons.[8] The handbook is designed for National Ozone Units, governments and fire protection communities. It identifies the types of standards and codes of practice that are relevant to the Montreal Protocol and provides step-by-step guidance on how to establish new, or revise existing, standards and codes of practice to promote the halon phaseout. The cases of China, India and Venezuela are good examples of standards and codes implemented in developing countries.

China

In 1993 the Executive Committee approved two projects to be implemented by the UNDP with the Tianjin Fire Research Institute for the revision of national fire codes and standards. These projects resulted in four revised national codes and the continuation of training programmes. Moreover, several technical assistance activities were funded under the World Bank-implemented halon sector phaseout plan to:

- revise and enhance standards for fire extinguishers and systems using CO_2 as agent;
- amend building codes for high-rise buildings and basements; and
- upgrade technical skills of fire officers through organized training programmes.

India

India was given financing to help revise national fire codes and related fire equipment standards. Changes and amendments to the National Building Codes are now under review by the relevant authorities. In total, about 30 international standards and codes of practice were reviewed, with 12 of these considered suitable to be new Indian standards.

Venezuela

Venezuela is an example of the difficulties which can arise when international standards are adopted without a proper understanding of all the implications. The local standard that required a minimum of 55 per cent of phosphate content for portable dry chemical extinguishers was replaced in 1998 by the International Standards Organization (ISO) standard, which has a fire extinguishing performance rating for the extinguishers, but does not prescribe the chemical content of the dry chemical powder. Following the ISO standard confirms the fire extinguishing performance of newly manufactured extinguishers, but does not ensure sustainable performance over time if the powder settles into a hard cake that is not expelled during discharge. Unscrupulous Chinese enterprises exploited this loophole and manufactured low quality ABC powders with 15 per cent phosphate content that were imported into the Venezuela market from China. This eroded the confidence in ABC dry chemical powder fire extinguishers, the usually preferred replacement for halon-1211 fire extinguishers.

Safety standards and quality control measures for sophisticated fire protection applications in the petroleum, airline and other sectors were developed by international organizations such as ISO and the International Civil Aviation Organization (ICAO) and were implemented nationally by fire authorities and insurance enterprises. The statute adopted in 2000 created a new standards organization, SENCAMER, to monitor consumer safety and labelling.

HALON BANKING, RECOVERY AND RECYCLING

Halon banking, recovery and recycling projects have played a key role in sustaining existing fixed halon fire suppression installations and, to a limited extent, existing portable fire extinguishers during the transitional period of adjustment from ODS to non-ODS firefighting, fire suppression and fire control agents and technologies.

The complete phaseout will depend on the sustainability, availability and adoption of the new non-ODS agents, new technologies and replacement cost. Without concerted effort by regulating authorities and governments, sectors such as civil and military aviation may procrastinate in selecting chemical or not-in-kind alternatives to halon and will be vulnerable to the uncertain supply of halon. Most airlines in developed countries maintain a 'young' fleet of aircraft typically 5 to 7 years old, and generally not exceeding 15. The 'aging' fleets are often taken over and operated by developing nations, which typically stretch their life to more than 20 years and up to 40. Once aircraft manufacturers implement halon-free technologies, developing countries may have to either service with halon or retrofit aircraft produced before the phaseout.

UNEP's role in facilitating contacts for halon trading

UNEP, in consultation with the Halons Technical Options Committee (HTOC), designed and launched a business-to-business web portal, the On-line Halon Trader (www.halontrader.org), in 2001. This website is designed for enterprises that use halons in critical applications, including owners, managers and/or operators of fire protection systems, fire control services, and other organizations related to fire protection. Through this free service, enterprises that need halon for critical applications ('halon seekers') are able to post listings of specific demand in a virtual marketplace. Companies or halon banks that can meet this demand with recovered, reclaimed or recycled halon ('halon providers') can respond or post their own listings about halons available for exchange. UNEP provides the platform for this exchange but does not in any manner become party to transactions between those who seek halons and those who provide them.

The main positive effect of halon banking, recovery and recycling (HBR&R) projects was to convince users that the halon phaseout is inevitable and accepted, and to show how recycled halons can still be available during a long transition period for critical uses and the servicing of installed capacities, particularly fixed installations and in aviation.

The following are the lessons learned from these projects:

- Government agencies may be able to legislate and regulate the operations, but often lack the necessary skills and operational flexibility to successfully operate and sustain HBR&R facilities.
- A private enterprise operating the project has to have the facilities, logistics, knowledge, background, contacts with clients and technical expertise to successfully operate and maintain the facilities. There must also be support from industry (related associations, the fire protection industry and fire equipment manufacturers). Although the HBR&R operation may be viewed as a competitor to the supply of newly produced halons, the terms and conditions must be made conducive to bring about cooperation.
- The operator may find it difficult or uneconomical to operate the HBR&R facility as an independent operation, particularly if newly produced halons at low prices are still available domestically or from imports. To be econom-

ically viable, it should form part of other fire protection or service operations, as in programmes in Venezuela and Brazil.

- The operator must have sufficient incentives incorporated in the pricing
  mechanism, such as buy-back schemes and trade-off. The pricing should be
  such that clients are encouraged to use the facilities. Brazil Airlines, for
  example, finds it cheaper and quicker to use fire protection services in
  Miami than those in Sao Paulo.

Halon Phaseout and Management in CEITs

The Global Environment Facility (GEF) helped reduce halon emissions in
countries with economies in transition (CEITs) by facilitating production
closures in Russia, funding technical assistance projects, and helping countries
establish recovery and recycling plans. The following section documents the
experience of CEITs in establishing halon banks and switching to alternative
fire extinguishing substances.

Technical assistance and training projects

The GEF provided technical assistance and training to the fire protection sector
in Belarus, Poland and the Ukraine. These programmes were intended to
educate stakeholders about the technical options for phasing out the use of
halons and to help them plan to manage existing halons during the transitional
period. One conclusion common to all workshops was the need to establish
systems to collect, recycle and recover halons. In the Ukraine, a Halon
Management Plan was subsequently developed, and the GEF funded an
additional investment sub-project that provided a modern halon recovery and
recycling capability. Unfortunately, a halon recovery and recycling system was
not originally proposed by Belarus and was excluded from the GEF project.[9]

Investment projects

Estonia, Latvia and Lithuania

By the time projects were implemented in Estonia, Latvia and Lithuania, it was
clear that drop-in replacements for halons were not a viable solution. Careful
halon management was needed in the interim period as the fire protection sector
moved to new technologies. Accordingly, the GEF helped to establish a
Regional Halon Management Bank to provide a basis by which access to halons
could be made available to service and maintain fixed flooded fire protection
systems and fire extinguishers in Estonia, Latvia and Lithuania. The project was
implemented as planned, and no problems were reported. Halon recycling and
storage equipment was installed at the Estonian Environmental Research Centre
(EERC) in May 2002, and the supplier provided training for EERC technicians.
The workshop was followed by halon alternative awareness workshops in all
three countries. A follow-up inspection, completed in November 2002, reported
that the bank was operating as planned, and the project was given a satisfactory
rating.

Box 9.1 Study of halon recovery and regeneration in Hungary

The halon recovery and regeneration project in Hungary was designed in response to provisions in Hungarian Ministerial Decree 22/1993, which stipulates, inter alia, that from 1997 halons must be recovered in a closed cycle system when appliances are serviced.

This project was implemented as designed. Additional equipment was required in the shape of a vacuum pump on the recovery unit, a pneumatic clamp to cope with different sized cylinders, an electronic scale, and a gas detector to ensure that air contamination in the reclaim centre was kept at safe levels. Unexpectedly high levels of contamination shortened the life of the recovery unit pump, which had to be replaced. Some of these items were purchased with unallocated project funds.

The contamination levels that have come to light through the operation of the halon recovery centre have revealed an unexpected level of illegal activity on the part of Hungarian enterprises that service portable halon extinguishers. According to the Fire Protection Association (FPA), unscrupulous enterprises have sometimes used sand and oil to make fire extinguishers appear fully charged when halon has leaked out. One regeneration filter was completely destroyed before this was realized.

The practice of adulteration is part of the lax attitude to fire protection that has developed in Hungary with the emergence of small and medium-sized enterprises struggling to survive in the harsh new economic climate. Initially, amounts recovered matched estimates, but the FPA was surprised by a sharp decrease in the second half of 1998. The reason for this is simply the high cost of halon regeneration when properly carried out.

Formerly, state-owned enterprises were subject to rigorous fire protection rules that were properly enforced. In theory, there still exists a legal requirement for all appliances to be checked at six-month intervals, but this regulation is frequently ignored. The FPA has responded to this by reporting the enterprises using their services to the Fire Protection Authority. Since only about 18 enterprises in Hungary service halon fire extinguishers, it will be easy for the FPA to detect those ignoring the regulation and act accordingly by withdrawing Permits to Operate.

Implementation of this sub-project thus had the added benefit of detecting enterprises that had ignored appliance-servicing regulations, thus improving the reliability of halon fire extinguishers and public safety. Previously 15 per cent of stocks were wasted when equipment was serviced; this has been reduced to less than 2 per cent with the new equipment.

The operation is still financially viable with the lower throughput. Enforcement of the appliance inspection rules by the FPA should result in the amounts handled by the FPA being restored to earlier levels. The operation is thus sustainable for the foreseeable future.

The FPA noticed that stocks have dwindled to about half their original level through firefighting, imperfect halon handling (leaks) and illegal export, principally to the Ukraine and Romania. Illegal export is a worrying trend that is being addressed by the Customs Department.

The FPA is pleased with the performance of the operation and feels that it has demonstration value for countries setting up these schemes. Future plans involve studying the possibility of establishing a halon 'bank'. This involves purchasing halons from enterprises whose fire protection requirements can be met by other means and storing it for use in critical applications. The FPA is responsible for ruling whether an application constitutes an 'essential use'. Such applications might include aircraft, children's hospitals or fine arts museums.

Source: World Bank (1999) 'Hungary project completion report'.[10]

Azerbaijan, Kazakhstan and Hungary

Halon management and banking programmes were also established in Azerbaijan, Kazakhstan and Hungary. While the projects in Azerbaijan and Kazakhstan were implemented as planned and rated satisfactory (despite implementation delays in the case of the latter), significant problems were encountered in Hungary (see Box 9.1).

Production closures

Russia was a special case among the CEITs in that it was the only halon manufacturer that was unable to phase out its production and consumption by 1994, as mandated by the Protocol. Other producers – the Czech Republic, Estonia, Hungary, Latvia, Lithuania, Poland, Slovakia and Slovenia – phased out before 1994 by themselves. As the largest donor to the special initiative to eliminate CFC and halon production in Russia, the GEF helped phase out halon production in three locations: JSC Halogen, Kirovo-Chepetsk Chemical Kombinat and the Russian Scientific Centre of Applied Chemistry.

JSC Halogen was a large chemical complex that produced a wide range of fluorocarbons, hydrofluoric acid, hydrogen fluoride, fluorocompounds and fluoroplastics. The plant had the capacity to produce 300 tonnes of halon-2402. Kirovo-Chepetsk Chemical Kombinat was a large chemical complex located in the northeast part of European Russia. It had the capacity to produce 1030 metric tonnes of halon-2402. The Russian Scientific Centre of Applied Chemistry was a research institute that also operated a small-scale production facility in St Petersburg. The facility had the capacity to produce of 80 metric tonnes of halon-1301 and 20 metric tonnes of halon-1211. It also has recycling capabilities. The Centre played an important role in the development of halogen chemistry in Russia and was an important player in the development of replacement and new halogen chemistry technology.[11] Production of halon in Russia stopped at the end of 2000, and the capacity for such production was permanently closed by mid-2001.

The GEF also planned to establish a halon banking programme in Russia due to the large stock of halon in existing equipment. The GEF was optimistic that the country's ongoing requirements for halon could be met by developing a halon bank. Demand for new halon was relatively low, and many existing fire protection systems were dormant or located in decommissioned military installations. The project, which cost US$1.5 million, paid for contract support to design the banking programme. Support was provided by local and foreign experts. However, 'the project did not succeed in initiating direct recovery and recycling of halons in the fire protection sector as intended'.[12] Project designers attribute these failures to the 'failure of the state-controlled enterprises in the sector to capitalize on the opportunities afforded by the project'. On a positive note, enterprises involved in halon recovery and recycling were reported to be developing halon management capabilities independently.

CONCLUSION

The phaseout of halons by developed countries in 1994 is an example of the highly innovative thinking of all those concerned with the implementation of the Montreal Protocol. In 1987, when the Protocol was signed, halons were considered irreplaceable for fire protection, but technical progress and subsequent amendments to the Protocol led to phaseout in less than a decade.[13] It is significant that the 1994 phaseout – two years earlier than for any other ODS – was first proposed by the TEAP and its Halons TOC and advocated by the military, electronics and aviation industries most affected by the decision. This astonishing early phaseout was accomplished by the industrialized countries by international cooperation and the invention and implementation of halon banking; technology transfer funded by the GEF and MLF helped CEITs and developing countries to phase out halons as well.

Technology Transfer to Phase Out ODSs in Solvents[1]

INTRODUCTION AND HISTORY OF THE USE OF ODSS AS SOLVENTS[2]

A solvent is a liquid that has the ability to dissolve, suspend or extract other materials, without unacceptable chemical or physical change to the material or solvent. Solvents are important because they make it possible to process, apply, clean or separate materials.

Carbon tetrachloride or CTC (CCl_4)[3] was commercially introduced in the early 1890s as a low-cost, low-odour, colourless and non-flammable liquid solvent to dissolve fats, oils and greases, and for metal and fabric cleaning. It is toxic when absorbed through the skin or when inhaled; it is a powerful narcotic that can result in unconsciousness and asphyxiation; and at high temperatures, it decomposes to form highly poisonous phosgene gas.[4]

Other chlorinated solvents introduced in the early 1900s with similar chemical characteristics, but without significant ozone-depleting potential, include methylene chloride or MC (CH_2Cl_2), chloroform $(CHCl_3)$, trichloroethylene or TCE $(CHCl=CCl_2)$ and perchloroethylene or PCE $(CCl_2=CCl_2)$. These solvents were synthesized in the first half of the 19th century but were not produced commercially, as solvents, for more than half a century.

1,1,1-Trichloroethane (CH_3CCl_3), also known as methyl chloroform or TCA, is an ozone-depleting solvent that is relatively moderate in toxicity. Its lack of chemical stability hampered commercial use until a multi-component stabilizer package was developed and introduced. It grew in popularity in the 1970s and 1980s as a metal degreaser and adhesive ingredient, often supplanting the less expensive but more toxic chlorinated solvents as they were controlled by health and safety authorities.

Until the 1970s, electronics and precision products were cleaned with alcohol or chlorinated solvents, with aqueous solvents first introduced in the mid-1970s. Once commercialized in the late 1970s, however, CFC-113 $(CF_2ClCFCl_2)$ quickly dominated most markets because it is a gentle solvent ideally suited to cleaning products with delicate components: it is compatible with a wide range of metals and plastics, dries quickly, has neither a flash point nor a flammable range, and has relatively low toxicity.

CFC-113, CTC and TCA once had large and economically important applications as cleaning solvents, but they are all ODSs. CFC-11 (CCl_3F) was also used in a few niche solvent applications. Ozone-depleting solvents were used in four major areas: dry-cleaning, electronics cleaning, metal cleaning and precision cleaning. Minor uses occurred in displacement drying, coatings, inks, adhesive bonding, aerosol product formulations and other miscellaneous applications.

ODS solvent use in dry-cleaning

Chlorinated solvents are conveniently used for dry-cleaning of fabrics because, unlike water, they do not distort fibres and because some soils are more soluble in a solvent. Water cleaning can shrink or stretch many materials and fabrics and is particularly unsuitable for garments with linings and interfacing made from dissimilar fabric.[5] In the early 1900s two chemicals were available to dry-clean delicate fabrics and leather: gasoline and CTC. The drawbacks of gasoline are self-evident. CTC, by comparison, was effective, non-flammable and low-cost (although more expensive than gasoline) and could be used to dissolve fats, oils and greases in metal and fabric cleaning.

In the 1910s the dry-cleaning industry was divided between flammable purified gasoline and the increasing use of CTC. In the late 1920s purified gasoline and, to some extent, CTC were replaced by Stoddart solvent, a mixture of aliphatic hydrocarbons with a much higher flash point than gasoline. However, CTC gained ground and was the main dry-cleaning solvent by the 1930s, except for delicate fabrics, suede and leather, for which Stoddart solvent remained popular.

Because of toxicity concerns, dry-cleaning with CTC was slowly replaced in most countries in the mid-1950s by PCE. A few countries, such as Japan, favoured the more expensive ODS, TCA, to replace CTC.

In the 1970s CFC-113 replaced solvents to clean certain delicate fabrics, leather and garments with decorations that are adversely affected by other solvents. By the early 1960s CTC was prohibited in most developed countries for cleaning applications due to its toxic effects on the liver and kidneys.

ODS solvent use in electronics manufacture and repair

Most of the ozone-depleting solvents used in electronics cleaning were applied to remove rosin-based soldering flux from electronics assemblies such as printed circuit boards and for developing dry film photoresists. Flux is a substance that facilitates soldering by chemically cleaning the metals to be joined. After the soldering is complete, corrosive or conductive flux residues must be cleaned off to assure the function and reliability of the electronic product. Other reasons for removing flux residue are that it ensures electrical performance, ensures adhesion of conformal coating, facilitates testing and inspection, prevents electro-migration, and makes products more attractive to customers. Prior to concern about depletion of the ozone layer, CFC-113 solvents mixed with alcohols were the solvents of choice for electronics cleaning. These blended solvents effectively removed flux residues without damaging solvent-sensitive components on electronic assemblies, were non-flammable and had low toxicity.[6]

ODS solvent use in metal cleaning

Metal cleaning is a surface preparation process that removes organic compounds such as greases and oils, particulate matter, and inorganic soils from metal surfaces. Stamping and grinding lubricants are good examples of such complex soils. Such soils can be an essential part of the production process as parts are prepared for subsequent operations such as assembly, painting, coating, electroplating, inspection, further machining and fabrication, and packaging. Traditionally, chlorinated solvents such as TCE, TCA, PCE, MC and CTC were used for metal cleaning and degreasing. CFC-113 had small uses, beginning in the 1970s with its use as a metal cleaner for complex avionic gyros, as concerns increased about the toxicity and the effects of long-term, low-concentration exposure to some chlorinated solvents.[7] However, CFC-113 is a poor general-purpose metal cleaner.

ODS solvent use in precision cleaning

Precision cleaning applications are characterized by the extremely high levels of cleanliness required to keep delicate instruments and surfaces operating effectively. Precision cleaning processes remove contaminants from metal, plastic and glass surfaces to a micrometre degree of fineness. Computer disk drives, gyroscopes, miniature bearings, medical equipment and supplies, and optical components are examples of instruments that require precision cleaning; this once depended almost entirely on CFC-113.[8]

TECHNICAL OPTIONS FOR PHASING OUT OZONE-DEPLETING SOLVENTS

Advantages and disadvantages of alternatives

By the time the Montreal Protocol entered into force in 1989, a majority of the solvents used were ozone-depleting. The transition to ozone-friendly alternatives required difficult choices, as each of the alternatives had advantages and disadvantages (see Table 10.1). The following sections describe the technical options for each solvent use: dry-cleaning, electronics cleaning, metal cleaning and precision cleaning.

Dry-cleaning

Dry-cleaners replaced their CFC-113 and TCA solvents with alternatives that include:

- PCE;
- hydrocarbons;

Table 10.1 *Advantages and disadvantages of alternatives to CTC, CFC-113 and TCA*

Alternative	Advantages	Disadvantages
Chlorinated solvents		
Methylene chloride (MC)	Effective cleaner	Toxic[9]
		Waste disposal
Trichloroethylene (TCE)	Effective cleaner	Toxic[10]
		Waste disposal
Perchloroethylene (PCE)	Effective cleaner	Toxic[11]
		Waste disposal
Aqueous cleaners	Effective cleaner	High energy consumption
		Waste disposal
		Process sensitive
Water only	No additional chemicals	Limited applications
		Waste disposal with some contaminants
		Ferrous metal rusting
		Water spotting
		Long process times
Hydrocarbon/Surfactant blends	Effective cleaner	Often combustible
	New formulations are not combustible	Toxic in some cases
		Waste disposal
		Expensive
Alcohols (C=1–4)	Effective cleaner	Flammable
		Waste disposal
		Some are toxic
		Costly in some countries due to duties on alcohol
No-clean technologies	Less or no residue to be cleaned	Not approved for some applications
		Work area must be kept very clean
nPB*	Effective cleaner	Ozone-depleting
		Relatively high toxicity
HFCs** and HFEs***	Effective for precision cleaning	Expensive
		Requires blending
HCFCs****	More or less 'drop-in'	Ozone-depleting, controlled
		Expensive
		Toxic[12]

Note: * Normal propyl bromide ($CH_3CH_2CH_2Br$); ** hydrofluorocarbons; *** hydrofluoroethers; **** hydrochlorofluorocarbons.

- silicone;
- glycol ethers;
- HCFC-225; and
- carbon dioxide.

PCE is widely used as a dry-cleaning solvent. Although it is still listed as a carcinogen,[13] recent toxicological and epidemiological evidence indicates that it may not be a human carcinogen and that indications in animal testing with other species are not relevant to humans. Modern PCE machines are extremely efficient and usually result in very low solvent emissions. Nonetheless, the use of PCE in dry-cleaning may be regulated in some countries, regions or localities. For example, the US Environmental Protection Agency (EPA) set national emissions standards for PCE in September 1993 that apply to both new and existing PCE dry-cleaning facilities.

The flammability of petroleum solvents effectively precludes their use in neighbourhood shops, although with proper precautions they can be a substitute for CFC-113 on many fabrics. In Australia, for example, a fabric labelling convention has been introduced that designates white spirit as a substitute for CFC-113 in the dry-cleaning of specific fabrics. Concern over the hazard of any liquid solvent has shifted a large portion of dry-cleaning to highly sophisticated cleaning facilities located in industrial areas. With this business model, neighbourhood shops wash cotton garments with water and act as collection and return locations for the remote dry-cleaning centres. The cost savings from economies scale of industrial cleaning usually more than pay for the complications of tracking and transportation. Neighbourhood air quality is improved, the ozone layer is protected, worker exposure is reduced and consumers pay less for cleaning, at the cost of a slightly longer lead time for the cleaned garments.

Recent improvements in dry-cleaning equipment to maximize recovery of cleaning solvents, while minimizing emissions, have resulted in increases in the use of flammable solvents. In addition, new petroleum solvents that have lower odour and toxicity are being marketed. Silicone solvents have recently been developed for use in dry-cleaning operations. While these materials work well, they must be used in equipment designed for hydrocarbon solvents, as they have flash points around 76°C. Glycol ethers were introduced as dry-cleaning solvents in the late 1990s. They can work in hydrocarbon machines under similar conditions.

Normal propyl bromide (nPB) was introduced in 2001 in an experimental dry-cleaning machine. This solvent is prohibited for such use in some European countries, however. It is not recommended because of uncertainty over its toxicity and because it is an ODS, albeit not yet controlled by the Montreal Protocol.

Speciality cleaning of delicate, intricate or animal-skin items can be achieved using hydrocarbon or HCFC solvents. Equipment for use with hydrocarbon solvents is available as a 'totally enclosed' design (where operators never handle fabrics saturated with solvent) or as an 'open transfer' design (where operators move solvent-saturated fabrics from the cleaning machine to a separate equipment that extracts the solvent and dries the fabric.) The lower cost of hydrocarbon solvents favours their use for specialist cleaning in developing countries.

Electronics

The alternatives available for the electronics industry are:

- 'no-clean' fluxing, with little or no subsequent cleaning;
- water-soluble chemistry, followed immediately by a straight water wash process;
- rosin or similar fluxes, followed by saponification and a water wash process;
- rosin or similar fluxes, followed by a hydrocarbon/surfactant (HCS, semi-aqueous) cleaning and water wash process combination; and
- hydrocarbon cleaning.

No-clean technology

The best choice is to avoid cleaning, whenever possible. The challenge of the Montreal Protocol caused industry to join together to find the most environmentally superior options, and nowhere was this leadership better demonstrated than in the elimination of solvents. Teams considered the need for various cleaning operations and invented not-in-kind solutions.

No-clean soldering technology was developed by flux manufacturers and evaluated by a team of experts organized by the US EPA and the Industry Cooperative for Ozone Layer Protection, with major technical contributions from AT&T, Ford, IBM, Motorola, Nortel and Texas Instruments and including many experts from the Solvents Technical Options Committee. The team perfected the technology, patented key aspects and then declared the technology to be in the public domain. No-clean soldering assembles newly manufactured components that are not oxidized using a non-corrosive flux that is not harmful if residue remains after soldering. Today, approximately 60 per cent of electronics products previously cleaned with CFC-113 are manufactured with no-clean processes. When first introduced, no-clean technology was described as an excellent compromise for most applications, except where a long product lifetime with ultra-high reliability is required. Some electronics products such as implantable medical devices and space hardware are still cleaned with CFC-113 to avoid 'out-gassing' from flux residues. The original disadvantages of a narrower operating window for soldering and an increase in faulty solder joints were rapidly resolved with better machines, new fluxes, controlled atmospheres and stringent process control. Today, this method is now used for the majority of electronics manufacturing, including military hardware. The Ford Motor Company and its electronics division Visteon estimate that no-clean soldering has saved hundreds of millions of dollars and made electronics virtually error-free.[14] In some 'high density' surface mount designs, solder ball removal is critical for proper end-use performance, however, and in these cases it is often simpler to clean than to remove all the solder balls manually, a time-consuming and painstaking process.

Water-soluble technology

This is the method of first choice where no-clean methods are not satisfactory and cleaning is required for any reason. It is the most frequently employed technology when assemblies are cleaned. Water-soluble cleaning technology provides excellent cleaning, but requires an expert to select the proper cleaning solution matched to the type of soil to be removed and the materials to be cleaned. In some cases, equipment operators must strictly control the cleaning process.

Saponification technology

Saponification is the conversion of an insoluble rosin, or similar flux or paste, into a soluble soap. The process consists of this chemical conversion followed by a water wash, as described previously for water-soluble technology. The chemical process is not always easy, especially where there are thick flux deposits. The correct choice of the combination of flux and saponifier is essential. The introduction of lead-free solders, with higher melting points and consequently harder flux residues, has led to new families of defluxing products using a combination of saponification and non-ODS solvent techniques.

Hydrocarbon/surfactant technology

This method uses a water-soluble or water-solubilized low-volatile hydrocarbon solvent to dissolve a rosin or similar flux. The solvent, with dissolved contaminants, is then washed off with water and rinsed, as described previously for water-soluble technology. This technology was initially popular in developed countries because it did not require a change in flux type, but is less in use today because it is a more complex process than water-soluble technology. There are two variants of this process: the first where the hydrocarbon-surfactant formulation is water insoluble, the second where the formulation is water soluble. The user will have to adapt operations to account for the difference in the waste stream, since the second formulation becomes part of the water waste stream and thus must undergo proper waste treatment.

Hydrocarbon technology

Straight hydrocarbon cleaning is rarely used today. All hydrocarbon solvents are flammable or combustible, and most hydrocarbon solvents are classified as volatile organic compounds (VOCs). The flammable ones, such as undiluted or blended isopropanol, present a risk that is unacceptable for most manufacturing applications and are frequently very poor in cleaning quality, because not all the thermally modified flux residues and organic metallic salts formed after soldering are soluble in these products. Some of the combustible blends are better in terms of cleaning quality but present the problem of being difficult to dry. Hydrocarbon solvents frequently require long drying times and four to ten times the energy to evaporate them than water.

Other solvents for electronics cleaning

Normal propyl bromide (nPB) is an ODS that was overlooked when Parties to the Montreal Protocol selected substances to control and has not been subsequently added to the Protocol. nPB is a potential candidate to replace ozone-depleting solvents with a higher ozone depletion potential (ODP), such as 1,1,1-trichloroethane, but is not recommended because of its likely toxicity and the impact on the stratospheric ozone layer if it were used in high volumes.[15]

HCFC-141b (CH_3CCl_2F) and HCFC-225 ($C_3HCl_2F_5$) blends are also still being used, but both these substances are controlled under the Montreal

Protocol and scheduled for phaseout. HCFC-141b was originally estimated to have a low ODP, but was later discovered to have an ODP comparable to TCA. Regrettably, many enterprises, particularly in Europe, initially replaced TCA (ODP = 0.1) with HCFC-141b (ODP = 0.11) in compliance with the Protocol, but with no benefit to the stratosphere. HCFC-141b use as a cleaning solvent is now almost obsolete in most developed countries.

HCFC-225 was the only hydrochlorofluorocarbon solvent to make a significant contribution to ozone layer protection that was invented after the signing of Vienna Convention.[16] HCFC-225 has a very low ODP (ODP = 0.03) compared to the CFC-113 it replaces (ODP = 0.8). The contribution of HCFC-225 was that it worked as well as CFC-113 anywhere CFC-113 was used in precision cleaning and electronics assembly. Enterprises not wishing to phase out CFC-113 could complain of the cost of HCFC-225, but could not claim that there was no choice, especially as low- or no-emission machinery became available to largely mitigate the difference in cost. HCFC-225 is also an excellent solvent for removal of perfluoropolyether-based greases and lubricants (Krytox®).

HFC and HFE blends are also used for some particular niche applications, but their performance, requiring long and complex cycles for effective total defluxing, is such that their price rules them out as suitable candidates for all but the most exceptional and delicate medical or defence-related applications, where very high cost is not a barrier.

BOX 10.1 JAPANESE DEVELOPMENT OF HFE (HYDROFLUOROETHER)[17]

Akira Sekiya, National Institute of Advanced Industrial Science and Technology, Japan

After the Montreal Protocol was signed in 1987, the best chemists in the world reconsidered how to satisfy every ODS application without the use of chlorine, bromine and iodine molecules that can deplete stratospheric ozone. There was fierce competition in developing HCFCs (hydrochlorofluorocarbons) and HFCs (hydrofluorocarbons) for CFC substitutes, but less work on the development of hydrogen-atom-containing fluorinated ethers and amines, including hetero atoms. Together with the National Institute of Advanced Industrial Science and Technology (AIST), ten private corporations in Japan have participated in the project proposed by the New Energy and Industrial Technology Development Organization (NEDO) and entrusted to the Research Institute of Innovative Technology for the Earth (RITE) to conduct research and development of fluorinated ethers and similar compounds.

Three types of refrigerants and several types of blowing agents and solvents have been selected as alternative compounds to CFCs, and Asahi Glass and Daikin Industries have commercialized some solvents. Research continues to reduce costs and to further investigate the uses of other fluorinated ether compounds.

Independent from the national project, AIST cooperated successfully with the Zeon Corporation to develop and commercialize the cyclic $C_5F_8H_2$ as a solvent with superior cleaning characteristics.

**BOX 10.2 INDUSTRY LEADERSHIP IN JAPAN FOR
PHASING OUT TCA**

*Akira Okawa, Japanese Industrial Conference for
Ozone Layer Protection (JICOP)*

The Japanese Electrical Manufacturers' Association (JEMA) member enterprises achieved the total phaseout of TCA by the end of 1995 through the leadership and hard work of the Technical Committee for 1,1,1-trichloroethane (TCA) which was established in 1990. This committee investigated the actual conditions of consumption of CFCs and TCA by major manufacturers and evaluated the technical and economic feasibility of substitution technologies. JEMA promoted the reduction of TCA and the substitution technology by publishing a guide book, *Japanese Case Studies in Successful Low-Cost Substitution*, which contained case studies collected from members, and by organizing seminars to disseminate the results. This hard work allowed JEMA member enterprises to achieve the total phaseout of TCA by the end of 1995.

Moreover, JEMA actively assisted smaller enterprises in phaseout by preparing the 'Field guide for non-TCA' and by organizing and teaching seminars for reduction of TCA. JEMA has also supported developing countries by sending technical experts to Southeast Asia and by preparing an English version of *Japanese Case Studies in Successful Low-Cost Substitution*.

Metal cleaning

Available metal cleaning alternatives are:

- 'no-clean' and 'keep-clean';
- mechanical cleaning (for example wiping or media blasting);
- steam cleaning;
- emulsion/hydrocarbon-surfactant cleaning;
- organic solvent cleaning (with solvents less toxic than non-ozone-depleting halogenated solvents);
- non-ozone-depleting halogenated solvents (TCE, PCE, MC, HFC and HFE, the last two often mixed with trans-1,2-dichloroethylene);
- organic solvent cleaning (with solvents more toxic than non-ozone-depleting halogenated solvents);
- volatile methyl siloxanes;
- HCFC-225;
- n-propyl bromide;
- HCFC-141b; and
- PFCs.

No-clean

No-clean solutions to the previous use of ODS solvents in metal cleaning included the invention of specially formulated cutting oils that had the added property of corrosion protection when left on the metal surface. Another not-in-kind option was ultrasonic forming equipment that avoided the use of

Box 10.3 Ozone-safe technical preferences for metal cleaning

Darrel Staley, Boeing Company[18]

Trichloroethylene, perchloroethylene and methylene chloride are the most obvious alternatives to 1,1,1-trichloroethane and carbon tetrachloride in general metal cleaning. They have similar physical and chemical properties; they are relatively inexpensive and require only minor modifications of the cleaning process. Recently, they have been scrutinized for their health and environmental impact, and most countries have established safe exposure levels. If existing equipment is utilized, the cleaning system must be upgraded to minimize worker exposure and potential emissions to air, soil and water.

Hydrocarbon solvents and their oxygenated derivatives are increasingly used in specialized equipment designed to prevent potential flammability and explosion hazards. These systems allow the use of a large number of solvents. However, hydrocarbon and oxygenated solvents are often regulated as VOCs. Many hydrocarbon solvents are toxic, especially those with an aromatic molecular structure. Some are carcinogenic or mutagenic. The toxicology of oxygenated derivatives is very complex, with some very toxic and others much less. Many solvent blends are available for manual cleaning and for cold immersion cleaning.

Not-in-kind technologies

Aqueous cleaners (composed of builders, surfactants and additives) have served well as substitutes in metal degreasing. The process involves washing, rinsing and drying. Aqueous cleaning uses immersion with mechanical agitation, ultrasonic or spray processes. Each requires different equipment with optional features such as heaters, dryers, automation equipment, filtration, recycling and water treatment. Proper waste disposal is important in aqueous cleaning, because some by-products are not biodegradable. Aqueous cleaning is more energy-intensive than solvent cleaning and thus may have a larger climate change impact where the energy is derived from fossil fuels. Ultrasonic cleaning is appropriate only for small parts and becomes inefficient if the water contains dissolved gases.

Emulsion cleaners and hydrocarbon/surfactants are used in semi-aqueous cleaning and have many different formulations. They are used with ultrasonic agitation and fluid circulation to clean metal parts. Their advantages are low vapour pressures, low evaporation loss, low flammability and high flash points. However, they have associated recycling and disposal problems and low contaminant-saturation capacity, and require special equipment. Furthermore, they are frequently highly alkaline, which necessitates strict health and safety measures.

Various mechanical cleaning methods have been used for metal surface preparation and proposed as possible alternatives for CFC-113 and 1,1,1-trichloroethane. Brushing, wiping with rags or sponges, use of absorbent materials, media blasting and pressurized gases are generally best suited for lower-grade cleaning requirements or as a pre-cleaning operation in the removal of solid and semi-solid soils. Pressurized gas (air, rare earth gases, carbon dioxide and nitrogen) may be used for removing particulate contamination. Compressed air typically found in workshops is totally unsuited for this application, however, because it is frequently heavily contaminated with entrained water and compressor oil, except if a suitable filter and dryer are used. Specific technologies under development include wheat starch, sodium bicarbonate and solid carbon dioxide blasting.

Thermal vacuum de-oiling can be used to clean parts after cutting, machining and stamping, and in preparation for coating and heat treatment. Oil is removed from parts by vaporization in a heated vacuum chamber to give a super-clean finish. The operation is simple, but contaminant-specific.

'No-clean' options for metal degreasing include water-soluble, emulsifiable machining and metal-forming lubricants (chlorinated or not). These are much easier to remove with aqueous cleaners and are not hazardous to workers. The true 'no-clean' methods are rarely used and consist of volatile pressing and cutting oils (so-called 'vanishing oils'). Dry lubricants and thin polymer sheeting which can be peeled from a metal surface after a metal-forming operation are also possible.

---

cutting oil that would need to be cleaned away by energizing the metal molecules to form complex shapes or to fit metal parts together.

Precision cleaning

Alternatives available for precision cleaning solvents include:

- HCFCs;
- HFEs;
- HFCs;
- chlorinated and other miscellaneous solvents;
- alcohols and ketones;
- aliphatic hydrocarbons;
- aqueous cleaning;
- hydrocarbon-surfactant (semi-aqueous) cleaning; and
- supercritical fluids.

HFCs and HFEs have successfully replaced CFCs, HCFCs and PFCs in certain precision cleaning operations, aerosols and carrier fluids, as well as in some niche industrial solvent applications. The unblended HFCs and HFEs and their properly formulated azeotropes and/or blends have zero ODP and mid-range global warming potentials (GWPs), as well as many desirable properties similar to CFC-113, such as low toxicity, good compatibility, excellent chemical and thermal stability, non-flammability, and selective solvency for the applications. Two HFC solvents currently commercially available are HFC-43-10mee ($C_5H_2F_{10}$) and HFC-365mfc ($CF_3CH_2CF_2CH_3$). Several manufacturers are working on other HFCs for applications that once used ozone-depleting solvents.

Chlorinated solvents, such as TCE, PCE and MC, are available in high-purity as well as commercial grades. They have been successfully applied in many precision cleaning applications, especially as a substitute for TCA.

Common organic solvents are alcohols such as ethanol, isopropanol, several glycol ethers (methyl, n-butyl and diethyl) and ketones, such as acetone and methyl ethyl ketone (MEK). Most of these solvents are selected for their high polarity and their very effective solvency, although they do have some level of toxicity. These substances are flammable, and care must be exercised. Some of these materials are also used for precision wipe cleaning. The inexpensive iso-propyl alcohol (also named 2-propanol) does not carry a high soil loading, and is thus best used as a finishing solvent rinse cleaner.

Aliphatic hydrocarbons, formulated for tighter control on composition, odour, boiling range, evaporation rate and so on, may be employed in some precision cleaning processes.

Aqueous cleaners use water as the primary solvent. Synthetic detergents and surfactants are combined with special additives.

Conservation and recovery procedures have proven to be effective in reducing the environmental impacts of industrial solvent usage. These proce-dures are extremely valuable, not only in cases where CFC-113 and TCA have not yet been replaced, but also where organic solvents must vaporize in the process of post-cleaning drying. Although solvents can be recycled, the recycled solvent is rarely used in precision cleaning applications due to uncertain solvent purity and the extraordinary value or importance of the parts cleaned. The solution to this problem is the individual recovery and handling of any solvents in use at a given location. Careful handling will help prevent the mixing of solvents, thereby allowing for treatment and potential reuse. Additionally, enter-prises may purchase small and inexpensive solvent reclamation equipment to offset the high costs of solvent disposal.

One application that required particular attention was the cleaning of perflu-oropolyether-based greases (Krytox®) in inertial guidance systems. CFC-113 was the only known readily available solvent for this product. Extremely expensive PFCs were usable but were not ideal, and they had very high GWPs, generally more than 10,000 times that of carbon dioxide. Later development proved that HFC-43-10mee was effective for this application. Supercritical fluid cleaning is best used for small, low volume parts where batch cleaning is not a problem.

Adhesive bonding

The preferred alternatives for consideration here are:

- 'no-solvent', hot melt radiation curable;
- aqueous emulsions;
- organic solvents (low ozone-depleting, low toxicity, low flammability solvents); and
- non-ozone-depleting conventional chlorinated solvents (TCE, MC and PCE).

Before TCA became readily available, most adhesives had a solvent base of either flammable solvents, such as xylene, toluene and ketones, or non-flammable methylene chloride (MC). Both types required good ventilation to prevent either the accumulation of explosive or toxic vapours (MC decomposes in the body to carbon monoxide). TCA was therefore welcomed as a solvent for adhesives, being less toxic and non-flammable.

While there is no drop-in replacement for TCA in adhesive bonding products, a variety of solvent-based and non-solvent adhesives provide high performance for specific applications.

Prior to the phaseout in developed countries, TCA was the major ODS used in adhesive formulations because it is non-flammable, dries rapidly, does not significantly contribute to photochemical air pollution and performs well in many applications, particularly foam bonding. While there are no drop-in replacements, there are alternative solvents that can be selected for specific applications. Where flammability is an issue, other halogenated solvents can be used as direct substitutes.

Aerosol formulations

Aerosol propellants – typically CFCs, HFCs, hydrocarbons or inorganic lique-fied gases – are used for cosmetic, convenience, medical and technical products. The active ingredients and any solvent or carrier are kept under pressure by the aerosol propellant in a metal container and dispensed in a controlled manner by activating a valve. The active ingredient is responsible for the effectiveness of the product (for example medicine, pesticide or shaving foam); the solvent or carrier dissolves active ingredients in the formulation to allow for a uniform dispensing of the product; the propellant expels the contents from the can. In some cases, the solvent is the active ingredient for manual cleaning applications. In some other cases, the solvent suspends the paint, medicine, pesticide or speciality chemical for uniform dispensing. In other cases, the propellant also acts as the solvent.

When HCFC-141b was mischaracterized by scientists in the late 1980s and early 1990s as a low-ODP substance, it emerged as a leading candidate to replace CFC-113, CFC-11 and TCA in aerosol products by enterprises in devel-oped countries because of the 1996 phaseouts. However, when it was determined that HCFC-141b has an ODP about equal to that of TCA, its use was promptly abandoned by responsible enterprises (particularly in the US and Japan) and subject to regulations discouraging or prohibiting most uses of it as a substitute for TCA around 2002 in both the US and the EU. The replacement solvents include petroleum distillates, water-based products, organic solvents, HFCs and HFEs.

Miscellaneous applications

There are a number of miscellaneous solvent applications, including oxygen systems cleaning, space vehicle manufacturing, carrier media, component drying, riveting and machining, aircraft hydraulic systems, leak detection, fabric protection and coating, laboratory testing, skin de-fatting prior to surgical procedures to prevent scalpel slipping, human blood plasma and whole blood freezing, cleaning of catheter and related plastic medical devices, hypergolic fuel line cleaning for satellite steering jets, mould release agents, motion picture film cleaning and critical military applications.

Primary ODSs used in these applications include CFC-113, TCA and CTC. HCFC-225, although ozone-depleting, is used to replace CFC-113 in some criti-cal applications where non-ODS alternatives and substitutes are not yet technically satisfactory and cost is not a barrier to implementation.

Parties to the Montreal Protocol granted a global exemption to the January 1996 developed country phaseout for laboratory and analytical uses; they periodically restrict that exemption based on technical findings by the Technology and Economic Assessment Panel (TEAP) that a specific application no longer requires ODSs. An exemption was granted to the US space programme for manufacture of solid-fuel rocket motors and other related miscellaneous uses.

SOLVENT TECHNOLOGY CHOICES MADE IN CEITS AND DEVELOPING COUNTRIES

This section examines the experience of enterprises in developing countries and countries with economies in transition (CEITs) in transitioning to new solvent technologies, paying particular attention to the suitability, sustainability and reasons for the selection of each given alternative. Information in this section is derived from project completion reports and other documents collected from the Global Environment Facility (GEF) and the implementing agencies of the Multilateral Fund (MLF) for the Implementation of the Montreal Protocol.

The phaseout of CFC-113 in developed countries had dramatic impacts in developing countries where multinational enterprises had manufacturing facilities or joint ventures. Foremost was the agreement negotiated by the US EPA and Industry Cooperative for Ozone Layer Protection (ICOLP) that environmental leadership enterprises would halt their ODS use in developing countries within one year of the corporate phaseout date in their home country, which was often well ahead of the January 1996 Montreal Protocol date. Second was the decision by multinational electronics enterprises to demand that suppliers in all countries certify their components, assemblies and products as 'ODS-free' in order to avoid the necessity of labelling products sold in the US as 'made with or containing ODSs', as required by the US Clean Air Act. Third was the market transformation that occurred when multinational enterprises halted their own ODS use and compelled suppliers to halt theirs – with the consequence that local suppliers of cleaning technology in developed countries abandoned ODS products and shifted to the next technology. Thus many electronics manufacturers in developing countries, such as Brazil, halted their uses early and never participated in the MLF or World Bank financing.

Table 10.2 provides insight into what alternatives are being chosen.

For applications not phased out by market transformation, aqueous cleaning was a major alternative among developing countries and CEITs, although other technologies were also represented. In CEIT projects funded by the GEF, no-clean, PCE, hydrocarbon and ethylene oxide were each used once, and two projects were initially planned using HFE but later switched to other alternatives. Despite some delays and technical setbacks, the solvent projects implemented in CEITs were largely successful, sustainable and suitable for the enterprises that selected them. In many cases, the alternative technologies improved worker safety and health by reducing exposure to toxic ODS solvents and associated chemicals and by modernizing equipment and codes of practice.

Table 10.2 *Solvents selected to replace CFC-113, TCA and CTC in China*[19]

ODS Solvent	CFC-113	TCA	CTC	**Total**
1992–1999				
Aqueous & semi aqueous	842	53	0	**895**
Miscellaneous non-ODP	85	5	0	**90**
Total*	**927**	**58**	**0**	**985**
2000–2002				
Aqueous	51	0	0	**51**
Chlorinated non-ODP	157	75	0	**232**
Hydro-carbon	266	127	0	**393**
HFE & HFC	56	0	0	**56**
HCFC-141b	469	7	0	**476**
nPB	370	19	23	**412**
Unknown	470	393	0	**863**
Total	**1839**	**621**	**23**	**2483**

Note: Units are ODP tonnes. * Total 1992–1999 figures do not include phaseout of 12 ODP tonnes from a rare application of CFC-11 as a solvent.

In developing countries, hydrocarbon and chlorinated solvents were second and third in popularity behind aqueous. No-clean and semi-aqueous cleaning technology was also used. HCFC-141b, HFCs and ethylene oxide were also selected for some special uses.

Some CEITs and developing countries were using massive quantities of CTC for dry-cleaning and spot-cleaning of garments and for cleaning agricultural machinery, electric motors, oxygen systems and storage vessels, copper pipe during manufacturing, energized electrical circuits, air-conditioner heat exchangers, iron castings, military equipment, and even bicycle chains in small shops. Most often, it was used without even the most elementary precautions to protect worker health and safety. Much of this consumption has ceased, but it still continues in many enterprises. A specific problem arises with CTC because it is easy and cheap to manufacture, even in quite small quantities, and in some countries the authorities had no idea how much was being made, imported, used or exported. As a result, reporting to the Ozone Secretariat was frequently very inaccurate. In many cases, CTC users were very reluctant to change to non-ODS TCE or PCE, which were 10–20 per cent more expensive, even though they could be considered as less toxic drop-in replacements. In some cases, TCE or PCE solvents were misrepresented by promoters of CTC as being more toxic than they were.[20] The worker health and safety hazards were sometimes exacerbated by the fact that the implementing agencies of the MLF could not economically undertake the work required to make the small modifications to existing machinery. The reporting problem for CTC was repeated, to a lesser extent, for the other ozone-depleting solvents, especially TCA. The reasons for this were twofold: first, in some cases the authorities were unable to track down the commercial chains of distribution, blending and labelling with a trade name of solvents; second, many countries have fairly open borders, allowing illicit trade to occur without the knowledge of the authorities.

Small and medium-sized enterprises

One of the biggest obstacles to implementing a phaseout of ozone-depleting solvents, especially in large countries, is the inability to trace small and medium-sized enterprises (SMEs). This means that many tens of thousands of enterprises had no technical help or financial aid. In some cases, the SMEs were suddenly confronted by an inability to obtain the solvent they had been using for years. Their suppliers were purely commercial and were unable to offer technical support or even an appropriate substitute. This led many enterprises into economic difficulties, especially if they were making a single range of products for a single client. In China, the Solvent Sector Plan (SSP) addressed this problem to some extent, and the annual progress reporting indicates that supply-side controls have enabled China to meet its annual reduction targets, but the effects of this at the enterprise level are not visible.

As shown in Figure 10.1, China also faced the difficult problem of managing orderly phaseout for its small enterprises. In the first ten years of China's phaseout efforts (including the first three years of the SSP), phaseout projects were completed or approved for fewer than 100 large solvent users. These large users accounted for approximately 2500 ODP tonnes, roughly half of the total phaseout effort required. Phaseout of the remaining 2500 ODP tonnes requires the cooperation of almost 2000 small users. The Chinese SSP considered this risk and tried several unique approaches, including a voucher system, gradual self phaseout and a reimbursement mechanism. The voucher system was the primary thrust for phasing out small users. This phaseout method utilizes chemical suppliers or other knowledgeable industry representatives to work as intermediate execution agents (IEAs), who identify enterprises

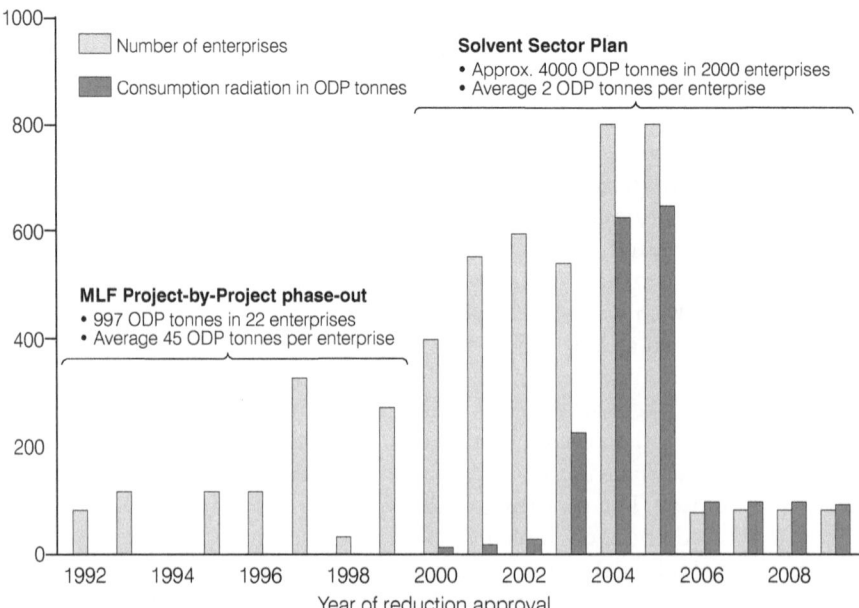

Figure 10.1 *Consumption reduction approved and number of enterprises using ozone-depleting solvents*

to apply for grant funding. Grants are paid in the form of a voucher or coupon to buy equipment and substitute chemical cleaning agents. The voucher's value is a function of the quantity and type of ODS solvent in use at the enterprise. IEAs are also responsible for verifying the claimed consumption level. This method has definitely helped identify more enterprises, though the ultimate level of success with this unique method remains to be seen.

The Chinese Solvent Sector Plan addressed the problems of identifying and assisting the large number of small enterprises using ODS solvent and by having a large number of 'designated local experts' trained by overseas solvent experts. The local experts and the enterprises chose from options judged by the international panel as technically suitable and environmentally acceptable. This approach greatly reduced the funding required for travel by overseas experts, leaving more in the budget to effect the conversions. In Indonesia, the overseas team of experts provided training and information to the various enterprises in seminars and workshops that were organized by the Indonesian Ministry of Environment.

EXAMPLES OF SPECIFIC TECHNOLOGIES IMPLEMENTED IN CEITS AND DEVELOPING COUNTRIES

The following sections explain in detail and by specific technology the experiences of enterprises and implementing agencies in phasing out ozone-depleting solvents in CEITs and developing countries.

Water-based (aqueous) technology

Most enterprises reported that they chose this technology because it is a non-ODS, non-flammable, non-toxic alternative. Had MLF or GEF funding not been available, many of these enterprises would probably have switched to a more toxic alternative. According to one CEIT enterprise, GEF funding allowed them to avoid using PCE as a solvent, which 'would have had adverse consequences for operators' health'.[21]

In comparison to baseline conditions, aqueous cleaning significantly reduced environmental exposure and safety risks. ODS emissions were eliminated. However, energy and water use were increased, which increased operating costs in some developing country firms. The main reason for the increased energy costs was thermal drying of parts with complex geometry (flat parts could be more effectively and economically dried by blasting them with high velocity air).

Many enterprises reported improvements in workers' health, safety and morale. By using water, firms eliminated the presence of harmful chemical vapours. Several enterprises located in Belarus reported that 'the delivery of modern equipment provided a psychological boost that improved staff morale at a time when production was dropping and the future looked uncertain'. In one case, staff members were so impressed by the improvements in the soldering station that they began pressing management 'to consider ways of improving conditions throughout the plant'.[22] International cooperation included many technical and training conferences in developing countries that included hands-on training and open-houses in facilities operated by multinational enterprises that had phased out ODSs. Top management of these multinational enterprises authorized cooperation on know-how and allowed their employees to give advice on which technology was most cost-effective. Japan hosted government and industry representatives for up to three or four weeks of training and other education. Multinational enterprises sometimes contracted for production the local enterprises they assisted with the phaseout because they were so impressed by their concern for the environment and the quality of their work. (The other side of this coin was that multinational enterprises sometimes abandoned work with local enterprises that were unable to implement phaseout, because they viewed that failure as an indication of technical capability.) The networks created by the protection of the ozone layer were naturally sustained when other environmental issues needed technical solutions. The value of this cannot be overstated.

Water-based technology often improved business. Developing country enterprises felt that the efficiency of operations had improved with the new technology and that the changeover was supported by 'excellent' technology transfer, technology support and operational procedures. Some developing country enterprises found that water-based technology lowered their operating costs. Others found that the new technology cleaned better than old processes – in one case, the improvement was a full 40 per cent. Other enterprises reported that product quality had improved. One enterprise located in the Philippines reported that the improvement in quality from the conversion earned several awards from customers as an Outstanding Supplier and Vendor. The conversion to water-based cleaning also paved the way for the enterprise to develop other innovative processes, such as no-clean manufacturing, and attracted new customers by allowing the enterprise to promote itself as environmentally conscious.

Enterprises located in CEITs also reported improved business after the transition to water-based technology. For example, the two enterprises producing televisions in Belarus typically purchased tuners from enterprises abroad, buying from the domestic enterprise only when they lacked foreign currency. The new equipment financed to protect the ozone layer allowed the domestic Belarus enterprise to improve quality enough to become the preferred supplier of tuners to both enterprises producing televisions. Another CEIT enterprise reported that conversion allowed the enterprise to compete on Western markets, and even set the stage for a beneficial joint venture with a German electronics firm.

In some cases, the new technology to replace ozone-depleting solvents had unexpected costs. For example, several firms did not foresee the added costs of treating contaminated water. In one case, three Hungarian firms implementing water-based solvent projects reported that the wastewater treatment requirements were not foreseen in the original project design. Two were able to obtain the needed wastewater treatment equipment with leftover funds; the other solved the problem by using a wastewater treatment facility nearby.[23] Some firms in developing countries reported that the increased costs of energy and wastewater treatment were offset by savings on chemicals, since water was much less expensive than other solvents.

Water-based cleaning projects were sustainable because old equipment was usually destroyed, leaving no chance of reverting to ODSs. In addition, the operating costs of using ozone-depleting solvents were higher than the cost of water-based systems.

One consistently occurring problem was that many of the aqueous cleaning lines used highly basic (alkaline) detergents. These should be handled only with adequate personal protection equipment (gloves, goggles, aprons and so on). In addition, eyewash and shower facilities should be available. Unfortunately, few of the MLF projects included these, nor were the beneficiaries warned of the danger – a splash of unwashed detergent in the eye may cause permanent blindness within 20–30 seconds.

Under an umbrella project in Argentina, nine enterprises replaced TCA by aqueous cleaning in the manufacture of carburettor parts, metal kitchenware, windshield wiper assemblies, automotive water pumps, silverware and stoves. The project also included manufacture of formal shoes, sports shoes and shoe soles. The umbrella project structure was used to reduce the cost of implementation. One large enterprise replaced TCA with a combination aqueous detergent plus mechanical process for sheet steel cutting.

In China, several enterprises converted from TCA for the cleaning of very small precision parts used in the fabrication electron gun assemblies for cathode ray tubes (CRTs) to aqueous processes, usually with ultrasonic agitation to ensure complete remove of stamping oils. The challenge of containing tiny stainless steel parts during cleaning was overcome by a local fabricator who designed rotating baskets.

The conversion of the largest single-point user of ODSs in China required the replacement of 15 cleaning machines, many of them so large they had to be cut up for removal from the enterprise buildings. Aqueous cleaning was preferred to replace ODSs in the high volume manufacture of inexpensive control circuitry used in domestic water heaters.

**BOX 10.4 OVERCOMING TECHNICAL DIFFICULTIES
AT TSVETOTRON BREST, BELARUS**

The conversion encountered some technical problems because the upgrade was carried out in two phases. In the first phase, alkaline stripping processing was introduced while the ODS-based photoresist developing process was retained. Once the firm acquired the new soda-based photoresist developing machine, the full conversion was completed. This two-step upgrade yielded a higher than normal rate of defective products, which impacted the enterprise's sales during that period. Additionally, after some time, adjustments were required in the wastewater system, requiring the installation of a new filter to bring effluent water quality up to national standards. The enterprise experienced an increase in energy-related operating costs, although these were offset by the higher quality and increased quantities of production.

Source: World Bank (2001).[24]

Semi-aqueous technology

Several Chinese enterprises were forced to relocate to new industrial sites outside the centre of Shanghai because the fire authorities had recently prohibited any flammable solvent ingredients, regardless of whether the mixed solvent was flammable or the drying system was safe. The enterprises eliminated the use of CFC-113 by replacing the existing vapour defluxer cleaning system with a batch semi-aqueous emulsion cleaner and stand-alone dryer. Drying is effected in an oven with proper safety features. The semi-aqueous cleaning process works very well for the printed circuit assemblies produced by the enterprise, without the effluent containing heavy metals that require water treatment. The new process equipment has a built-in closed-loop water system to minimize environmental impact. The enterprise already has an aqueous cleaner on site, so any added equipment needed to comply with environmental regulations should already be in place.

A Turkish enterprise manufacturing military telecommunications equipment replaced its CFC-113 use with a process involving the use of a diglycol ether solvent followed by a water wash in a series of open tanks and an automatic hoist of the type often used in electroplating lines. The water purification system consisted of a still in a closed circuit, so that there was no wastewater. The measured ionic contamination level on the cleaned assemblies was about one-fifth of that which had obtained using the CFC-113 installation.

No-clean technology

No-clean technology was popular in Southeast Asia, where many enterprises built electronic components for consumer electronics products. For example, one enterprise changed first to water and then to no-clean because it was difficult to sufficiently dry the air-conditioning tubes and because the water discharge regulation was very stringent in the green area where the factory was located. The no-clean system avoided both product drying and wastewater disposal. The same experience was found in Brazil.

One enterprise located in Hungary converted to a residue-free soldering process that did not require cleaning. Environmental risks were significantly reduced, and no new health and/or safety risks were created. This project was particularly interesting because the no-clean process eliminated the solvent cleaning equipment, but itself required no new equipment. The project was implemented as planned, and no problems were encountered. Although costs increased as a result of the new technology, the project was rated as sustainable by the World Bank upon completion.

Hydrocarbon technology

In many developing countries and CEITs, hydrocarbon formulations replaced TCA in the production of correction fluids and were also used to eliminate CFC-113 in dry-cleaning. Since hydrocarbons are flammable, precautions had to be taken to reduce fire risks. In the production of correction fluids, such precautions included explosion-proof equipment, external ventilation systems for filling lines and mixing rooms, inert gas firefighting systems, and explosion-proof fixtures for electrical and lighting installations.

Fire protection for hydrocarbon dry-cleaning used a nitrogen blanketing system with sufficient pressure to assure explosion-proofing. Additional precautions were required to prevent dyes dissolved by hydrocarbon-based solvents from staining subsequent batches of clothing. The solution was an effective dye filtering system in the recirculating lines.

Slovenia's largest dry-cleaning enterprise chose to replace CFC-113 and CFC-11 used to clean fine garments with aliphatic hydrocarbons. The enterprise chose replacement technology manufactured in Japan and then modified in Germany to address safety issues. The German safety equipment monitors the hydrocarbon/air concentration in the equipment and automatically shuts down as the lower explosive limit (LEL) is approached.[25]

No reversion to ODSs is foreseen. Operating costs were reduced, and project implementation proceeded smoothly. The project did not cover the cost of any special safety training, but the equipment manufacturer provided some basic training during commissioning. The project implementation report said that 'there has been virtually no change in operating procedures, except that operators are now required to wear gloves when removing garments from the cleaning drum and placing them in the dryer'.

However, not all enterprises were adequately warned of the fire hazard precautions that were necessary. An example of an accident waiting to happen was an Egyptian enterprise manufacturing garden furniture which was using hot kerosene in a spray machine to remove mould release compound from the products. The kerosene tank was heated by a tube carrying the flue gases of an open-flame oil burner situated under the tank. Furthermore, none of the pump or switchgear electrics were intrinsically safe or explosion-proofed, and only one small fire extinguisher was in direct proximity to the machine.

Ethylene oxide mixtures

Mixtures of ethylene oxide (EO, chemical formula C_2H_4O) and CFC-12 (CCl_2F_2) were occasionally used in CEITs and developing countries to sterilize medical instruments. The CFC ingredient reduces the fire and explosion hazards encountered in using pure ethylene oxide.

India used CO_2/ethylene oxide technology to replace CFC/ethylene oxide sterilizers and to replace nozzles for siliconization of syringes by an alternative to CFC-113. There were no major problems or delays. In Argentina, no major problems were experienced throughout the implementation of the project, except that portable ethylene oxide and LEL sensors with accurate sensitivity were difficult to obtain on the market.

In Mexico, several different methods were approved to implement sterilization alternatives under an umbrella project. Certain of these were preferred to provide sterile goods for deployment on ambulances, where in situ methods would not be practical.

One Hungarian enterprise produced ethylene oxide sterilizers for the medical industry. The enterprise originally chose transitional HCFC-124

(CF_3CHClF) to replace CFC-12. This HCFC choice proved both unsustainable and unacceptable because HCFC-124 was a solvent to the plastic medical devices and left an oily surface. The enterprise solved this problem by developing an HFC-134a (CF_3CH_2F)/ethylene oxide mixture that inerted the flammability and had no undesirable solvent effects. The HFC-134a/ethylene oxide mixture proved to be more sustainable in both an environmental and a business sense: it is a non-transitional substance and has allowed the business to continue operating. According to the World Bank, 'their solution appears to be novel and thus has valuable demonstration potential outside Hungary'.[26]

Chlorinated solvents (MC, TCE and PCE)

In the Democratic People's Republic of Korea, PCE was used to eliminate the use of TCA and CTC in cleaning operations prior to the plating of metal parts at a factory producing televisions. The choice of PCE-based cleaning chemicals required drying aids, a stabilizer additive to prevent acidification of the solvent, and vapour solvent recovery and distillation to minimize losses from liquid drag-out and slower drying due to potential for liquid entrapment in parts. The new equipment maintained product quality. The use of PCE cleaning agents requires careful handling and monitoring of workplace exposure, but is generally safer than the use of other flammable or toxic substances. A spill containment pan was needed for each chlorinated solvent installation to prevent any groundwater contamination in the event of a spill or leakage.

In India, TCE was used to eliminate the use of CTC in cleaning. The new technology, a closed-type degreasing cleaner, has considerably improved the health protection of operators at Sapna Coils. TCE was not used in CEITs.

In Hungary, one enterprise converted its dry-cleaning processes to use PCE instead of CFCs. The enterprise chose PCE because it is non-flammable and cleans better than hydrocarbons. Environmental risks were reduced, but health risks may have increased as a result of using PCE. To mitigate these health risks, four additional pieces of equipment were purchased to allow the residual sludge from the machines to be removed without exposing the operators to PCE vapours. As a result of the new technology, the enterprise now uses equipment that is completely hermetic. Hungarian environmental authorities have classified the plant as a 'non-emitting' process and have removed the requirement for annual inspection. Because the equipment was irreversibly converted and because the new system has lower operating costs, there is little chance of reverting to the old process. However, PCE is not entirely satisfactory because it is more aggressive toward delicate fabrics like silk. The enterprise did not rule out any hydrocarbon process, but favoured water-based reagents as a solution.[27]

In China, one enterprise replaced its TCA installation for air-conditioner compressors with TCE, using two 'near-zero-emission' vapour degreasers. This was very successful: although the capital cost was high, this was offset by the very low consumption of the low-cost solvent and proved a very satisfactory solution to a problem. The solvent consumption was so low that the project included the necessary materials to ensure solvent stability.

Many of the projects lacked adequate health and safety equipment when using halogenated solvents, including perfluoropolyether. If a major leak were to occur, it is essential that operators have access to filtered gas masks. Furthermore, all installations using toxic solvents should have means to monitor the levels of solvent vapours in the operators' workplaces. This was rarely provided in the projects. This concern was illustrated by an Egyptian plant moulding shoe soles. It replaced its TCA installation by a very sophisticated 'zero-emissions' machine using PCE. However, the operators were opening the machine to remove the parts before the solvent had been fully evacuated, creating excessive vapour exposure levels throughout the workshop, which also employed juveniles and women of child-bearing age. This example shows the importance of correct operator training as part of a project.

HCFC-141b

HCFC-141b was not used in CEITs. It was used in Colombia to replace CFC-113 in the production of silicon-coated hypodermic needles and catheters at Laboratorios Rymco. There were technical problems in the implementation of the needle coating process. Difficulties were experienced in achieving quality control standards, and there were safety, health and environment risks related to the volatility of HCFC-141b. To mitigate these risks, the storage conditions are controlled to avoid evaporation and workplace exposure. Detection devices, extraction systems, training of personnel and safety devices were provided through this project to control and minimize the related risks. While HCFC-225 would have been a 'drop-in' replacement for CFC-113 in this application, the cost due to 'total loss' of the HCFC-225 during deposition was deemed unacceptable.

HFCs and HFEs

HFCs were used in China to dry metal parts cleaned in aqueous solutions and to dry circuit boards cleaned in semi-aqueous solutions. The new processes use HFC-43-10mee as the basis for various formulations. HFC-43-10mee is a non-flammable, low toxicity solvent used in vapour-tight equipment; therefore the overall working environment has improved, and the risk to personnel of exposure is minimal. The enterprise was satisfied with the cleaning results of the new process and is committed to protecting the environment, so the risk of the enterprise returning to use of CFC-113 is minimal. However, the enterprise expressed some concern regarding the high operating costs using the current ODS-free solvents and may actively look to find other cheaper ODS-free alternatives.

Similarly, a Chinese enterprise manufacturing thick-film hybrid circuits had its CFC-113 installations replaced by sophisticated machines for use with HFC solvents. Because of the very high cost of these solvents, the machines were designed for minimal losses. The project was very successful. Unfortunately, the enterprise, desirous of cutting costs in a competitive market, reverted to using HCFC-141b in the same machines, in spite of the fact that the MLF project covered the cost of the more expensive solvent for a number of years. Even though the losses of HCFC-141b were very small, there was therefore a minimal negative effect on ozone depletion as a result of this decision, made on purely economic grounds.

A Russian enterprise replaced CFC-113 for the cleaning of various precision metal space parts with an azeotrope of HFE, trans-1.2-dichloroethylene and ethanol. (This solvent is commercially available as HFE-71DA.)

HFCs were not used in CEITs.

BOX 10.5 JAPAN/INDIA BILATERAL CTC PHASEOUT PROJECT

Tsutomu Odagiri, Japan Industrial Conference on Cleaning (JICC)

The Japanese Government is currently supporting a carbon tetrachloride conversion project in India within the framework of the MLF. The JICC organized the project and provides technological experts who supervise the investment in new cleaning facilities on the basis of international tenders.

Project objective

The Indian Government had been planning for some time to phase out early and totally the use of about 4150 ODP tons/year of carbon tetrachloride for metal cleaning by the major steel enterprises and the copper pipe manufacturers, including the phaseout of:

- 225 ODP tons of CTC used at SAIL (Steel Authority of India Limited) as a cleaning solvent for high voltage switchgear, transportable and stationary electrical motors, and oxygen-producing equipment, piping and storage vessels;
- 40 ODP tons of CTC used at two plants of WEC (Western Engineering Company) as cleaning solvent in the manufacture of components such as copper tubes/coils, evaporators and air-conditioners;
- 100 ODP tons of CTC used at NCPL (Nissan Copper) as cleaning solvent in the manufacture of copper tubes/coils; and
- 50 ODP tons of CTC used at HMT (Hind Metal and Tubes) as cleaning solvent in the manufacture of copper and brass tubes.

The Japanese Government proposed the cooperation for the project, and in 2003 the MLF Executive Committee approved the execution of the project on the basis of the bilateral fund. The Japanese Government contributed US$5,000,000 and the United Nations Development Program (UNDP) was the implementing agency in the conversion from carbon tetrachloride to trichloroethylene of 26 cleaning units at nine plants owned by three enterprises.

Japanese experts, including Mr Hiroo Kitamura (Planning Chairman of the Conference) and experts from an Indian branch office of Mitsui Bussan Plant and Project acted as engineers and intermediaries to the team of experts from the Indian enterprises. The Japan/India team selected the technology, and enterprises from several countries participated in the international tender. A Japanese manufacturer (Cleanvy Company) was the successful bidder for the cleaning units, which were introduced in 2006 to accomplish the total phaseout of carbon tetrachloride in 2007.

The phaseout of CTC in the four enterprises in India started in August 2006.

SOLVENT TECHNOLOGY COOPERATION
OUTSIDE THE FUNDING MECHANISM

Chapter 4 of this book described how ozone-depleting solvents, with the exception of CTC, were substantially eliminated by actions outside the funding mechanism. Leadership enterprises like Seiko Epson and Nortel made dramatic and daring pledges to halt all use. Partnerships like ICOLP and JICOP and the TEAP Solvents Technical Options Committee combined the talents of the best engineers in the world. These experts identified, developed, and perfected alternatives and shared the information and even the intellectual property worldwide. Private enterprises opened their factory doors to suppliers and competitors in developing countries. Nortel set up expert networks in Mexico and Turkey to guide technology choice, train equipment operators and implement new technology. Motorola organized industry in Malaysia and Indonesia to eliminate CFC-113, and then expanded the project to regionally manage waste from electronic product manufacture. Minebea developed an elegant combination of aqueous, hydrocarbon and chlorinated processes for precision cleaning, and set up a technical exchange in Thailand. Seiko Epson developed a centralized cleaning facility and an infrastructure for moving parts from suppliers through cleaning and then to the next assembly process.

CFC taxes and labelling laws accelerated the market transition and commercialized alternatives to achieve full economies of scale and equal or lower cost than the ODSs replaced. This aggressive effort reduced the incremental cost of eliminating ozone-depleting solvents and accomplished a large portion of the phaseout in developing countries without financing.

CONCLUSION

Canada's Nortel and Japan's Seiko Epson were the first multinational companies to declare a goal of the complete phaseout of CFC-113 and other ODS solvents. Joined by other leadership companies, the electronics and aerospace sectors were among the first to organize a global search for environmentally superior technology. Global public–private partnerships perfected no-clean soldering, aqueous cleaning and a dozen other solutions tailored to each unique cleaning challenge. Once technology was available, multinational companies undertook a rapid global phaseout, often implementing the most innovative technology first in developing countries where they were expanding production facilities. They demanded that suppliers halt ODS use and they cooperated fully worldwide. The positive experiences in CEIT and developing country sectors using solvents were:

- market forces and industry cooperation led to phaseout of most ozone-depleting solvents worldwide without financing;
- HCFC-141b, TCE and HFCs were avoided by most companies supplying multinational companies and in all GEF projects; and
- worker health and safety was improved by many MLF and GEF projects.

Negative experiences in the sectors using solvents were:

- many small and medium-sized enterprises were left to fend for themselves; and
- inadequate training after conversion to toxic chlorinated solvents sometimes jeopardized worker health and safety.

The transition to ozone-friendly technology in the solvent sector reveals three important lessons: first, corporate leadership can transform global markets at lower total cost and with little financing; second, partnerships of industry and governments can pick environmentally superior technology and reduce its cost though information sharing and economies of scale; and third, agencies implementing technology change must be vigilant in identifying and resolving issues of worker health and safety.

Technology Transfer to Phase Out ODSs in Pest Control

Melanie Miller and Marta Pizano[1]

INTRODUCTION

Following methyl bromide's listing as a controlled ozone-depleting substance (ODS) under the Montreal Protocol in 1992, many countries have undertaken activities to develop, transfer and implement alternative technologies. As a result, the consumption of methyl bromide has been reduced substantially in all regions and in all uses controlled by the Protocol. This chapter provides an overview of the major barriers and challenges that arose, and identifies the main methods of technology transfer and change that occurred.

MONTREAL PROTOCOL CONTROLS

Methyl bromide is a poisonous gas which has been used primarily as a pesticide since the 1930s.[2] It has several minor uses as a laboratory and analytical agent and as feedstock. In the past it was also used as a halon fire extinguishing agent.[3] Its versatility as a broad-spectrum pesticide, particularly in high-value crops, led to increased sales in many countries. During the 1980s global sales increased at the rate of about 5 per cent per year.[4] Methyl bromide came to the attention of the Montreal Protocol in 1991 when the Montreal Protocol Scientific Assessment Panel reported that anthropogenic (human-made) methyl bromide had a significant detrimental impact on the ozone layer.[5] The Parties convened a special Scientific Assessment Panel (SAP)/Technology and Economic Assessment Panel (TEAP) assessment on methyl bromide alternatives and ozone impacts, compiled by TEAP experts, methyl bromide users, agricultural researchers and atmospheric scientists.[6] Following these assessments in 1992, methyl bromide was added to the list of controlled ODSs under the Montreal Protocol (Copenhagen Amendment), and a freeze date was established for developed countries. At the 1997 Meeting of the Parties in Montreal, the phase-out schedule was set for 1 January 2005 in developed countries (non-Article 5 parties) and 1 January 2015 in developing countries (Article 5 parties).[7] The phaseout schedules covered all uses of methyl bromide as a pesticide, except for quarantine and official pre-shipment (QPS), which were exempted. Like other ODSs, any methyl bromide used entirely as feedstock in the manufacture of other chemicals is also exempt from phaseout because the methyl bromide is converted to other substances and is not emitted.

Impact of Protocol controls
on worldwide methyl bromide consumption

In the 1990s methyl bromide was produced in at least fourteen facilities in eight countries. By 2006 production was reduced to eight facilities in four countries, primarily in the US, Israel and Japan.[8]

Annual global consumption of the methyl bromide uses controlled by the Protocol was about 64,418 tonnes in 1991, and had fallen to about 20,752 tonnes by 2005. Consumption for controlled uses is predicted to decline to about 13,500 tonnes in 2007 if recent trends continue.

Methyl bromide has been consumed for controlled uses by 135 out of 188 countries that have reported data to the Ozone Secretariat since 1990. Fifty-six per cent (75 of 135) of these methyl bromide user countries no longer consume methyl bromide. By 2008 all but seven developed countries will complete their methyl bromide phaseout.

Critical use exemptions (CUEs) are exceptions from the scheduled methyl bromide phaseout date permitted in specific cases where alternatives are considered to be technically or economically infeasible (Decision IX/6 of the Protocol). In 2003 developed countries initially requested CUEs amounting to 18,704 tonnes of methyl bromide for 2005, raising concerns that CUEs would allow substantial quantities of anthropogenic methyl bromide to remain in use. However, actual methyl bromide consumption (production and imports) reported for CUEs in 2005 was about 11,468 tonnes, 39 per cent lower than the quantity originally requested. The methyl bromide consumption (production and imports) authorized for CUEs in 2007 is less than 6560 tonnes, which is less than 12 per cent of 1991 baseline consumption. In parallel, the number of individual CUEs was reduced from 134 in 2005 to 46 or fewer in 2008. While there is no room for complacency, it is clear that countries have made progress in reducing CUEs.

It is therefore clear that the Montreal Protocol has led to substantial reductions in the controlled uses of methyl bromide. Scientists have reported recently that the concentration of methyl bromide in the atmosphere has decreased significantly since 1998, correlating with the reductions achieved by the Protocol. In the northern hemisphere, the methyl bromide concentration in the atmosphere fell dramatically from about 10.5 parts per trillion (ppt) in 1999 to about 8.8ppt in 2003.[9]

However, QPS treatments are not controlled by the Protocol. After falling during the 1990s, the use of methyl bromide for QPS has recently increased, due to a new International Plant Protection Convention agreement aiming to control pests in raw wood pallets and packaging used in international trade. The agreement requires methyl bromide or heat treatment of raw wood pallets,[10] or, as an alternative, processed wood or plastic pallets can be used. Many operators are choosing to use methyl bromide because it tends to be cheaper or more convenient than the other options.

MAJOR USES OF METHYL BROMIDE

Since the 1930s methyl bromide has been used in many different pest control applications because it is a versatile pesticide product and provides a relatively rapid treatment, which users value. It is convenient to use and diffuses well in soil, thus reaching the target pests. Due to its high toxicity, methyl bromide is capable of eliminating a very broad spectrum of pests, from fungi to insects to mammals. Methyl bromide has been used mainly for controlling pests in soil before planting high-value crops such as strawberries, tomatoes, cut flowers and nursery plants. To a lesser extent, it has been used for controlling pests in flour mills, food processing facilities, stored products (for example cocoa beans and grains) and timber. Table 11.1 lists the main applications where methyl bromide is used as a pesticide (also called a fumigant since methyl bromide is a gas). Methyl bromide treatments are called fumigations.

In addition to its use as a pesticide, methyl bromide is used for several minor purposes: for laboratory and analytical purposes and for feedstock in the manufacture of other chemicals. In the distant past, methyl bromide was also used as a halon-1001 fire extinguishing agent.

Because of methyl bromide's versatility, many methyl bromide users have regarded it as an unparalleled pesticide product. However, methyl bromide does have several disadvantages. For example, US quarantine authorities recommend against the use of methyl bromide on products such as automobiles, electronic equipment, machinery with milled surfaces, products containing magnesium (subject to corrosion), goods made of natural rubber (for example foam rubber in furnishings), items with a significant sulphur content, baking powder, bone meal, high protein flour (for example wholewheat flour or soybean flour), and fat products (butter, lard and other fats).[11] Dutch government inspectors recently reported that methyl bromide altered the chemical composition of some pharmaceutical products when shipping containers were treated; recipients of the medicine would not know of such alteration.[12] Methyl bromide gas has significant toxicity to plants (phytotoxicity), which makes it an effective weed treatment but shortens the shelf-life of cut flowers and perishable commodities when methyl bromide is used as a post-harvest treatment.[13]

The use of broad-spectrum pesticides such as methyl bromide is increasingly regarded as an old-fashioned approach to pest control, because many non-target organisms are affected. Many farms and food enterprises have discarded broad-spectrum pesticides in favour of integrated pest management (IPM) or other approaches that aim to control specific pest species while avoiding damage to non-target species. Concerns about severe poisoning of

Table 11.1 *Main uses of methyl bromide pesticide products (fumigant)*

Items treated with methyl bromide	Examples of crops or commodities treated	Main target pests	Montreal Protocol requirements
Soil fumigation. Approximately 60% of total methyl bromide fumigant in 2005	Soil treatment prior to planting high-value crops, e.g. strawberries, tomatoes, sweet peppers, aubergine, melons, flowers, seedlings, nursery plants	Soil pests: nematodes, fungi, weeds	**Non-QPS:** subject to phaseout schedules (Articles 2 and 5); Critical use exemptions may be authorized for specific uses (Decision IX/6); Trade and licensing controls (Article 4); Data reporting requirements (Article 7)
Structural fumigation of buildings, transport. Approximately 4% of total methyl bromide fumigant	Flour mills, pasta mills, food processing facilities, empty ship holds	Stored product insects, rodents	
Durable products post-harvest fumigation. Approximately 33% of methyl bromide fumigant	Stored grains, cocoa beans, coffee beans, nuts, dried fruit, timber, wood products; Stored, imported or exported products	Stored product insects	
		Quarantine and officially regulated insects	
Raw wood pallets and packaging materials that are imported or exported. Included in durables (above)	Raw wood pallets and packaging materials that can carry wood-destroying pests in international trade	Quarantine and officially regulated insects	**QPS:** exempted from phaseout schedules; Data reporting requirements (Article 7)
Perishable commodities post-harvest fumigation. Approximately 3% of methyl bromide fumigant	Fresh fruit, fresh vegetables in international trade	Mainly for quarantine and officially regulated insects	

agricultural workers, water pollution and food residues resulting from the use of methyl bromide led some governments to prohibit or restrict methyl bromide use during the 1980s.[14] Methyl bromide is toxic to the nervous system, and hundreds of cases of methyl bromide poisoning have been reported due to accidental exposure of workers or bystanders.[15] Methyl bromide was one of the two pesticides most often involved in serious (systemic) occupational poisonings during a 40-year period in California.[16] Recently, a study of 55,330 men who work as pesticide applicators identified a strong correlation between occupational exposure to methyl bromide and prostate cancer.[17]

IDENTIFICATION OF ALTERNATIVE TECHNOLOGIES AND MAJOR CHALLENGES

When replacing methyl bromide, alternative technologies need to be technically effective in controlling the target pests, affordable, accommodated within the production timetable, and acceptable to regulators and customers. In 1992 a number of methyl bromide manufacturers, fumigators and users claimed that there were no viable alternatives, and this view is still held by some today. However, by 1994 the Methyl Bromide Technical Options Committee (MBTOC) made a detailed assessment and concluded that technically feasible alternatives existed or were at an advanced stage of development for more than 90 per cent of methyl bromide use, including some quarantine uses.[18]

The MBTOC Assessment of 2002 concluded that alternatives existed for more than 93 per cent of methyl bromide consumption (excluding QPS) and estimated that only about 3200 tonnes of methyl bromide (7 per cent of controlled uses) lacked alternatives at that time.[19] By 2002 there were many documented examples of alternative technologies used in commercial practice, and documented cases of adoption of alternatives that provided satisfactory levels of pest control and crop yields comparable with methyl bromide; a few types of alternatives in fact provided better yields than methyl bromide.[20] These reports confirmed that alternatives were not generally able to control the same very wide spectrum of pests that methyl bromide controls, so a simple substitution of one for one was not feasible. Users often resolved this problem by using several compatible techniques together, enabling the necessary spectrum of pests to be controlled – for example by using a product or technique that controls nematodes plus a product or technique that controls fungi. Application methods were also being improved. This mixing and matching of alternatives in appropriate combinations is inevitably more complicated than the traditional use of methyl bromide. However, identifying relevant combinations was made manageable by focusing on the target pests that needed to be controlled. This paralleled the approaches used in integrated pest management (IPM) programmes already adopted in some agricultural sectors, which often reduced production costs in the longer term.

Recently, the MBTOC Assessment of 2006 concluded that technical alternatives exist for almost all remaining controlled uses of methyl bromide (including critical use exemptions), although these alternatives may be more difficult to adopt in certain pre-plant sectors (for example certain types of strawberry nurseries, some orchard replant industries and control of broomrape weeds in certain locations) representing about 1136 tonnes of methyl bromide use. The MBTOC noted that economic constraints, regulatory issues and the period of time required for uptake of alternatives affected the phaseout rate of these remaining uses in developed countries.[21]

TYPES OF TECHNOLOGIES ADOPTED

The MBTOC Assessments of 2002 and 2006 noted that the following types of alternative technologies have been adopted:

- In the soil sector, for crops such as strawberries, tomatoes and flowers:
  - chemical products – combinations of 1,3-dichloropropene (1,3-D), chloropicrin, metham sodium, dazomet, nematicides, fungicides, herbicides and barrier sheets; and
  - non-chemical techniques such as substrates, steaming, grafted plants, resistant plant varieties, solarization, biofumigation and crop rotation.
- In the post-harvest sector, for grain, stored commodities, mills and food processing facilities:
  - chemical products such as phosphine, sulphuryl fluoride, propylene oxide, ethyl formate and other pesticides; and
  - non-chemical techniques – combinations of IPM (for example altering structures to prevent pest entry, cleaning programmes and pest monitoring), heat, cold, pressure, vacuum-hermetic treatments, low-oxygen atmospheres, irradiation and other techniques.

Each term above – 'steam', 'substrates' or 'phosphine', for example – in fact covers many different types of products, equipment and application methods, some of which are much more effective than others. As a result, training or technical assistance is important for transferring the effective methods and achieving successful results. Case studies have shown that the adoption of alternatives is not intrinsically difficult; however, access to the relevant know-how is essential.[22] Most of the alternatives adopted today were listed in the 1992 report of the SAP/TEAP, and several new methyl bromide alternatives were identified after 1992, such as vacuum-hermetic treatments and biofumigation. It is expected that alternatives will continue to be improved and optimized as more experience is gained, in the same way that the application methods for methyl bromide itself were considerably improved from the 1960s.

Alternatives do not deplete the ozone layer, but some still have undesirable toxic properties like methyl bromide or pose other environmental problems, such as waste disposal from the use of plastic films. Such problems have been addressed by setting up plastic collection/recycling schemes, for example. Some sectors have adopted a strategy of using fumigant alternatives (for example phosphine or chloropicrin) as short- and medium-term alternatives, while continuing to invest in more environmentally-friendly methods for the longer term. This two-stage strategy was adopted in a regional CEIT/GEF (country with economy in transition/Global Environment Facility project), for example.

Not-in-kind alternatives

Alternative technologies can be divided into two broad groups: in-kind alternatives and not-in-kind alternatives. The latter generally require greater changes on the part of users. Two examples are described below.

Source: Photography courtesy of C. Spotti, SIS Fumigation Company, Italy

Figure 11.1 *Example of equipment used for applying alternative pesticide products via modified drip irrigation systems in Italy*

Soil fumigants are often considered to be in-kind alternatives. Like methyl bromide, they are pesticide products applied to soil and active in gaseous form. Soil fumigants have been developed in new formulations, application equipment and methods have been improved, and barrier sheets have been developed to retain gases in soil and reduce the doses necessary to control pests.[23] The improved methods can control fungi, nematodes and many weeds and produce crop yields that are comparable to those obtained using methyl bromide.[24] When necessary, additional weed control methods are also employed. In some cases, existing methyl bromide injection equipment has been adapted, but in many other cases, different equipment is used for applying alternative fumigants, for example through drip irrigation systems (watering systems for crops). These in-kind alternatives have been adopted in many regions of the world for crops such as strawberries, flowers, melons, tomatoes, and peppers and other vegetables.

Substrates provide an example of not-in-kind technologies. Substrates are soil-like materials, based on the same principle as potted plants. Crops are grown in pest-free materials such as peat, peat-substitutes, coconut fibre, rice hulls, pumice, pine bark and/or local clean waste materials and contained in pots, bags, trays, gullies and various other types of containers. Due to their high crop yields and other commercial benefits, substrates have been adopted for strawberries, tomatoes, peppers, other vegetables, flowers and tobacco in all areas of

Source: Photography courtesy of M. Barel.

Figure 11.2 *Production of carnation flowers in bags filled with local substrate materials in Kenya*

the world (including Europe, North America, Japan, China, Kenya and South Africa), often as a result of general economic developments in agriculture. Some substrate systems are high-tech, requiring substantial capital investment and intensive inputs, whereas others are low-cost, simple systems which use relatively inexpensive materials.

Technologies suitable for service providers or end-users

Alternative equipment can be divided into two general groups, with different implications for technology transfer depending on whether it is suitable for service providers or individual end-users (although some equipment is suitable for both groups):

1   **Technologies for pest control service providers.** Some technologies are used by fumigation enterprises or contractors to provide commercial pest control services to farmers or other end-users, in the same way that fumigation enterprises have provided commercial methyl bromide fumigation services to end-users. In such cases, one item of equipment can serve many end-users, as and when needed. Examples include mobile equipment for applying soil fumigants by injection or drip-irrigation, mobile steam equipment, and heaters for carrying out heat treatments in structures. Pest control service providers can also utilize stationary equipment, such as chambers where end-users bring products for treatment (with, for example, low-oxygen atmospheres, heat, cold, pressure or chemicals), examples being

---

Box 11.1 Rapid adoption of not-in-kind technology: Example of tobacco seedlings

1992

Tobacco seedlings were among the five main soil uses of methyl bromide in 1992, important in Brazil, Greece, Japan, South Africa, Spain, the US, Zimbabwe and many other countries. The US alone used around 1680 tons of methyl bromide for tobacco seedbeds in 1992, accounting for 8 per cent of methyl bromide use in the soil sector.

1994/1995

Tobacco experts and methyl bromide users reported that there were no viable alternatives and very few potential alternatives were on the horizon.

Mid–late 1990s

Substrates (floating tray systems) were widely adopted in the US and were introduced in Brazil, Argentina and Zimbabwe.

2002

Within a few years, substrate systems had been adopted for the majority of tobacco seedling production in many countries.

Commentators have recently assumed that phasing out methyl bromide in the tobacco sector was easy. But in fact the changes implemented in the tobacco system were much more fundamental than those generally needed for crop systems that adopt methyl bromide alternatives. Substrates required many changes in technology, infrastructure, management and the production chain:

- large initial investment (partly covered by Multilateral Fund (MLF) funds in developing countries);
- large technology transfer effort to achieve adoption by many thousands of farms;
- new know-how required by everyone in the technology and information supply chain: farmers, extension personnel, agricultural consultants, suppliers of agricultural materials and tobacco enterprises;
- entirely new production methods from the system of seedling production to transplant operations in the field; and
- new manufacturing facilities and distribution chains for the new materials and equipment.

The rapid and substantial change in the 1990s in this sector resulted from a strong willingness to phase out methyl bromide. In this sector, a major motivation for change was the desire to avoid public criticism for ozone depletion in an industry already facing strong public concern about the health hazards of smoking.

B-Cat or ECO2 facilities in Europe and Asia, which are often located at ports. Greenhouse equipment and materials for mass-production of grafted plants (for example International Nursery, Grow Group and Agrimatco in Morocco) or tobacco seedlings (for example large-scale floating tray production in Argentina) provide further examples of technologies that are suitable for enterprises that can provide a service (in this case seedlings) for many end-users.

2   **Technologies for end-users.** Alternative technologies such as fixed steam boilers, small mobile boilers, substrate systems, hand-held application equipment and IPM systems are more suitable for end-users such as farmers and food enterprises.

When equipment is used by service providers, the necessary training and transfer of know-how can be targeted primarily at those service providers. Training service providers (who are relatively few in number) requires less time and fewer resources than training hundreds of end-users. In cases where end-users need to make adaptations to accommodate the new technologies (for example adapting existing farm irrigation systems to carry treatments), the service providers can also be trained to give end-users the relevant technical assistance. This type of technical support was provided by several major soil fumigation enterprises (pest control service providers) in Italy and Spain, for example, as illustrated in Box 11.3.

The capital cost of equipment can of course present a substantial barrier, particularly in developing countries; however, this problem has been largely overcome by the MLF, which provides equipment for end-users or service providers.

Proprietary technologies and trademarks

Some alternative technologies are proprietary products subject to trademarks or proprietary restrictions; examples include certain resistant plant varieties, Cocoons^{TM} for vacuum-hermetic treatments, and chemical products such as ProFume®, ECO2Fume®, Vapormate^{TM}, Telone®, Agrocelhone^{TM}, Basamid® and Nemasol®. However, proprietary status has not prevented widespread use, and the trademarked products are exported and supplied to many countries. Moreover, many alternative products and equipment are non-proprietary generic items that can be manufactured by any enterprise that wishes to do so; examples include many types of substrate materials and containers, a number of IPM materials, steam equipment, heating and cooling equipment, equipment for low-oxygen atmospheres, vacuum pumps, and chemicals that are no longer covered by patents.

In some markets, enterprises that imported and distributed methyl bromide gained exclusive rights for the distribution of major methyl bromide alternative products. This business arrangement appears to have been counterproductive in a number of cases, where the methyl bromide importers continued to give priority to selling methyl bromide and put little or no effort into selling or promoting the alternative products. The arrangement was more successful, however, in countries where national policy changes reduced the availability of methyl bromide substantially, thereby opening the door for sales of alternative technologies.

TECHNOLOGY TRANSFER AND CHANGE IN DEVELOPED COUNTRIES

This section examines some of the components and pathways of technology transfer and change, identifies divergent approaches, and provides an overview of how major barriers and challenges were addressed in developed countries.

Methyl bromide consumption patterns in developed countries

Developed countries used more than 56,000 tonnes of methyl bromide in 1991, falling to 6560 tonnes of authorized consumption in 2007 (less than 12 per cent of the 1991 level). However, individual countries achieved technology transfer and change at different rates. The reasons for these differences are discussed in later sections.

In 1991 the US, the EC, Japan and Israel together used 97 per cent of the total methyl bromide consumed in developed countries. Figure 11.3 shows the trends from 1991 to 2007. US consumption (in other words production and imports) fell sharply from 1998 to 2002, then increased until 2004 as the scheduled phaseout date (2005) approached and enterprises stockpiled methyl bromide. In 2007 US consumption finally fell to pre-2002 levels. Consumption in the EC, the second biggest consumer in 1991, has shown a steadier downward trend since 1999, falling to a low level of consumption in 2007. Consumption in Japan and Israel started at a lower level in 1991 and decreased from 1998, levelling off recently.

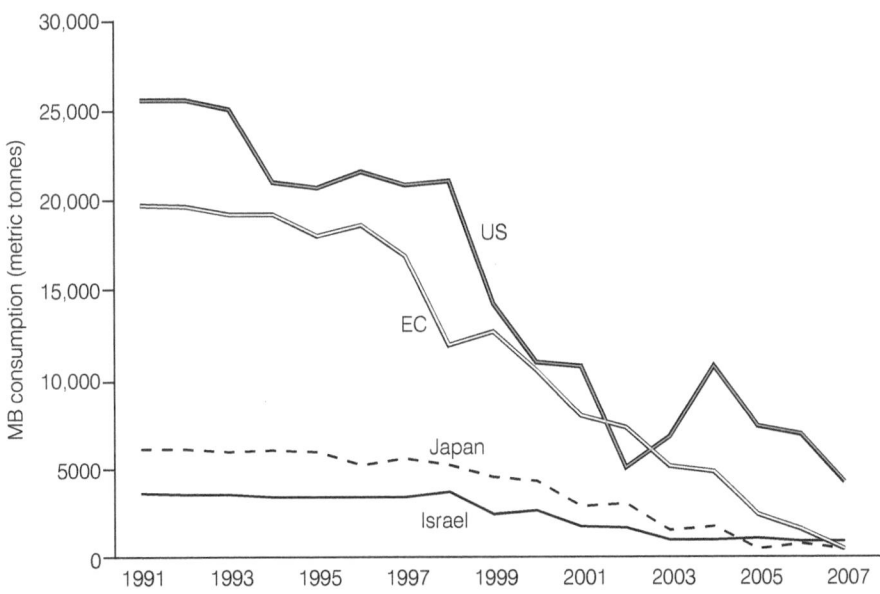

Source: MBTOC (2007)[25]

Figure 11.3 *National methyl bromide consumption in the US, the EC, Japan and Israel, 1991–2007*

Box 11.2 Technical leadership by post-harvest fumigation enterprises

In the 1990s Fumigation Services and Supply (FSS) and Insects Limited were among the first enterprises to develop and adopt methyl bromide alternatives in post-harvest applications. FSS carried out methyl bromide fumigations and other pest control services in mills, food processing facilities, grain stores, shipping containers and other facilities. At an early stage, the enterprise owner, David Mueller, made a policy decision to protect the ozone layer by phasing out methyl bromide from the enterprise's commercial operations. FSS made substantial investments and worked with other enterprises to test alternatives and develop them to a commercial scale, particularly alternatives based on insect biology and combination treatments. They developed and commercialized ECO2Fume® (phosphine plus CO_2) with BOC Gases and Cytec Industries, combination fumigations using heat, and integrated pest management techniques based on cleaning and pest monitoring.[26] FSS also worked with Dow AgroSciences LLP and food enterprises in the testing, registration and commercial development of ProFume® (sulphuryl fluoride). FSS and its sister company Insects Limited organized national and international conferences and conducted many training courses in methyl bromide alternatives for fumigators and pest control operators. They also assisted developing countries in adopting methyl bromide alternatives in MLF projects.

FSS announced to its customers that it would phase out controlled uses of methyl bromide by the end of 2004. It made substantial technical changes in its operations and indeed has not used methyl bromide for uses controlled by the Montreal Protocol since 31 December 2004. On the issue of technology change, David Mueller comments that:

> the old adage is correct: where there's a will, there's a way. Viable and economic methyl bromide alternatives are available for mills and similar post-harvest uses, but each enterprise needs to be willing to embrace change.

Mueller also reports that:

> We still hear comments like 'alternatives are too expensive', 'food production will be impossible without methyl bromide' or 'methyl bromide is necessary for us to survive'. But the reality is that many thousands of former methyl bromide users in the US and around the world have acted environmentally responsibly by eliminating methyl bromide, and they are still operating successful businesses as before. In fact, many of us have found out that alternatives offer different and greater advantages than methyl bromide.

Research and development

Initial research examined alternatives treatments one by one, with little success. Progress was made primarily when more innovative enterprises, farmers and researchers focused on combining several treatments and made improvements in equipment and application methods. Boxes 11.2 and 11.3 provide examples of fumigation enterprises that showed leadership in the technical development and trialling of alternative technologies. A Canadian report in 1998 noted that research and commercial demonstration projects in Canada had resulted in the

---

BOX 11.3 TECHNICAL LEADERSHIP BY
SOIL FUMIGATION ENTERPRISES

A major Italian methyl bromide fumigation enterprise, SIS, worked with farmers and other enterprises in the 1990s to transfer and adapt alternative products from other countries, develop equipment, and register new application methods (via drip irrigation). They built nurseries to produce grafted plants for tomato, aubergine and other vegetable crops, which could be used in combination with solarization, IPM or an alternative fumigant. They also established some greenhouses using substrates. SIS used its national network of extension personnel to provide necessary technical assistance to farmers, enabling them to adapt existing irrigation systems and make other changes so that alternatives to methyl bromide could be used. SIS carried out 465 treatments using alternatives in 2002, this rising to 4000 by August 2004. They found that the new products generally competed favourably in terms of cost with methyl bromide; however, additional weed control was sometimes needed, and the waiting period was longer than with methyl bromide, requiring changes in production timetables or earlier treatments to avoid delays. Moreover, the work of extension personnel was found to be essential in changing the rather fixed habits of farmers.[27]

In Spain, another major methyl bromide fumigation enterprise, Agroquimicos de Levante (AQL) registered Agrocelone™ soil fumigants initially in Spain, then in Morocco, Lebanon, Chile, Cuba and other countries. It developed suitable application equipment for different types of application, for example broadacre injection, raised bed injection and application thorough drip lines. AQL promoted these alternatives by collaborating with researchers and growers in trials and demonstrations and providing technical support to farmers.[28]

---

availability of many alternative products, technologies and services which have proven advantageous to the Canadian agriculture and agri-food sector.[29] In general, much of the cost of R&D was borne by individual enterprises, but a number of governments also provided funding for research programmes. An innovative approach to research and development funding was adopted in Australia, where methyl bromide users agreed to place a levy on methyl bromide imports (about Aus$0.40 per kg methyl bromide by 1998), this revenue being matched by government funds.[30]

Agricultural extension activities

Alternatives to methyl bromide have often required substantially different skills and know-how. In order to control pests and achieve satisfactory results, it is necessary for alternative technologies to be applied using correct methods, and this requires relevant know-how and thorough training. Some developing countries organized and implemented special large-scale training and extension programmes for fumigators and many thousands of methyl bromide users. In contrast, developed countries did not organize large-scale training programmes; they did, however, carry out other types of extension activities. In the US and Italy (Box 11.4), for example, university extension departments worked with fumigation enterprises to carry out demonstrations on a variety of crops in a number of geographic locations. And in Australia, for example, agricultural

---

Box 11.4 Information dissemination in Italy

In 1996 a survey was carried out among Italian farmers to identify problems relating to soil disinfestation in the vegetable and ornamental sectors and the percentage of farmers willing to adopt new solutions. Eighty per cent of the farmers interviewed used methyl bromide, and almost all farmers were sceptical about applying alternative technologies. Most farmers did not accept the idea of reducing methyl bromide doses, but almost 80 per cent supported the use of virtually impermeable film (VIF) sheets. According to the survey results, there was little interest in adopting substrate systems, a technique which had proven to be a good methyl bromide alternative in Italy. However, the survey identified that the farmers trusted the extension service as an important means of diffusion for innovative technologies. The survey confirmed that the fear of costs and loss of yields and the lack of know-how and information were the major factors preventing the widespread use by farmers of the methyl bromide alternatives already available at that time.[31]

To address the low level of awareness, training and information activities were carried out by the Ministry of the Environment in collaboration with agricultural institutes, universities, fumigators and several farmers' associations. Technical seminars, meetings, workshops and open days were held at farms and research centres to present the results of field demonstrations, often at a time when the crops could be observed in the field. Scientific and technical publications, leaflets and videos were produced, aiming to reach farmers, technicians, consumers and politicians. As a result, after 1999 the number of farmers in Italy who adopted alternatives increased.

---

researchers held many meetings with local groups of growers to demonstrate and provide technical assistance on the most relevant methyl bromide alternatives, and published guidelines on alternatives and regular updates for growers on results of regional trials and demonstrations.[32] In several countries, fumigation enterprises played a leading role in providing technical assistance and/or training, as illustrated in Boxes 11.2 and 11.3.

Governments in some developed countries depended largely on market mechanisms to provide the diffusion of alternative technologies. This market approach worked rather slowly in situations where there was a large supply of methyl bromide and little immediate pressure or incentive for users to change.

Economic considerations

The adoption of methyl bromide alternatives required from modest to large initial investments, depending on the technology and situation. For example, fumigation enterprises which provide services to farmers purchased equipment such as mobile boilers for steam treatments or new injection equipment for soil fumigants. Changes in operating costs also arose from changes in materials, labour inputs and timetables.

At first glance, some alternatives appeared unreasonably costly because they required a large initial investment in training or equipment. However, when investment and operating costs over several years were compared with methyl bromide, some alternative technologies were actually more cost-effective.[33] The net revenue (profitability) over several years is the key parameter affecting end-users, and alternatives with higher costs than methyl bromide, such as substrate systems, can actually be as profitable as methyl bromide as they provide higher crop yields and quality. Similarly, an alternative that provides lower yield, such as biofumigation plus solarization systems, can be as profitable as methyl bromide if the operating costs are sufficiently lower.[34]

Another factor accounting for differences in the rate of technology change observed between regions is that methyl bromide prices have varied from one location to another. The law of supply and demand means that the price of methyl bromide tends to decrease in locations where stocks are plentiful and to increase in locations where the methyl bromide supply has been restricted, making the relative prices of alternatives less or more attractive respectively. Moreover, as alternative products and materials became more widely available in a region, their prices tend to become more favourable due to economies of scale and/or increased competition among suppliers.

Regulatory approval of alternative technologies

Pesticides and fumigants, like methyl bromide, normally have to be registered by the government authorities responsible for pesticide safety before they can be used, so the availability of particular chemical products varies from country to country. Phosphine, for example, is registered in many countries, while ethyl formate is registered in only a few. The process of applying for a new pesticide registration is usually very expensive in developed countries. Registration is normally carried out by the enterprises that wish to sell a pesticide product, and such enterprises tend to focus on sectors where they can expect a large market, ignoring small markets. However, despite these substantial barriers, a number of alternative products have been registered as methyl bromide alternatives.

On the other hand, many non-chemical alternatives (for example substrates, steam, grafted plants and biofumigation) do not face such barriers because they do not require registration by pesticide authorities, so they are accessible immediately to users.

In addition to issues related to the registration of chemical products, other regulatory issues sometimes limit or affect the choices made by users, such as disposal/recycling of plastic waste, restrictions on uses (for example buffer zones) and regulations existing in the countries to which treated products are exported.

Market demands and certification standards

Whether methyl bromide or alternatives are used, agricultural products have to be acceptable to purchasers (supermarkets, wholesale purchasers and the final consumers) and have to meet commercial expectations relating to visual appearance and uniformity. Purchasers, from supermarkets to individual consumers, became increasingly concerned about pesticide residues and the environmental impacts of agriculture during the 1990s. As a result, some regions adopted IPM and environmental programmes. Several agricultural associations in the Almería region of Spain, for example, established IPM standards which identified

---

**BOX 11.5 TECHNOLOGY TRANSFER ACTIVITIES
AND LEVIES IN AUSTRALIA**

Australia used 679 tons of methyl bromide in 1997 for products such as grain, strawberries, tomatoes, peppers, melons and cut flowers. Methyl bromide use was reduced to about 55 tons in 2006. Substantial technology change has been accomplished in the last decade as a result of various initiatives.

In 1994 meetings were held in agricultural regions of Australia to consult methyl bromide users and identify methyl bromide usage patterns and sectors that would be most affected by the Montreal Protocol phaseout. The meetings identified existing and potential alternatives and helped growers to identify targets for methyl bromide reductions in each region. Affected horticultural sectors were encouraged to make a commitment to a coordinated national phaseout of methyl bromide.[35]

A Methyl Bromide Consultative Group was set up in 1995, with representatives of methyl bromide users, researchers, and national and regional government. This group coordinated the development of a national strategy for methyl bromide reductions and phaseout.[36] An Alternative Research Coordination Committee was also set up in 1995, comprising researchers, government representatives and key methyl bromide users' representatives, to coordinate research and trials of methyl bromide alternatives and promote communication and education for farmers about alternatives and the phaseout. Trials were conducted on the main horticultural crops, different soil types and climates (temperate, subtropical and arid). In the following years, substantial changes were made in the application methods for alternatives, particularly techniques that improved dispersal in the soil.

A key to progress in Australia was the introduction of a voluntary levy on methyl bromide imports. This provided funds for the alternatives research and information programme. In 1998 the levy raised approximately Aus$250,000, which was matched by funds from the national Horticultural Development Corporation, giving about Aus$500,000. Environment Australia also required methyl bromide importers to purchase import licences and charged activity fees.[37] This increased methyl bromide prices, making alternatives more attractive for commercial use.

A communications network was established, comprising contact people in each region for disseminating information. A newsletter, *National Methyl Bromide Update*, was launched in 1997 to provide growers with news about the ozone layer and Montreal Protocol, research results, progress in the implementation of alternatives, and case studies on farms that used alternatives.[38]

---

acceptable pest control methods and prohibited the use of methyl bromide (Box 11.7). And supermarkets in Europe established 'good agricultural practice' (GAP) standards called EUREP-GAP, requiring fruit, vegetable and flower producers to introduce more environmentally friendly practices. EUREP-GAP standards for cut flowers and ornamentals do not permit the use of methyl bromide.[39] Other agricultural certification programmes, such as the Milieu Programma Sierteelt (MPS), the Kenya Flower Council, FlorVerde and the Flower Label Programme, have also had a significant impact in eliminating the use of methyl bromide from many farms around the world that produce cut flowers and ornamentals. These standards have led to substantial technological changes in the participating farms.

BOX 11.6 TECHNOLOGY TRANSFER AND REGULATORY ACTIVITIES IN JAPAN

Japan used 5336 tons of methyl bromide in 1998 (excluding QPS), primarily for crops such as strawberries, tomatoes, peppers, melons, watermelons, cucumbers and flowers. Methyl bromide use was reduced to about 546 tons by 2005 as a result of the following activities.

Campaigns with farmers

In 1992, within a year of methyl bromide being listed as an ODS, the Japanese Government started campaigns to encourage farmers to reduce methyl bromide use and emissions and encourage alternatives. The Ministry of Agriculture, Forestry and Fisheries (MAFF) issued a notice with guidance to farmers as follows:

- farmers are encouraged to make efforts to reduce use of methyl bromide;
- farmers are encouraged to apply alternatives as far as possible; and
- it is strongly recommended to cover the soil surface with sheets in the event of unavoidable methyl bromide use; this is strictly required in greenhouse facilities.

MAFF established a methyl bromide reduction panel, which consisted of all divisions of MAFF concerned with vegetable production, pest control, pesticides and pest control research, to promote alternative technology transfer.

Technology transfer activities

- Leaflets were produced to explain alternative techniques.
- Farmers were provided with new equipment and trained in alternative application technology by experts from local technology extension centres.
- Pesticide enterprises were encouraged to develop new pesticides and formulations for soil diseases and insects.
- Pesticide enterprises were encouraged to expand the registration of existing pesticides to cover the same spectrum of pests as methyl bromide.
- National, local and industrial agriculture research institutes were encouraged to develop alternative technology with financial incentives from government.
- When new alternatives were submitted for registration, they were given preferential and intensive review by the pesticide registration authorities to allow earlier registration than normal.
- Enterprises manufacturing pesticide application equipment were encouraged to improve the necessary alternative equipment and materials.
- Demonstrations of alternative technologies were made on farms using government financial support for materials, equipment and renting fields.
- When technically and economically feasible alternatives became available, the relevant methyl bromide uses were removed from approved pesticides lists managed by local technology extension centres.
- Alternatives were adopted for crops such as tomatoes and strawberries. Tray cultivation systems using substrates were adopted for strawberry fruit and plant production, for example.[40]

Box 11.7 Impact of agricultural production standards in Almería, Spain

The Almería region of Spain produces peppers, aubergine, tomatoes, melons and other cucurbits, accounting for approximately 70 per cent of the production and export of fresh vegetables in Spain. In 1995 methyl bromide was used on about 1430 hectares (3534 acres).

During the 1990s several growers associations, the Association of Harvesters and Exporters of Fruit and Vegetables (COEXPHAL) and the Association of Commercial Suppliers of Horticultural Products of Almería (ECOHAL) worked with certification bodies and other groups to develop IPM standards (norm UNE 155001) for the production of fresh vegetables.

Initially, the associations merely requested and encouraged growers not to use methyl bromide. Later, the production standards did not permit methyl bromide to be used, and COEXPHAL publicly expressed support for eliminating methyl bromide.

Throughout this period, the associations and technicians provided farmers with information and technical assistance about alternatives. Afterwards, the associations commented that information and training of growers were important factors in the successful adoption of alternatives.

As a result of this work, farmers made substantial technological changes, and methyl bromide was virtually eliminated throughout the large region of Almería by 2002.[41]

Specialists reported in 2002 that the alternatives met the high standards demanded by supermarkets and enabled growers to compete successfully in international markets despite fierce competition from producers with lower labour costs.

Regulatory frameworks and policy-led approaches

A small number of developed countries established national action plans which aimed to phase out most uses of methyl bromide within a short period of time. Examples are provided by The Netherlands and Denmark (Boxes 11.8 and 11.9), which carried out activities such as:

- review of methyl bromide uses and identification of relevant alternatives;
- amendment of national regulations or policy measures, including restrictions on specific uses of methyl bromide;
- providing technical assistance for fumigators and methyl bromide users; and
- identification of the responsibilities of government departments and stakeholder groups.

Promotion of new business opportunities

Methyl bromide phaseout provided opportunities for new businesses to flourish by making or supplying necessary alternative products or services. Since most countries import methyl bromide, there are opportunities for import substitution and local job creation.[42] The Canadian Government noted in 1996 that ODS phaseout provided an opportunity to develop new businesses and industries, at the same time as benefiting the environment.[43] A Canadian indus-

Box 11.8 Early phaseout of methyl bromide in soil sector in The Netherlands

The Netherlands is a major horticultural producer of crops such as strawberries, tomatoes, peppers and cut flowers. The country relied heavily on methyl bromide in the 1970s and was the largest user of methyl bromide in Europe at the time, using about 3000 to 5000 tons in 1980–1981. This was a substantial quantity of methyl bromide for a relatively small country. In this same period, The Netherlands experienced serious methyl bromide poisoning incidents among agricultural workers, methyl bromide drift in residential areas, and methyl bromide residues in water and vegetables. In response to these concerns, the government carried out a review of methyl bromide uses and alternative options and decided to phase out methyl bromide in soil fumigation. The government thus took the following steps with regard to the use of methyl bromide:[44]

- reviewed all methyl bromide uses and identified suitable alternatives;
- prohibited use of methyl bromide for crops where alternatives were available;
- required permits for each methyl bromide fumigation, issued case-by-case, based on evidence of soil-borne pests and lack of suitable alternatives;
- required fumigation sheets to be left on soil for a longer period (ten days);
- increased safety restrictions on methyl bromide, such as larger buffer zones around sites;
- provided technical assistance to help farms to adopt alternatives;
- provided financial grants for a short time to assist farms in purchasing steam boilers;
- improved existing alternatives and developed additional alternatives as a result of work by research institutes, fumigators and others; and
- encouraged industry–government cooperation.

A major soil fumigation enterprise, Marten Barel BV, used methyl bromide for many crops until 1981, but rapidly switched to providing mobile steam treatments and supplying substrates as an alternative business. They reported that:

> *methyl bromide phaseout led to a number of technical innovations in horticulture. For example, we developed more efficient steaming systems that used less fuel and reduced costs. The methyl bromide phaseout led to modernization and great improvements in crop production, with many benefits for growers as a result.*

Horticultural production is of considerable importance to the Dutch economy. The value of the sector in 1980 was DFL11,185 million, with employment of about 87,400 people. At that time, the government estimated that the phaseout would have considerable impact on businesses, increasing greenhouse costs, for example, by DFL13,500 to 30,000 per hectare.[45] As a result of the national plan, however, most methyl bromide use was phased out within a year – methyl bromide was no longer permitted for uses where alternatives were available, and a permit was required for each methyl bromide fumigation – but the agricultural production statistics showed that crop production was maintained during the phaseout period and actually increased in subsequent years, largely due to leading alternative technologies (substrates). The production value of the relevant crops doubled from DFL5955 million in 1980 to DFL12,757 million in 1991.[46] The government and other observers later concluded that the removal of methyl bromide had benefited horticulture in The Netherlands because it acted as a catalyst for the development and widespread adoption of new and improved production practices.[47]

**BOX 11.9 EARLY ACTION FOR OZONE PROTECTION
IN DENMARK**

Soon after the Montreal Protocol placed controls on methyl bromide in 1992, Denmark identified methyl bromide uses throughout the country and carried out a technical review of existing and potential alternatives. The review was extended to other countries in Scandinavia and published as a report of the Nordic Council of Ministers in 1993.[48] The report concluded that alternatives were available for most uses of methyl bromide and identified specific problem areas where alternatives were not available or required further development. The report recommendations stated that:

> *The political decisions concerning a phasing out of methyl bromide as a pesticide should not be delayed by demands for further research into alternatives [...] In our opinion, suitable alternatives already exist for the different areas of application that we have studied in this report. [...] During our work with this report, we have been able to identify factories, pest control enterprises, horticulturalists and so forth who are handling or have handled similar problems without applying methyl bromide.[49]*

Subsequently, the Danish Government approved a regulation in 1994 which established individual phaseout dates for the various uses of methyl bromide, including for tomatoes, flowers and other crops, flour mills, warehouses for grain, seeds, herbs, spices, nuts, dried fruit, tobacco, furs, extermination of wood boring beetles, wood packaging, museum objects, historic buildings, ships, and aircraft.[50] Denmark successfully completed national methyl bromide phaseout by January 1998.

**BOX 11.10 POLICY DRIVING CHANGE:
FUMIGATION OF HOMES IN CALIFORNIA**

Safety restrictions placed on methyl bromide in a specific sector – fumigation of homes in California – led to rapid technical changes:

- In 1990 California used around 2300 tons of methyl bromide for fumigation of homes to control pests.
- Many fumigators in 1990 believed it would be very difficult to find or adopt other pest control methods.
- Following regulatory restrictions on methyl bromide due to accidental poisonings in homes, other products were widely adopted within three years.
- In 1992 California used around 430 tons of methyl bromide for domestic dwellings,[51] representing an 81 per cent reduction in two years.

---

try–government group, for example, produced a booklet publicizing methyl bromide alternatives and Canadian enterprises that could supply alternative technologies and know-how.[52] An EC report in 1997 also pointed out that methyl bromide phaseout provided opportunities for new or expanded industries, such as manufacturing equipment for pest control treatments, making substrates from waste materials or recycling agricultural plastics. The report also

BOX 11.11 POLICY-LED TECHNOLOGY CHANGE IN THE EUROPEAN COMMUNITY

Tom Batchelor, Director, TouchDown Consulting, formerly European Commission specialist on ozone layer protection

The EC was once the world's second largest consumer of methyl bromide, but today it is one of the smallest consumers, with almost 98 per cent of its 1991 consumption eliminated (excluding QPS). The EC regulation on ozone-depleting substances (Regulation No 2037/2000) placed several restrictions on methyl bromide that were more stringent than required by the Montreal Protocol:

- from 1 January 2001 methyl bromide consumption for uses controlled by the Protocol was restricted to 40 per cent of 1991 consumption, as indicated in Figure 11.4 below;
- from 1 January 2003 methyl bromide consumption was reduced to 25 per cent of the 1991 level;
- from 1 January 2006 the use of stocks of methyl bromide was banned except for authorized CUEs and QPS;
- small disposable cans of methyl bromide were prohibited, making it less accessible to untrained farmers;
- methyl bromide leakage in soil fumigation was reduced by requiring the use of virtually impermeable film (VIF) or equivalent; and
- a quantitative limit was set on the amount of methyl bromide used for QPS.[53]

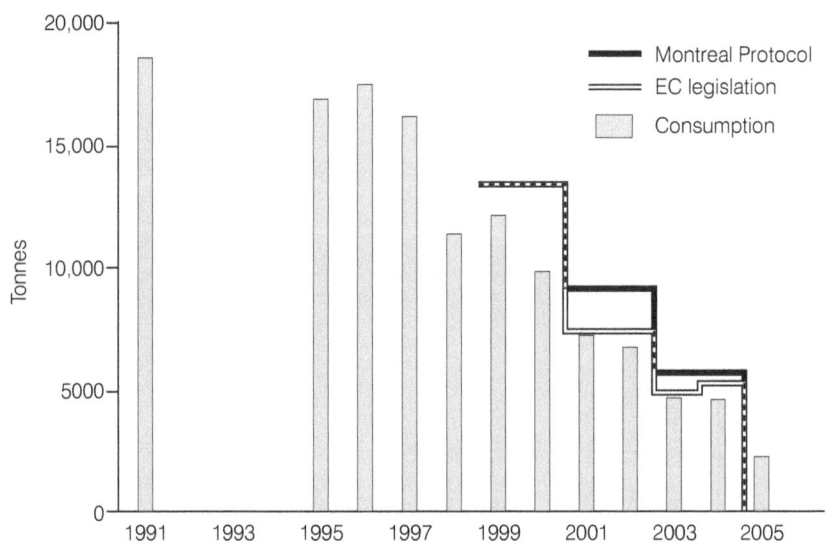

Note: The graph shows consumption as reported to the Ozone Secretariat. Consumption in the period 1992–1994 (not shown in the graph) was probably in the same range as reported in 1991.

Figure 11.4 *EC consumption of methyl bromide, indicating limits established by the Montreal Protocol and EC regulations, 1991–2005*

The EC slightly increased the quantity of permitted methyl bromide slightly in 2004, when additional countries joined the EC, although in practice the additional quantity was not used.

In the 1990s methyl bromide users claimed that eliminating methyl bromide would reduce crop yield, damage profitability and increase unemployment; however, many methyl bromide users successfully adopted alternative technologies in the EC. As the phaseout date of 2005 approached, some methyl bromide users requested critical use exemptions (CUEs). The EC discouraged applications for CUEs allowing methyl bromide, by establishing detailed evaluation procedures. A team of experts worked with the European Commission to assess whether alternatives were available, suitable and economically feasible, and identified measures that could reduce the use and emissions of methyl bromide. The European Commission adapted the existing ODS licensing procedures and issued quotas for methyl bromide imports/use to specific registered fumigation enterprises for each individual CUE sector.[54] This meant that fumigation enterprises had to specifically make a request before any methyl bromide could be imported or produced, in contrast to the previous licensing system which allowed methyl bromide importers and manufacturers to put methyl bromide onto the market right up to the quantitative limit permitted by the regulation, without regard for actual need. In 2005, 127 fumigators were registered in the database, but this number was reduced significantly in successive years as alternatives were adopted and methyl bromide was no longer needed.

The Commission and EC countries developed a European Community Management Strategy (ECMS) for phasing out CUEs, as required by the Protocol.[55] The ECMS contained, among other information, an extensive database of methyl bromide alternatives available in Europe. This database was an important tool leading to reductions in the amount of methyl bromide required for critical uses.

The above policies, together with the efforts made by methyl bromide users, reduced the CUEs as indicated in Figure 11.5, leading to a rapid adoption of alternative technologies in the EC.

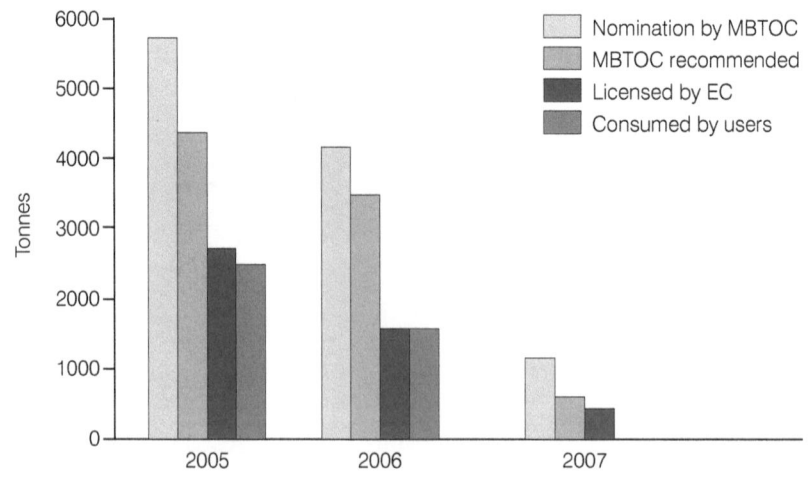

Source: TouchDown Consulting, Belgium.

Figure 11.5 *Nominated and authorized methyl bromide consumption for CUEs in the EC, 2005–2007*

identified examples of opportunities for new service industries such as pest identification and monitoring services, training, agricultural software, and mobile disinfestation services.[56]

Levies and taxes

A small number of countries placed levies or taxes on methyl bromide as a disincentive for using it. Often the resulting funds were used for the promotion of alternatives. From 1996, for example, Environment Australia (formerly the Australian Environmental Protection Agency) required methyl bromide importers to purchase import licences at a cost of Aus$10,000 for each two-year period. In addition, the agency charged an activity fee of Aus$90 per tonne of imported methyl bromide. The revenue was placed in a fund and used to support activities for phasing out ODSs.[57] Australian methyl bromide users also decided to introduce a levy – a type of voluntary tax – on methyl bromide imports, to generate funds for trialling and developing alternatives. The levy was collected at wholesale level by the methyl bromide importers. The levy funds were matched by funds from the government (the Horticultural Development Corporation), giving about Aus$300,000 per year for the development of alternative techniques and communication about alternatives with farmers.[58]

The Czech Republic's ozone protection legislation placed taxes on producers and importers of ODSs, and the revenue is used by the state Environmental Fund for Ozone Layer Protection. From January 1996 this tax was applied to methyl bromide.[59] A regulation in Slovakia placed a fee on all ODS imports, including those of methyl bromide. The fee increased the price of methyl bromide compared to alternatives, encouraging users to shift away from methyl bromide.[60] Similar levies and taxes have been used in other environmental areas. Poland, for example, established a fee on the emission of pollutants to the atmosphere (of US$0.02 per kg), which could act as a disincentive to use pollutants such as methyl bromide.[61] And in Sweden, an environmental levy of approximately US$1.50 per kg of active pesticide ingredient raised about US$3 million per year for research, development and extension in non-chemical and IPM methods.[62]

Targeted grants and loans

Latvia operates an Environment Investment Fund which provides loans for enterprises to implement environmentally friendly technologies. In the agricultural sector, a number of national and regional governments promote agricultural innovation and exports by providing grants for activities such as advisory services, training and pilot projects. These give important economic signals to farmers and can help determine their choice of pest control methods.[63] The regional government of Ragusa, in Sicily, for example, organized a programme in the 1990s to promote new agricultural technologies. It subsidized the purchase of plastic sheets for solarization (25 per cent of cost reimbursed) and machinery to lay sheets for open-field solarization (13 per cent reimbursed). Irrigation systems were also subsidized.[64]

TECHNOLOGY TRANSFER AND CHANGE IN CEITS AND DEVELOPING COUNTRIES

In 1991 annual methyl bromide consumption was about 8340 tonnes in developing countries, a relatively low level compared to developed countries. However, consumption in developing countries increased to more than 18,100 tonnes in the 1990s before falling to 9285 tonnes by 2005. Consumption is expected to be about 7000 tonnes in 2007 if recent trends continue. The majority of developing countries achieved a freeze on methyl bromide consumption in 2002 and a 20 per cent reduction in consumption by 2005, as required by the Montreal Protocol. Many developing countries have reduced methyl bromide well beyond the scheduled Protocol requirements. In 2005, 80 per cent (115 of 144) developing countries consumed less than 50 per cent of their national baselines.

The Montreal Protocol's MLF and the GEF provide technical and financial assistance to developing countries and CEITs, assisting the process of technology transfer and adoption and thus enabling large volumes of methyl bromide to be phased out. A number of methyl bromide demonstration projects have also been funded from other sources, by the developing countries themselves (projects in China being examples) or through bilateral assistance from the Governments of Australia, Germany, Italy, Canada and Spain. In many developing countries, farmers, institutes, producer/exporter groups and others have also carried out their own trials to test or adapt alternatives and have adopted alternatives using their own financial resources and efforts. The following sections describe the main types of activities that have contributed to technology transfer and change in developing countries.

MLF demonstration projects

Initially, priority was given by the MLF to identifying, evaluating and demonstrating methyl bromide alternatives (Decision IX/5). As a result, the MLF funded more than 40 demonstration projects that aimed to transfer, evaluate and adapt relevant technologies from regions that already used alternatives. Trials were carried out at farms and institutes to compare the efficacy, crop yields and costs of methyl bromide and several alternatives, and to customize alternatives to suit local conditions. In 2002 the MBTOC reviewed the technical results of MLF demonstration projects and concluded that, with two exceptions, 'for all locations and all crops or situations tested [...] one or more of the alternatives have proven comparable to methyl bromide in their effectiveness'. (The two exceptions involved ginseng root rot in China and high-moisture fresh dates in North Africa, where limited trials had failed to identify suitable alternatives.) The MBTOC noted that, in many cases, combined techniques provided more effective results than individual techniques. Particular attention needed to be paid to appropriate, effective application methods. The MBTOC concluded that the demonstration projects showed that the tested alternatives could be introduced into developing countries and adapted successfully within two to three years; in some cases, this time period even included the registration of pesticide products.[65]

MLF *phaseout projects*

The MLF has funded many projects to assist developing countries achieve early phaseout of methyl bromide by building local capacity, helping methyl bromide users to install alternative equipment and products, and providing training in the use of alternatives. This access to technical know-how and financial resources has had a large impact on the rate of technology change. The first MLF project aiming to phase out methyl bromide was approved in 1998. By December 2006 there were more than 46 projects, with approved funding of more than US$77 million. The MLF projects in total are scheduled to eliminate 9105 tonnes of methyl bromide in developing countries; by December 2006, 5245 tonnes had already been phased out. Most developing countries have achieved the scheduled methyl bromide reductions in MLF projects. In some countries there were delays, and schedules were renegotiated, but in other countries projects achieved the methyl bromide reductions faster than scheduled.

The MLF Executive Committee guidelines for methyl bromide projects recommend that stakeholders should be fully involved in the development of projects. This includes relevant government departments, farmers, farmers' associations, extension agencies, enterprises and other groups who have an interest in methyl bromide use and alternatives.

MLF projects commonly included three main components for promoting and accomplishing technology transfer: policy development; training, technical assistance and information dissemination; and agricultural equipment and materials.

Policy development

The MLF guidelines require countries to develop a package of policy measures so that the methyl bromide phased out by a project will not be reintroduced after the project is completed. Governments are expected to monitor and introduce limits on the quantity of methyl bromide that can be imported and produced. They are encouraged to undertake other activities which discourage the use of methyl bromide and promote the use of alternative technologies.

Training, technical assistance and information dissemination

The projects often provide capacity-building for relevant agricultural institutions. Extension specialists and experts develop training programmes and extension materials in local languages. Trainers and extension personnel are normally trained first by regional specialists or international specialists (who thus transfer relevant skills and know-how from other regions). The local trainers then train farmers and others, often in field workshops. Following this, trainers often make two or three follow-up visits to each farm to continue the training and ensure that farmers will be able to use the alternatives properly. Additional farm advice is provided if necessary. Some projects train very large numbers of farmers. A project in Malawi, for example, trained more than 400,000. The training is often organized in batches so the work can be done in stages in each region. The five-year vegetable project in Lebanon, for example, aimed to train 10 per cent of farmers in the first year, 20 per cent in the second year, 30 per cent in the third and fourth years, and 10 per cent in the final year. However, the rate and method of training varies from one country to the next, reflecting different agricultural and social needs. Some projects also provide publicity and general extension information so that students and farming communities become aware of methyl bromide problems and available solutions.

The MLF projects have shown that very large numbers of farmers can be trained relatively quickly, even for not-in-kind technologies.[66] In the Macedonia MLF project, for example, about 12,500 farmers, located in different regions, were trained in all aspects of the selected alternatives in the period from 2002 to 2004, using 'train the trainer' methods, and phaseout was achieved much earlier than expected.[67] And a project in Argentina established a well-organized programme which trained more than 2700 farmers in the first year and continued to train further thousands in each subsequent year (Box 11.12).[68] This was particularly impressive because many of the recently trained farmers in Argentina are in economically deprived and inaccessible regions. It is notable that many were able to adopt successfully an alternative technology which requires many new skills compared to the traditional methyl bromide application method. The organization of training programmes for hundreds or thousands of methyl bromide users has been common in MLF projects in developing countries; by contrast, large training programmes on methyl bromide alternatives were rarely organized in developed countries.

As part of the training programmes, many MLF projects have produced practical booklets or manuals in local languages to assist the adoption of alternative technologies by farmers and others. The booklets and guidelines describe how to apply alternatives correctly for crops such as tomatoes, peppers, strawberries, melons and tobacco. In general, developing countries have carried out much more awareness raising and information dissemination about the ozone layer, methyl bromide and alternatives than developed countries.

Agricultural equipment and materials
The choice of alternative technologies in projects depended either on the results of demonstration projects or on experiences in similar crops and situations. The MLF expects countries to introduce the cheapest effective alternative technologies. The project proposals normally identified the value of any methyl bromide-related equipment currently used by farmers and calculated the funds necessary to cover the incremental capital costs of equipment and materials necessary to set up alternatives on the premises of methyl bromide users. The operating costs of methyl bromide and alternatives were often compared over a period of about four years or so, and the MLF provided funds for incremental operating costs if these were considered significant. Farmers or service providers who participated in the projects' training programmes received necessary equipment. In projects for methyl bromide phaseout in seedbeds, for example, the MLF typically provided a proportion of the funds needed for seed trays, materials for building shallow pools, small microtunnels (for individual farmers) and/or large plastic greenhouses (for seedling service providers). The quantity of essential materials and equipment was normally calculated on the number of hectares on which methyl bromide was used.

BOX 11.12 TECHNOLOGY TRANSFER
IN ARGENTINA AND BRAZIL

In Argentina, more than 268 tons of methyl bromide were used for tobacco seedbeds by about 24,000 farms. Following a MLF demonstration project which trialled several different alternative technologies, a MLF phaseout project started in 2002 in Argentina with technical assistance from the United Nations Development Programme (UNDP). The following technology change programme was carried out:[69]

- Information sessions with stakeholders were held regularly in all regions.
- Commissions were created for every province and were put in charge of making the main project decisions.
- Methyl bromide-free crop production protocols were developed.
- Distribution of alternative materials to farms and technical assistance were provided to 16,000 growers in each year of the project.
- Continuous awareness-raising activities were held in rural primary and secondary schools, where most students are sons or daughters of tobacco growers and farm employees.
- In addition to these targeted campaigns, awareness-raising activities were undertaken with the help of the media.
- 9055 growers and 189 technicians were directly trained through seminars, field days and different kinds of meetings.
- Leaflets, manuals and other diffusion materials were produced and distributed in each region. The project team participated in several rural fairs, where it demonstrated alternative technologies to more than 30,000 people.
- The project team developed a pine bark substrate that can be successfully used with the trays. It also designed trays and seeding machines and devised fertilization and other techniques which are well suited to Argentinean conditions.
- Agreements were signed with the governments of the seven provinces where tobacco production takes place.
- Nearly all tobacco organizations signed a firm commitment to phase out methyl bromide by 2007.
- By 2005, 73 per cent of the planted area (62,200 hectares) had switched to alternative technologies. The main alternatives adopted were substrates/floating trays and metham sodium, with a small proportion using heat or steam.[70]

Brazil is the second largest producer of tobacco in the world. The sector involves about 140,000 growers and their families. In 1998 tobacco was the highest methyl bromide-consuming sector in Brazil. Total methyl bromide imports in 1998 were reported at 1414 tons, and 703 tons were used for tobacco seedbeds.

The demonstration project for tobacco in Brazil was implemented with the United Nations Industrial Development Organization (UNIDO) from 1998 to 2000, in coordination with EMBRAPA, the Brazilian Institute for Agricultural Research. The aim was to test the economic and technical feasibility of several different seedling production technologies in order to identify the most appropriate. Alternatives were also tested by the Association of Tobacco Growers of Brazil (AFUBRA) and by the Association of Tobacco Manufacturers (SINDIFUMO). The conclusion was that substrates – floating tray systems – could be adopted.[71]

A phaseout project was implemented in 2001–2002 in cooperation with AFUBRA and SINDIFUMO, who also co-financed the project. The project first trained the technicians of SINDIFUMO and AFUBRA with the assistance of international consultants. The technicians then transferred the technology and necessary knowledge to 140,000 farmers by meetings, training videos and visits to farms. The project gave farmers part of the equipment needed for putting the floating systems in place. Initially, the substrate system cost substantially more per hectare than using methyl bromide, but by the end of the project it cost only about 28 per cent more, and this difference was compensated for by the higher survival rate of substrate seedlings.[72] The project carried out other activities to promote technical change and help the farmers to eliminate their dependence on methyl bromide.

By the end of 2004 only about 10 per cent of the tobacco growers used methyl bromide; by 2005 methyl bromide consumption in the tobacco sector was reported as zero and methyl bromide use for the crop was banned.[73]

GEF projects

The GEF has also provided financial assistance for several methyl bromide projects for countries that are not eligible for MLF projects. In 1999 the GEF approved a regional demonstration project to assist eight CEITs in demonstrating alternatives, raising awareness and developing national policies.[74] UNEP requested individual national demonstration activities for all eight participating countries, but a demonstration was funded only in one (Poland), and consequently the results were not specifically relevant to the others. Nevertheless, the participating CEITs examined policy options and developed national plans for promoting methyl bromide phaseout, as illustrated by Box 11.13.

BOX 11.13 NATIONAL PLANS AND POLICY DEVELOPMENT IN CEITS

Methyl bromide was used in Central and Eastern Europe as a soil fumigant for crops and for stored products. A number of CEITs examined the following issues when developing policies to promote technology transfer and methyl bromide phaseout:[75]

- **Review of existing regulations.** Most CEITs already had legislative controls on ODSs; these were extended to cover methyl bromide. Under pesticide regulations, methyl bromide is classed as a toxic substance and is restricted to a list of permitted uses or specific crops. Some countries also have use restrictions, such as limits on the frequency of soil fumigation in Croatia or the requirement for a safety buffer zone between fumigation sites and residential areas in Bulgaria.
- **Identification of alternatives.** Surveys were carried out to see if alternatives were already used for key pests and methyl bromide uses. Some countries already used substrates, grafted plants, pesticides and other alternative technologies.
- **Barriers to adoption of alternatives.** Several CEITs identified the main barriers to adoption of alternatives and potential solutions such as practical research to increase efficacy or make alternatives easier to use. Lack of knowledge about how to use alternatives could be addressed by training trainers, while lack of local availability of alternatives could be improved by encouraging local enterprises to produce or supply alternatives.
- **Raising awareness.** Several CEITs produced publicity materials. Bulgaria, for example, produced several booklets about the methyl bromide issue, such as

Without Methyl Bromide, But How? Approaches to Methyl Bromide Phase Out in Bulgaria (see Figure 11.6). This informed the public about the reasons why methyl bromide needed to be phased out, its main uses, alternative products already available in the country, obstacles to adopting alternatives and activities that would help to overcome the obstacles.

- **National action plans.** National action plans for methyl bromide phaseout were developed by a number of CEITs. Hungary and Poland, for example, made the following plan in 2000:
    - Year 1: Assess methyl bromide use and collect data;
    - Years 1–2: Identify alternatives via workshops and demonstration projects;
    - Years 2–3: Stakeholder activities, such as encouraging local production of alternative products, holding seminars for stakeholders, and encouraging voluntary reductions by methyl bromide users;
    - Years 2–3: Establishing a policy framework by preparing/modifying legislation on ODSs, setting a clear phaseout schedule;
    - Years 2–5: Information and awareness: inform growers about effective alternatives, distribute technical manuals and leaflets for farmers and other methyl bromide users, provide information for newspapers and magazines, and set up ecolabels for agricultural products that do not use methyl bromide; and
    - Years 2–5: Implement alternatives: carry out training of trainers, disseminate information about successful alternatives, and hold regional training workshops.
- **Regional information exchange.** CEITs held regional workshops and meetings for discussion of methyl bromide alternatives and policy approaches among government officials, methyl bromide users and experts in alternatives. As a result, countries have been able to compare regulations, share their experiences, and learn about successful cases of alternatives and methyl bromide phaseout.
- **Economic incentives and disincentives.** Countries reviewed national policies to remove economic incentives that promoted the use of methyl bromide, and considered disincentives. A regulation in Slovakia, for example, placed a fee on all ODS imports. Poland established a fee on the emission of pollutants to the atmosphere (US$0.02 per kg), which acts as a disincentive to use pollutants such as methyl bromide. Latvia runs an Environment Protection Fund which provides grants for research on alternatives, and enterprises can receive loans from an Environment Investment Fund to implement environmentally-friendly alternatives.[76]

Figure 11.6 *An example of public information materials*

Subsequent GEF projects focused on methyl bromide phaseout and included regional CEIT methyl bromide phaseout projects in Bulgaria, Hungary, Latvia, Lithuania and Poland (with the UNDP and UNEP) and in Ukraine (with the World Bank), with a project under preparation in South Africa (World Bank).

The GEF projects generally have the same general components as MLF projects, being based on the MLF model: policy development, training and technical assistance, provision of part of the necessary equipment and materials (to cover the incremental costs), and full involvement of stakeholders.

ADDITIONAL ACTIVITIES THAT AFFECTED TECHNOLOGY TRANSFER AND CHANGE

This section identifies some additional activities that promoted technology transfer in developing countries, aside from the activities commonly carried out in MLF and GEF projects.

Comprehensive plans and regulatory frameworks

Several developing countries, such as China, the Philippines and Thailand, drew up very comprehensive and detailed national action plans for phasing out methyl bromide, while this occurred rather rarely in developed countries. The types of regulatory reviews and regulatory frameworks adopted in some developing countries were also more innovative and comprehensive than the regulatory changes made in many developed countries.

Local suppliers and manufacturers

Some MLF projects actively encouraged local businesses to import or manufacture alternative technologies, as part of the strategy of technology transfer. In some cases, this was done to persuade methyl bromide importers to import alternative products so that they would not lose business and would thus not oppose methyl bromide phaseout. In other cases, local enterprises were encouraged to supply alternative products in order to build up a market and infrastructure to support the continued use of alternative technologies after the project ended. Local enterprises were also encouraged by projects to manufacture alternative equipment or products locally – in cases where products were not patented or protected by intellectual property rights – in order to make the alternatives more affordable to local users. In Argentina, several enterprises were encouraged by the MLF project to manufacture trays for seedlings, in order to reduce prices paid by farmers and to ensure the economic viability of the alternative technology. In Kenya, methyl bromide users and researchers identified local substrate materials, such as pumice and coconut fibre, to avoid having to pay for more expensive imported materials. In Chile, steam boilers were made locally after several machines were imported; the boiler design was based on locally available parts and components.

Box 11.14 Technology transfer activities in Jordan

In 1998 Jordan used about 325 tons of methyl bromide for crops such as tomatoes, cucumbers, peppers and strawberries. Jordan carried out a MLF demonstration project, followed by a phaseout project with assistance from GTZ (Gessellschaft für Technische Zusammenarbeit), the technical cooperation agency of Germany, and by 2005 national methyl bromide consumption was down by 67 per cent.[77] The project included a number of policy-related activities aimed at promoting technology change:[78]

- **Regulatory framework.** Using the existing pesticide control structures, the pesticide registration committee issued regulations to reduce the imports of methyl bromide over time.
- **Creating awareness.** Numerous press interviews were given by the National Ozone Officer, the secretary general of the Ministry of Agriculture, expert farmers and project staff to create public awareness of the problem of methyl bromide.
- **Socio-economic assessment.** A survey of farmers found they were concerned about the high price of methyl bromide and the fact that they could not buy it on credit like other agricultural materials. They cared little about the acute toxicity of methyl bromide to humans or negative effects on soil, water, air and the ozone layer. The survey identified only two ways to gain farmers' interest in methyl bromide alternatives in Jordan: opportunities to save money and the fact that methyl bromide would not be available in future.
- **Training.** The project provided farmer-to-farmer training and technical assistance. A survey found that 87 per cent of the farmers trained in the Jordan Valley adopted alternatives in 2000.
- **Testing of consumer labels.** The project developed a consumer label to identify products that have been grown without the use of methyl bromide. The labels were tested on shipments of strawberries to a European supermarket. The supermarket gave positive feedback and encouraged the producer to continue labelling products in this way.
- **Economic assessment of methyl bromide suppliers.** Methyl bromide was imported by three enterprises. These enterprises sold many types of agricultural inputs, so did not rely on methyl bromide alone for their business and thus had a relatively relaxed attitude towards methyl bromide phaseout.
- **New business in alternatives.** The project identified opportunities for enterprises in Jordan to make or sell alternative technologies. One Jordanian enterprise, for example, entered into a joint business agreement with a foreign company to register and market an alternative fumigant. Another Jordanian enterprise planned to produce grafted plants, following the successful introduction of grafting technology in Morocco.
- **Monitoring.** The project established a monitoring system to measure progress. Participants surveyed the use of methyl bromide on farms, as well as practices used by farmers who had been trained in alternatives, to detect any reversal to methyl bromide due to problems. From these data, the project designed follow-up field activities and adjusted training programmes.

Development of novel technologies in developing countries

In a number of cases, enterprises, farmers or researchers in developing countries have developed novel alternative technologies. Innovative equipment was developed by the Fosfoquim enterprise in Chile, for example: Fosfoquim's

Horn Generator enabled phosphine to be produced and released much more rapidly and provided faster post-harvest treatments. The same Chilean enterprise also developed and adapted pure phosphine so that it could be used for many more purposes than in the past, such as on fresh grapes for export.[79] In CEITs, researchers developed combinations of dazomet and Trichoderma for vegetable crops.[80] In Colombia, cut flower producers developed many novel technologies, such as compost seeded with beneficial micro-organisms and low-cost substrate systems such as beds of burned rice hulls.[81] In Morocco, growers adapted grafting technologies and created cost-effective methods that enabled grafting to be adopted on a large-scale. And in Kenya, gravity-fed irrigation systems were developed so that small-scale farmers would be able to use substrates and other alternatives in rural areas where there was no electricity.

Leadership activities in developing countries

A number of farmers and enterprises in developing countries showed leadership in adopting alternative technologies at a relatively early stage, using their own financial resources. Several enterprises in Brazil, for example, actively encouraged the adoption of methyl bromide alternatives from 1996, providing technical assistance to farmers and disseminating technical information. And Colombia provides another example. Columbia is one of the largest cut flower producers in the world. Although the country imported methyl bromide from 1991 to 1996 (390 tonnes of methyl bromide in 1996), it was largely used by the banana sector; the flower industry did not become dependent on methyl bromide and developed successful alternative technologies based on substrates, steam, alternative fumigants and IPM. Colombian flowers are produced to a high quality and generally attain higher prices in world markets (€6.96/kg) than

---

**BOX 11.15 TECHNOLOGY TRANSFER
IN CHINA'S STRAWBERRY SECTOR**

- Mancheng was the largest methyl bromide-consuming region in China, using about 400 to 500 tons per year. Strawberry production has become the most important source of income for many farmers in Mancheng, where the cultivated area is about 5000 hectares.
- After substantial trials, chloropicrin (pic) was registered in China in 2002 and has gained in popularity with farmers. Chloropicrin manufacturers have also increased their extension network, and the Ministry of Agriculture has supported demonstration activities under a MLF project.
- Two methods of applying pic were developed – injection equipment, and novel capsules which can be placed in soil and allowed to dissolve and release the fumigant – which addressed the problem of its strong unpleasant smell during application.
- Use of pic expanded relatively quickly because it is cheaper than methyl bromide; it has already replaced 40 tons of methyl bromide (10 per cent of initial consumption) in 2006, and an additional 100 to 150 tons are expected to be phased out in 2007.[82] Other more environmentally friendly alternatives are under development with the aim of providing environmentally sound solutions for the longer term.

flowers produced in countries that have relied on methyl bromide, such as Ecuador, Kenya and Israel (€6.34/kg, €3.87/kg and €3.58/kg respectively).[83]

Environmental certification programmes

Industry environmental standards and certification programmes have assisted the adoption of methyl bromide alternatives in developing countries. International environmental certification programmes for flowers, such as MPS, FlorVerde and EUREP-GAP, do not permit the use of methyl bromide in the production of flowers and ornamentals. More than 5000 farmers have implemented the MPS programme in 22 countries, including Kenya, Zimbabwe, Zambia, Costa Rica and Ecuador.[84] Many of these farms have made the necessary technical changes using their own resources, without assistance from international bodies like the MLF or GEF.

EXPERIENCES AND BARRIERS IN TECHNOLOGY TRANSFER PROJECTS

Some methyl bromide manufacturers, methyl bromide importers and fumigation enterprises told methyl bromide users that methyl bromide did not really deplete the ozone layer, that no viable alternatives existed and that thousands of agricultural jobs would be lost if methyl bromide was phased out. Many methyl bromide users believed these false claims and, as a result, some MLF projects faced an uphill struggle in trying to counter misinformation at the grass-roots level.

Involvement of key stakeholders from the beginning is an accepted principle of MLF projects but was not always fully applied. Lack of sufficient consultation sometimes led to delays and blockages later. On the other hand, consultation also sometimes enabled methyl bromide users who did not want methyl bromide to be phased out to block project progress. In some cases, farmers and enterprises have taken strong and progressive action to adopt new technologies, while in other cases, they have prevented technological change.

In some sectors, enterprises and farms that participate in project activities are very open to sharing technical know-how with other enterprises or farms in the locality, even though they are competitors. However, in other cases, such as the melon sector in Central America or specific cut flower sectors, producers are very reluctant to share their advanced technical information because of the intense competition and lack of government extension services.

Information materials (for example manuals and newsletters) that focus on local needs (such as local crops, pests and circumstances) were found to have greater impact on methyl bromide users than general information. Farmers and other methyl bromide users need to know where they can find local suppliers of alternative equipment, products and services, so contact lists of enterprises and suppliers are therefore a useful resource.[85]

An important goal of MLF projects is to achieve a sustainable phaseout that will be maintained after the projects have been completed. One aspect of this is the selection of suitable technologies for each situation. A recent study conducted by the MLF found that technologies promoted in projects were generally chosen on the results of demonstration trials, following discussion with key stakeholders and information about commercial adoption taking place in the same country or in similar regions and sectors being provided. However, there were instances where advanced technologies had been implemented or equipment delivered without a solid examination of their technical or economic sustainability. Examples of this were steam equipment for field crops like strawberries and tomatoes, and electronic meters that cannot be calibrated in the country. This may have arisen partly following suggestions by bilateral and implementing agencies and/or its consultants, but may have also occurred at the request of governments, farmers or enterprises who wanted advanced technologies.

The viability of maintaining a prohibition on methyl bromide imports in countries that have now phased out methyl bromide is problematic when neighbouring countries continue to use methyl bromide. Ozone Officers have often reported that containers of methyl bromide are easily being transported, unreported, across borders.[86] Regional meetings have considered the feasibility of creating 'buffer zones' or other measures that could help to prevent illegal trade. The MBTOC produced a short report on harmful trade in methyl bromide in response to a Montreal Protocol Decision (Decision Ex. 1/4 (9a)) and identified several options that would help to prevent harmful trade in methyl bromide:[87]

- stronger systems of licensing methyl bromide imports could be implemented;
- the countries manufacturing methyl bromide could ensure the prior informed consent of government authorities in importing countries before allowing shipments; or
- developing countries could inform the Montreal Protocol about their actual need for methyl bromide so that the quantity manufactured for basic domestic needs could be lowered by the Montreal Protocol if it exceeds need.

A barrier that arises in technology transfer with respect to more modern chemical alternatives is the need for registration as a pesticide. In developed countries, registration procedures are normally very slow and extremely expensive. However, in contrast to expectations, registration did not turn out to be a substantial barrier to making progress in developing countries.[88] Several developing countries have now registered alternative pesticide products, while others have been able to achieve substantial methyl bromide reductions using other types of alternatives that did not require registration.

Another factor beyond the control of MLF projects is the price of methyl bromide. Fluctuating prices and aggressive marketing of methyl bromide have influenced the economics of alternatives. In Iran and Turkey, for example, methyl bromide prices increased, making alternatives more attractive to users, while in Ecuador methyl bromide is an inexpensive fumigant, and the cost per treatment is lower than alternatives. In other countries, such as Costa Rica, prices of methyl bromide were reported to be comparable to those of chemical alternatives. However, prices can fluctuate greatly over time.

A recent study on the results of MLF methyl bromide projects indicated some lessons that were apparent from all projects that have been completed to date:[89]

- Technically effective alternatives to methyl bromide have been found for almost all pests and diseases.
- Projects need to make more effort in documenting their economic viability and overall sustainability.
- Successfully evaluated alternatives can be introduced into developing countries within periods of two to three years. In fact, activities related to demonstration projects have led larger or more technically prepared growers to adopt alternatives at their own initiative.
- The capability to adapt to site-specific conditions is essential to the success of any alternative.
- Project implementation and follow-up are better when growers' associations, growers' cooperatives or large enterprises take part in them.

A number of MLF and GEF projects suffered delays for technical, administrative and/or political reasons. Several countries were not able to reduce methyl bromide use according to the timetable initially agreed with the MLF and found it necessary to adopt a revised phaseout schedule. This problem arose as a result of scepticism towards alternative technologies on the part of growers, lobbying campaigns to retain methyl bromide, and large CUEs granted to developed countries for crops such as melons and strawberries.[90] One such example occurred in Argentina, where initially strong progress was made in phasing out methyl bromide. By 2004 Argentina had phased out 51 tonnes more than the amount committed to in the MLF project agreement, in other words methyl bromide reductions were achieved faster than the project schedule.[91] However, the project later encountered reluctance from the strawberry producers, and it was not possible to meet the original project schedule.

Nevertheless, overall, developing countries have made very substantial technical changes in the methyl bromide sector, and the result can clearly be seen in the large methyl bromide reduction achieved by many. Most (136 of 144) developing countries achieved the 20 per cent reduction in methyl bromide consumption in 2005 required by the Montreal Protocol. In fact, most countries reached this target several years before the target date.[92] However, eight countries did not achieve the 20 per cent reduction and therefore failed to comply with the Montreal Protocol. A recent study conducted by the MLF[93] identified the following reasons why a small number of countries had not complied with the freeze or 20 per cent step:

- political and economic transformation processes in the country, implying radical structural changes;
- late ratification of the Montreal Protocol (after 2000) and/or its amendments;

- late preparation and implementation of country programme and/or phase-out projects;
- delayed approval and implementation of ODS-related legislation;
- weaknesses of the National Ozone Unit (late start, delayed implementation, frequent staff changes, and/or communication difficulties within the environment ministry and/or with other ministries);
- reluctance of stakeholders to actively cooperate in the ODS phaseout process or lack of sufficient involvement of key sectors or stakeholders since the onset of the project's activities;
- low methyl bromide consumption baseline due to exceptional circumstances (for example war or economic recession) or insufficient data collection; and
- expansion of the main sector using ODSs – particularly methyl bromide – after the baseline years.[94]

The majority of countries were able to meet their commitments, and when problems occurred, they were addressed in many cases. The remaining barriers in developing countries generally appear to be more political than technical in nature. Through its ability to assist countries in overcoming barriers and problems, the MLF has provided an example of international partnership by transferring ozone-safe technologies and technical know-how to many parts of the world.

REMAINING CHALLENGES

Both developed and developing countries have achieved substantial methyl bromide reductions, but specific sectors and regions remain reluctant to change, and a number of challenges remain in the technology transfer process:

- Technical R&D is needed to develop and improve alternatives in several specific areas (for example high-moisture fresh dates and strawberry plant nurseries).
- Some methyl bromide users and researchers continue to try to 'reinvent the wheel', rather than learning from users who have successfully adopted alternative technologies.
- Ineffective methods are sometimes employed. In such cases, insufficient effort has been put into transferring the necessary know-how.
- The process of registering alternative pesticide products is often slow and expensive in developed countries and difficult where markets are small; however, developing countries have been able to register products in a more timely manner.
- There is a large global stockpile of methyl bromide, which has reduced methyl bromide prices and led to dumping in some developing countries. The use of methyl bromide is strongly promoted by some enterprises.
- There are difficulties in tracking methyl bromide imports, and occurrences of unreported/illegal trade.
- QPS is exempted from Montreal Protocol controls, and methyl bromide

increases have occurred recently due to an International Plant Protection Convention standard for raw wood pallets and packaging.[95] Although an alternative heat treatment and the use of processed wood pallets are both feasible, the use of methyl bromide is cheaper or more convenient, and there is thus no incentive to use alternatives.

FACTORS THAT ASSISTED ADOPTION OF ALTERNATIVE TECHNOLOGIES

Based on experiences in diverse countries, this chapter has identified a number of activities and factors that assisted technology transfer in the methyl bromide sector, helping to overcome technical, financial, legal, policy and other barriers to change. The factors can be summarized as follows:

- Development of regulatory frameworks and policy support, such as:
  - high-level political support for ozone layer protection and ODS phase-out;
  - limits on the national supply (production/imports) of methyl bromide, as required by the Montreal Protocol;
  - de-registration of methyl bromide for crops and situations where alternatives are available;
  - issue of permits for each methyl bromide fumigation (case-by-case, requiring evidence of pests and lack of alternatives);
  - increased restrictions on the use of methyl bromide, such as larger buffer zones for safety;
  - requirements for minimizing methyl bromide use and emissions, such as sealing standards in fumigated structures or barrier sheets laid on soil for ten days; and
  - taxes, levies or other measures that increase the price of methyl bromide relative to alternatives.
- Technology transfer and building supportive infrastructure, such as:
  - R&D and demonstration trials, adapting alternatives to local needs;
  - financial assistance to overcome barriers of investment and technology change;
  - training and technical assistance programmes in the field;
  - providing information relevant to local circumstances and production systems, such as case studies, user manuals, contact details of local suppliers of alternatives, and news and updates on alternatives for methyl bromide users;
  - involvement of stakeholders, including local enterprises that can potentially supply alternative products;
  - building local infrastructure to support alternatives, encouraging local enterprises to manufacture or supply alternative products and services;
  - fast-track registration of alternative products;
  - use of market signals, such as international certification programmes, to promote adoption of alternative technologies; and

- fostering leadership and constructive activities when technical change is viewed as difficult and inconvenient by end-users.

CONCLUSION

The Montreal Protocol control measures have been an important catalyst for technological change. Methyl bromide use has increased recently in the quarantine and pre-shipment sectors which are not controlled by the Protocol. This is in marked contrast to the sectors that are controlled by the Protocol, where about 80 per cent of methyl bromide has been phased out to date and further reductions are under way. The change to non-ODS techniques was achieved by the efforts of many individual farms, companies and specialists with an interest in integrated pest management and more environment-friendly production. The process has demonstrated that technology change for the protection of the global environment is feasible in agricultural sectors in many different cropping systems, climates and circumstances.

However, some countries achieved technological change much earlier than others, particularly where key actors (such as farmers' groups, companies, projects or governments) strongly promoted change and alternative techniques. Activities that were more effective in achieving technological change included: training programmes to allow large numbers of users to learn how to use alternatives; banning uses of methyl bromide for which alternatives were available; and the introduction of ODS taxes or other economic measures. Successful agricultural sectors that never became reliant on methyl bromide also provided useful models or examples for similar sectors in other countries that were dependent on methyl bromide. Governments in developing countries tended to be more proactive than in industrialized countries, often as a result of MLF or GEF projects which provided encouragement and support for policy development. With some notable exceptions, many industrialized countries relied primarily on market diffusion of alternative technologies, leading to a slow pace of technical change because the existing infrastructure supported the continued use of methyl bromide.

Overall, technological change in the methyl bromide sector was more protracted than in other ODS sectors because the topic became very contentious in some agricultural regions. Methyl bromide manufacturers and their distributors and customers questioned the ozone science for more than a decade after other ODS producers and their customers had conceded the harmful effects of chlorinated and brominated ODSs. There was little or no public awareness or concern about methyl bromide and the impacts of ozone depletion. In contrast to other ODS sectors, large or conspicuous international companies rarely made public statements or pledges endorsing methyl bromide phaseout. Furthermore, methyl bromide producers and users came to believe that phaseout was more technically and economically difficult than in other sectors and that it was less important to do their part to protect the ozone layer because methyl bromide emissions were a smaller portion of the total ODSs that had been already discharged into the atmosphere. Many also believed that alternatives needed to be less expensive and better performing than methyl bromide in order to be considered, although this was not an appropriate or justified approach. An additional barrier to technical change in developing countries was that, without a Protocol control measure beyond the 20 per cent reduction step, the MLF Executive Committee considered the completion of methyl bromide phaseout to be a low priority. Also, the relatively large CUEs requested in industrialized countries slowed the pace of change in developing countries. Nevertheless, in spite of such barriers, technology transfer and adoption of alternatives have progressed greatly, resulting in very large reductions in methyl bromide consumption worldwide. Successful examples of alternative technologies have now been reported in many sectors and are available for those who choose to take further steps to protect the ozone layer.

Barriers to Technology Transfer Faced by CEITs and Developing Countries

INTRODUCTION

Ever since the global environmental issues occupied the centre stage in international discussions in the 1980s, there have been several debates in the United Nations and the World Trade Organization on determining what the 'barriers' to the transfer of environmentally sound technologies are, particularly to developing countries. Barriers have been identified as institutional, social, political, technical and economic factors that prevent or slow the adoption of environmentally superior technologies. However, while it is true that technology transfer can be challenging, it is wrong to label every challenge a 'barrier' to technology transfer; with a little effort, many barriers turn out to be surmountable. This chapter will demonstrate that in the case of technology transfer for protection of the ozone layer, many barriers that were feared failed to materialize, and that the concerted efforts of industry, governments, implementing agencies and non-governmental organizations helped to solve the problems that did arise.

TECHNOLOGY TRANSFER PRIORITIES: OUTCOMES OF THE 1992 EARTH SUMMIT AND THE 2002 WORLD SUMMIT ON SUSTAINABLE DEVELOPMENT

In 1992 and again in 2002 the world community came together for two historic environmental conferences: the Earth Summit in Rio de Janeiro and the World Summit on Sustainable Development in Johannesburg. Technology transfer was a major subject of debate. In the course of recent decades, experts and academics had identified dozens of potential barriers – particularly intellectual property rights – that might stand in the way of technology transfer for environmental protection (see Box 12.1).[1]

These conferences resulted in action plans to help developing countries overcome barriers to technology transfer. The 1992 Earth Summit led to the development of Agenda 21, a comprehensive blueprint of action to be taken globally to minimize human impacts on the environment in the 21st century.[2]

Chapter 34 of Agenda 21 dealt specifically with the transfer of environmentally sound technology, cooperation and capacity-building. The 2002 summit resulted in the Johannesburg Plan of Implementation of the World Summit on Sustainable Development, which called upon the global community to:

> *Promote, facilitate and finance, as appropriate, access to and the development, transfer and diffusion of environmentally sound technologies and corresponding know-how, in particular to developing countries and countries with economies in transition on favourable terms, including on concessional and preferential terms, as mutually agreed, as set out in Chapter 34 of Agenda 21.*

This plan of implementation was supported and elaborated by the Bali Strategic Plan for Technology Support and Capacity Building developed in 2004.

The international debate on technology transfer and support has now moved into a wide arena that encompasses a whole set of international and

---

BOX 12.1 INTELLECTUAL PROPERTY AND
ACCESS TO TECHNOLOGY

Some scholars believe that the intellectual property regime is the primary obstacle restricting developing country access to better technologies. Many believe that multinational enterprises are overcharging developing countries for access to new technologies, imposing unacceptable conditions, or short-changing developing countries in other ways.

The intellectual property debate has evolved over time, partially in response to the perceived failure of the 'regulatory' approach to technology transfer, in which some developing countries attempted to screen technology transfer agreements between foreign and domestic enterprises and discourage what they saw as unfair practices, such as export restrictions for products produced under licence imposed by foreign enterprises. The goal of regulatory approaches was to encourage transfer of technology from multinational corporations to locally owned businesses, using regulation to remedy any inequalities that might result.[3] Now, instead of limiting intellectual property rights, many developing countries are increasing protection for intellectual property, in the belief that strong intellectual property protection will attract foreign direct investment (FDI) which will spread technology across borders (see 'Pathways of technology transfer' on page 12).[4]

The debate continues to this day. Some experts argue that strong intellectual property rights could encourage the transfer and diffusion of environmentally sound technologies; others remain firm in their assertion that intellectual property undesirably and significantly increases costs in the adoption of such technologies.[5]

The Montreal Protocol experience is that almost all ozone-safe technologies were successfully diffused throughout developing countries without any restriction on intellectual property; that some technologies important to stratospheric ozone protection were protected by intellectual property but were transferred under fair and favourable conditions negotiated by the Global Environment Facility (GEF) and Multilateral Fund (MLF); and that only in two conspicuous cases were the owners of intellectual property important to ozone protection unwilling to allow use of their technology. Some developing country enterprises ultimately developed their own technical solutions or found solutions from other suppliers.

national activities needed to make technology change and transfer a success.

What is not apparent from these debates is the prioritization of the activities. What are the barriers most often faced? What are the steps that will most contribute to the introduction of environmentally sound technologies? The technology change already achieved to implement the Montreal Protocol offers valuable lessons. What were the important obstacles, and how frequently did they occur? What were the solutions?

TECHNOLOGY TRANSFER PROBLEMS THAT AROSE DURING THE IMPLEMENTATION OF THE MONTREAL PROTOCOL

The following sections describe the technology transfer problems of implementing the Montreal Protocol, based on expert observations and on an analysis of over 1000 completed technology transfer projects financed by the MLF and GEF. Challenges to technology transfer in countries with economies in transition (CEITs) and developing countries fall broadly into the following categories:

* poor infrastructure and utilities (power, communications and transport);
* inadequate/poorly implemented laws and regulations;
* shortage of trained technical and managerial personnel;
* cumbersome government and banking agencies;
* weak local supporting industry, especially in the production of parts and components;
* high cost of certain technology agreements and inappropriateness of technology transferred; and
* overall political and macroeconomic instability in some countries.

CEITs and developing countries overcame these and other problems in order to develop or obtain technologies to protect the ozone layer. The problems revealed by this analysis fall into five different categories:

1   problems at the enterprise level, due to lack of knowledge and training, ineffective functioning, or financial problems;
2   problems at the country level, posed by the delays caused by and ineffective functioning of the local institutions and inadequate infrastructure;
3   problems at the international implementing and financing agency level, due to policies and procedures;
4   problems created by suppliers of equipment; and
5   problems gaining access to ozone-safe technologies.

It should be noted that many projects were executed without problems, and that this study relates only to the projects that mentioned specific problems in their completion reports.

Enterprise-level problems

The key barriers that occurred at the enterprise level are summarized below.

Financial problems

Some of the enterprises faced bankruptcy while implementing alternative technologies. A number of enterprises were unable to raise the necessary financial support to meet the counterpart costs (costs over and above the grants from the MLF or the GEF). Some enterprises incurred huge losses in the first few years after introducing the alternatives, despite the availability of incremental operating costs as part of MLF project grants.

> *It is quite remarkable that the phase-out is being accomplished – virtually without exception – to the great satisfaction of developing country governments and enterprises, despite the challenges of simultaneously undergoing painful industrial modernization, reform and rationalization unconnected with the ozone protection regime!* (Tony Hetherington, former Deputy Chief Officer, MLF Secretariat)

Particularly in CEITs, but also in some developing countries, many firms were rejected outright due to financial problems or because of the perception of the relevant implementing agency that they would not survive the transition to a market economy.[6] A number of selected projects ran into financial difficulties during project implementation that caused delays or resulted in project cancellation. Other enterprises remained viable, but could not or did not raise the required counterpart funds.

MLF and GEF policies did not typically provide additional funds as incentives to avoid transitional technologies or to select technologies that were both ozone-safe and environmentally superior with respect to climate, human health or other co-benefits. As a result, some enterprises could not afford to purchase their first choice in equipment, and settled for the second best.

Inadequate technical and administrative capacity

Many of the enterprises lacked in-house technical resources to deal with change, were not familiar with the procedures of the MLF, the GEF or the implementing agencies, and took a long time to agree with the equipment specifications. Most small enterprises are not accustomed to estimating, engineering or implementing anything more than relatively minor plant modifications and are not equipped to manage a sudden change to more complex equipment and manufacturing processes.

In a few CEIT enterprises, problems related to lack of experience arose. In one case, inexperience with alternative technology resulted in damage to property and significant delays. In several cases, extensive outside assistance and training helped enterprises acquire the skills and capacity to use the technology. Some firms had trouble understanding or complying with implementing agency procurement procedures. Other firms criticized themselves for lack of planning during initial stages. Several submitted applications for funding, only to realize halfway through the project that they had omitted a necessary item. It was then necessary for the firm to spend extra money or to make cuts in other project areas in order to obtain the needed products or services.

Managerial problems

Many enterprises were family-owned or family-dominated. This created problems whenever the owner was incapacitated or otherwise unable to personally attend to project implementation of the change. In some enterprises, the management changed several times. In more than one case, family feuds between those who stood to inherit the business resulted in delays.

Convincing the top management

In a number of enterprises, the technical staff were easily convinced about the merits of the alternative technologies, but the top management was overly cautious about the new technologies, and delayed implementation.

Country-level problems

Country-level barriers that affected technology transfer for ozone layer protection were extremely varied. Among the extremes were terrorist incidents, economic recessions, political turmoil, licensing delays, disease outbreaks, and lack of or inadequately enforced policies and regulations.

Political turmoil, war, terrorist incidents and health risks

Terrorist incidents from 1998 (the year of the bombing of the US Embassy in Nairobi) resulted in many industrialized countries issuing travel advisories against travelling to certain countries. Projects were delayed when experts were prohibited from travelling or cautioned not to travel to certain countries or necessary meetings. In a few countries, occasional political turmoil resulted in delays. After the break-up of Yugoslavia in 1991, the civil war in Bosnia and Herzegovina and Croatia had devastating impacts on many businesses. In China and Southeast Asia, the health situation due to SARS discouraged the suppliers of equipment from visiting.

Economic recessions

In the late 1990s unfavourable economic conditions delayed projects in many countries such as Indonesia, Brazil and Malaysia. In Zimbabwe, the deteriorating economic situation led to huge foreign exchange problems and economic slowdown. During the 1999 recession in Thailand, many suppliers stopped providing credit and favourable terms to their customers, and required cash in advance. The financial crisis in the CEITs led to the loss of markets and decreases in sales. In some cases, macroeconomic conditions in the country caused implementation delays. In one case, the local bank acting as the financial intermediary was forced to close; in others, project financiers were not willing to release funds until economic conditions were stable; and in others, restructuring delayed projects or brought the continued viability of enterprises into question.

Delays by governments

Projects were delayed in many countries due to slow approval of permits and licences by town planning authorities, fire departments and other local bodies. This was particularly true for alternative technologies that required relocation and rebuilding of facilities. In some countries, the approval process was also slow for technology transfer contracts, tax exemption for customs duties and the signing of grant contracts. Although ODS-free technologies were supposed to be exempt from taxes, several enterprises in at least one country were asked to pay them nonetheless. Customs clearance for imported machinery was also a problem in several countries. In a minority of cases, a variety of problems were attributed to overall 'weak institutional capacity' at the national level. In two cases, complaints were voiced that country programme coordination was poor, missing key opportunities to phase out ODSs in project or entire sectors, or delayed excessively. In other cases, lack of 'institutional capacity' resulted in poor coordination between environmental, enforcement, customs and tax officials, hampering ODS phaseout.

Banking and infrastructure delays

In many countries, the banking procedures for money transfers were very long. In others, such as Nigeria, insufficient telecommunications systems and severe banking problems created many delays. Some enterprises in Nigeria requested that all payments be transferred to personal accounts in banks in the UK, which was unacceptable to the UN implementing agencies.[7]

Weak local supporting industry also created challenges for some firms, particularly small and medium-sized enterprises (SMEs), which often rely on technology suppliers to help them or supply parts if something goes wrong with their equipment. Firms often wanted certain technologies, but went with those available from local suppliers instead. For example, in the foam sector, several enterprises expressed the desire to use water-based foam-blowing technologies, but could not find local suppliers and ended up choosing hydrochlorofluorocarbons (HCFCs).

Lack of policies and regulations

In many countries, there were large imports of second-hand appliances containing CFCs, and this discouraged conversion to alternative technologies. The availability of cheap CFCs and non-availability of alternative refrigerants were constant concerns for developing country enterprises. Many countries delayed the introduction of policies and regulations to encourage ozone-friendly technologies, sometimes due to lack of awareness on how best to create these policies and regulations, and sometimes in order to appease importers of second-hand refrigerators and other ODS-dependent technology.[8]

Implementing agency-level problems

'Bottlenecks' at the level of the implementing agencies and/or bilateral partners were frequently experienced. These were often attributed to:

- inadequate flexibility;
- poor sourcing of technical expertise;
- slow disbursement of funds;
- poor appreciation of local conditions by some consultants and bilateral partners; or
- competition between implementing agencies and bilateral partners, which sometimes caused confusion.

In CEITs, many ODS-using firms were denied assistance converting to non-ODS technology because they were considered too risky. The result was that larger, better-established firms were converted, while many small or medium-sized firms were left to themselves. Implementing agencies caused delays in many cases and for many reasons. In some cases, implementing agencies were underfunded or lacked sufficient staff. Non-investment projects aimed at building administrative and institutional capacity remedied this to some extent. In other cases, the complaint was that staff were not proactive in coordinating activities. In others, high turnover delayed projects while new employees were brought up to speed. And in still other cases, the delays were the result of general administrative inefficiency, frequently exacerbated by the placement of the Ozone Unit at a low level in the bureaucracy. On several occasions, enterprises complained that implementing agency procurement and other procedures were excessively complicated, and that training was not sufficient.

MLF-level problems

The MLF has been widely praised for its mammoth effort to change technologies in favour of the ozone layer at the global level in many countries, including some where the environment for technology change has been very adverse. However, there were some shortcomings, such as frequent changes in policies and guidelines; inflexibility and rigid adherence to rules; and sometimes under-funding conversion costs, which resulted in the choice of cheaper, inappropriate technologies by developing countries.

Problems created by suppliers of equipment

There were no complaints from the great majority of projects regarding the supply of equipment. However, a few complaints were voiced. Enterprises in Argentina, Brazil, China, Colombia, Egypt, India, Romania, Syria and Turkey experienced delays in equipment supply and installation; the absence of the supplier's technicians at crucial stages created problems in some projects in Egypt and India. Thailand also noted a lack of local representatives of suppliers. In China, Egypt, India, Morocco, Pakistan, Sudan and Thailand, trials revealed product defects. Inadequate documentation by the suppliers created some problems in China, Egypt and Thailand, and in some cases there were linguistic difficulties that created problems. In more than one case, instruction manuals and other documents were not translated into the local language. At the height of the ODS phaseout, global demand for alternative technology was quite high, and in some technology sectors there was a limited number of suppliers to meet this demand. For many CEIT enterprises, this meant long waiting lists and delivery delays that hurt business. Although contracts with these suppliers contained clauses that would have allowed recipients to collect damages, few took advantage of this opportunity. As one firm explained, legal action would have delayed implementation even further and cost too much. In several cases, firms that provided technologies failed to address technical defects later discovered. As one enterprise put it, 'the supplier seems to have disappeared from the map and is not available to deal with the exceptional level of mechanical failures experienced by his equipment'. Another reported that 'several of one US supplier's machines have broken down, and the manufacturers will not reply to requests for help'. These experiences made other firms in the region, particularly those from the country where the 'problem enterprise' was located, reluctant to adopt alternatives.

Problems posed by lack of access to ozone-safe technologies and intellectual property protection rules

The technologies for the majority of projects in aerosol products, foams, refrigeration and air-conditioning were in the public domain; there was no need for the recipient to sign licence agreements. The few exceptions competed against public domain technologies and were selected only if benefits were evident, sometimes by recipients paying differences in costs because money approved by the MLF was not adequate for paying for the technology. There were a few cases – in China, Indonesia and Thailand – where there were problems in obtaining technologies from some suppliers, but the problems were ultimately solved by going to others (see Box 12.2 for examples).

While alternative technologies may have imposed additional cost on enterprises, the alternative chemicals used, however, may be manufactured by the enterprises using patented/licensed processes (for example HFC-245fa or HFC-365mfc).

Transfer of some technologies proved impossible

Intellectual property protection rules kept some countries from using specific process technologies to manufacture particular ODS alternatives. Two disputes that arose were over the transfer to India and South Korea of the technology to manufacture HFC-134a and the transfer of technology for the manufacture of fire suppressant FM-200.[9] These are discussed in detail below.[10]

The case of HFC-134a[11]

> *There are only five or six enterprises that control the patents and trade secrets related to HFC-134a. One of these enterprises offered an unlimited licence for the technology to Indian enterprises for a reported US$25 million and the alternative offer of a joint venture with majority stake or export restrictions on HFC-134a produced in India. Both options were unacceptable to the Indian producer.*

BOX 12.2 QUOTES FROM PROJECT COMPLETION REPORTS INDICATING INTELLECTUAL PROPERTY RIGHTS- RELATED PROBLEMS

HFC-134a for refrigeration in China

The major problem in project implementation was the difficulty in identifying an appropriate technology supplier and negotiating a technology transfer contract. Chongqing General Machinery Factory (CGMF), with assistance from the former Ministry of Machinery Industry, had contacts and discussions with several potential technology suppliers in 1994, but York and Trane (US enterprises), and Hitachi and Ebara (Japanese enterprises) declined. Only one enterprise expressed willingness to sign a technology transfer contract, but it charged a higher price and kept the 'impeller strength analysis software' and technology support service for the second and third years out of the contract.[12]

Hydrocarbon solvents in India

The enterprise licensed and procured technology from Widetech, Berhad in Malaysia. The technology transfer fee and the licence fee were the most expensive items in the service contract – constituting 57 per cent of the total contract value.[13]

The MLF adopted policies not to fund experimental research and development of indigenous technology unless the country agreed not to demand additional financing if the research failed to provide a solution. Such a commitment has been considered too strong by developing countries seeking funds from the MLF for this purpose, as evidenced by a lack of applications to the MLF for this purpose.

Confident that they could develop a unique manufacturing process at a lower cost, two Indian producers of CFCs commissioned the Indian Institute of Chemical Technology (IICT) in 1993 to develop bench-scale technology for the synthesis of HFC-134a indigenously. The research was considered successful, and in 1998 this privately-funded project entered its second phase, with a pilot plant to scale up the technology with the aim of making it ready for commercial use by the end of 1998. Nine years later, it was still not commercialized.

The MLF provided at least US$17 million to China to help the country develop domestic capacity to produce HFC-134a at the Xi'an Jinzhu Modern Chemical Industry Company.[14] China is now a producer of HFC-134a.

The case of HFC-227ea[15]

HFC-227ea, commercially known as FM-200, is generally considered the closest substitute to halon-1301 for a small number of specialized fire protection applications. Although FM-200 is not ozone-depleting, it is at least 40 per cent less efficient than halon-1301, is more toxic and has a high global warming potential. In addition, the cost of FM-200 is about three times that of halon-1301. The

Box 12.3 South Korea intellectual property RIGHTS CASE STUDY

Korean Trade Promotion Agency[16]

The Republic of Korea and the Montreal Protocol

In South Korea, most patent applications for chemical substitutes for ODSs have been submitted by foreign firms, mainly for the production of HCFC-141b and HFC-134a. Most of these patents will expire after the phaseout deadline for South Korea, implying that the local producers of HCFC-141b and HFC-134a would have to pay for the rights to produce locally or would have to pay world prices to import the chemicals. South Korean firms are of the opinion that the concession fees demanded by technology owners represent an expression of a lack of intention to transfer the alternative technology.

In addition to the high prices to be paid for acquiring alternative technology, South Korean businesses are also at a disadvantage due to the various conditions unfavourable to South Korea in technology transfer agreements with their foreign partners. Among 168 Japanese technologies introduced into South Korea in 1994, 15 (8.9 per cent) were not allowed to be consigned to a third party and 13 (7.7 per cent) were granted on a non-exclusive basis and on the condition that improved technologies should be shared between two parties during the contract period. Seven (4.2 per cent) were prohibited to be used for export products and three (1.8 per cent) were granted on the condition that the licensee not deal in competitive products or technologies. Among the 209 US technologies introduced to South Korea in the same year, 16 (7.7 per cent) were conditional upon the sharing of the improved technologies, 12 (5.7 per cent) were granted on a non-exclusive basis and 10 (4.8 per cent) were not allowed to be consigned to a third party. Such conditions have been reported to inhibit the effective transfer of technology, and have been considered unreasonable at times by South Korean firms.

Although there is the option of applying for a compulsory licence in the International Agreement on Trade-Related Aspects of Intellectual Property Rights (TRIPs) where the owners of patented technologies refuse to transfer their technologies, this recourse is generally very time-consuming due to the many stringent procedures and conditions. For instance, if the technology for the primary alternative substance (for example HCFC) is required for the production of the secondary alternative (for example HFC), the country could resort to compulsory licensing, but one of the requirements for this procedure is that the invention claimed in the second patent (HFC technology) should involve an important technical advance of considerable economic significance in relation to the invention claimed in the first patent (HCFC technology). However, there are no specific guidelines for defining 'important technical advance' or 'considerable economic significance'. Therefore, in practical terms, it is difficult to utilize the compulsory licence clause of the TRIPs agreement, especially for SMEs, which have limited negotiating abilities. Thus, to improve the prospects of a technology transfer, the varied and stringent conditions for the compulsory licence described in Article 31 would have to be modified.

Patents make the development of indigenous technology development more difficult. A case in point was a local refrigerator manufacturer which spent US$16 million over three years in a bid to develop a new compressor for an alternative refrigerant. The firm encountered a conflicting patent from a Japanese compressor maker and had to extend its project for another year to ensure that the new compressor developed by the enterprise did not conflict with this patent. Moreover, considerable time, effort and finances have to be spent to check for any conflicting patent right. Depending on the specific case, it costs between US$1300 and US$1500 and takes between two and three weeks to get such information. Even then, obtaining sufficient information can be a problem. Thus the process of developing the manufacturer's own compressor was an arduous and an expensive one.

To reduce the emissions of ODSs and encourage indigenously developed alternative technology, the Government of South Korea authorized the Korea Specialty Chemical Industry Association (SCIA) to tax ODS production and import of specified ODSs and products containing them. The money goes to a fund that has been primarily used for development of CFC substitute substances and for loans to enterprises wishing to transition to ozone-friendly production. The loans are conferred for activities such as technology development/equipment installation for the use of alternative substances; equipment installation for the production of alternative substances; technology development/equipment installation for reduction in the use of specified substances; equipment installation for destruction, recycling and reuse of ODSs; and equipment installation for upgrading efficiency in using specified substances. South Korea considered the local development of alternative technology imperative considering that transfer of such technology from foreign sources has not happened in accordance with the provisions of the Montreal Protocol.

The Korea Institute of Science and Technology (KIST) has been assigned the task of developing alternatives indigenously. This institute is supported by the Ministry of Trade, Industry and Energy and the Ministry of Science and Technology. In addition to KIST, private businesses such as Ulsan Chemical Co participated in the first-stage project under the supervision of the Ministry of Science and Technology and the Ministry of Trade, Industry and Energy.

patents for the production of HFC-227ea have expired, but in 1995 a US enterprise filed methods and composition patents with 20-year protection in a number of countries, including China, South Korea and Russia. Indian law did not until recently allow methods and composition patents. According to industry sources, China and Russia have successfully developed the process for FM-200 through indigenous research and development, but will be prevented from marketing the final product in countries where the patent is binding.

The US owner of the FM-200 patent requires that licensed fire protection systems satisfy the relevant design, component and inspection requirements of the Underwriters Laboratories (UL) or Factory Mutual (FM) and any additional requirements of the National Fire Protection Association (NFPA) standard 2000. Only three enterprises, in the US, the UK and Australia, have satisfied the required approvals. So far, no Indian firm has received permission to sell the relevant hardware.

The owners of the patent have offered joint ventures with a majority share holding, but are not interested in licensing the technology to wholly domestically owned enterprises. Indian firms currently marketing halon systems would prefer to purchase the technology outright and not be bound by UL, FM and NFPA standards.

India's experience is that where the alternative technology is easily accessible, commercially viable and not covered by intellectual property rights, the transition has been smooth. However, where the technology or process is restricted under intellectual property rights, with only a few technology owners, the experience has been negative.

WHAT EXPERIENCE SHOWS ABOUT ACTUAL BARRIERS TO TECHNOLOGY TRANSFER

The firm-level problems encountered during the implementation of the Montreal Protocol are consistent with the barriers hypothesized by the Intergovernmental Panel on Climate Change, the Technology and Economic Assessment Panel and the World Trade Organization,[17] such as shortage of trained workers, financial problems, limited access to finances, lack of awareness of alternative technologies and lack of supporting infrastructure. The experiences of the CEITs and developing countries with the ODS phaseout have brought new barriers to light, though, such as convincing top management and resolving the unique problems of family-run enterprises.

What is notable, however, is that fewer problems were encountered in practice than were put forth in theory, and they rarely created barriers to ODS phaseout. This is perhaps evidence of the effectiveness of technology transfer support provided by the MLF, the GEF and UNEP. Financial barriers, one of the most commonly debated barriers to technology transfer, were overcome through access to GEF and MLF grants. Lack of skills and capacity, another commonly debated barrier to technology transfer, was overcome through capacity-building projects sponsored by implementing agencies, as well as the training provided by technology suppliers and financed by the MLF and the GEF. Information-based barriers to technology transfer were eliminated by the successful efforts of UNEP (see 'Information exchange' in Chapter 13).

Many country-level barriers were either problems of governance or unexpected events (for example government delays, banking delays, delays caused by the SARS outbreak in Southeast Asia and political turmoil in Africa). Lack of policies and regulations, a barrier to the transfer of environmentally superior technology often debated at the international level, was successfully addressed under the Montreal Protocol. By becoming party to the Protocol, all countries committed to the phaseout schedules agreed by the Parties. Furthermore, implementing agencies of the MLF and the GEF often required documentation of country laws and regulations prior to providing technology transfer assistance. For example, the MLF required countries to have refrigerant management plans or to control imports of ODSs and ODS-based equipment as a precondition for the implementation of refrigeration recovery and recycling sub-projects. In addition, the good work of the International Network for Environmental Compliance and Enforcement (INECE), the Environmental Investigation Agency and others helped to ensure that lack of enforcement did not constitute a barrier to technology transfer at the country level.

One unanticipated country-level barrier to technology transfer in developed countries was the existence of numerous laws and policies that mandated the use of ozone-depleting substances. Fire safety laws for weapons systems,

BOX 12.4 LESSONS LEARNED BY IMPLEMENTING AGENCIES

Policy-related lessons

- Awareness of the Montreal Protocol and phaseout schedules is essential.
- It is necessary to provide alternatives and options.
- National Ozone Units need to understand technology assessment principles.
- It is necessary to collaborate with other agencies to get projects off the ground.
- Networking seems restricted to governments, and needs to be broadened.
- It is essential to use national experts.

Technology-related lessons

- New chemicals and products often have problems of flammability, toxicity, contamination and safety.
- Alternative technologies require careful implementation to mitigate inferior technical performance (thermal insulation, for example).

Cost-related lessons

- Old alternatives to ODSs tend to have lower investment and operational costs than new alternatives.

Market-related lessons

- It is essential to understand the influence of external and local factors on the market.

How implementing agencies can help overcome barriers to technology transfer

- Build a combination of cross-linked incentives and disincentives.
- Obtain grants for ODS-consuming enterprises.
- Mount public awareness campaigns and implement eco-labelling.
- Introduce legislation and enforce ODS phaseout and product bans.
- Exert influence in industrial associations to set technical standards.
- Encourage governments to impose local sales taxes on ODS-based equipment.

Source: UNEP OzonAction Programme[18]

aircraft, ships and racing cars, for example, often mandated the use of halon fire suppression systems. Quarantine standards often encouraged or required methyl bromide for certain pests and products. CFC refrigerants were required by law in some areas, and both industry and military cleaning standards required cleaning with CFC-113. In these cases, enterprises and governments seeking change organized strategic projects to remove specific barriers.

The implementing agencies have learned many lessons about how to improve technology transfer and minimize barriers (see Box 12.4).

It is notable that intellectual property rights did not constitute as large a barrier to technology transfer as was feared. In many cases, the technologies needed to phase out the use of ozone-depleting substances were in the public domain. In a few cases, there were problems in obtaining technologies from some suppliers, but the problems were ultimately sorted out by going to others. In the two cases where intellectual property considerations constituted significant barriers to technology transfer, solutions were ultimately found. In the case of the transfer of technology for the manufacture of HFC-134a, developing country enterprises developed their own processes to avoid paying licensing fees or accepting conditions they considered to be unacceptable. In the case of HFC-227ea, China and Russia successfully developed the process through indigenous research and development.

CONCLUSION

Fewer barriers were encountered in practice than were put forth in theory, and barriers rarely slowed the ODS phaseout. Many enterprises experienced no problems transitioning to ozone-safe technologies. The main problems encountered in technology transfer to developing countries were related to the need for effective governance in many areas, such as banking, finance, customs, and policies and regulations, and the need for more efficient functioning of the international agencies entrusted with the task of assisting the developing countries with technology transfer investment projects. Intellectual property protection did not constitute a major barrier to ozone layer protection, because many of the most important technologies were in the public domain, because the MLF and GEF paid licensing and other technology transfer fees when necessary, and because in the few cases where intellectual property constituted barriers to technology transfer, some developing country enterprises ultimately developed their own technical solutions or found solutions from other suppliers.

Awareness and Capacity-Building

Rajendra Shende and Tilman Hertz[1]

INTRODUCTION

Technology comprises the processes, equipment, substances and procedures that are put into practical application for the benefit of society. Technology transfer for environmentally sound technologies includes both technology and know-how, and organizational and managerial procedures.[2] This implies that human resource development and local capacity-building should also be addressed for technology transfer to be successful.

This chapter illustrates the role of the information and networking services, or the 'software', of technology transfer to phase out ozone-depleting substances (ODSs) under the Montreal Protocol through the United Nations Environment Programme's (UNEP's) OzonAction Programme and other implementing agencies of the Multilateral Fund (MLF). The chapter also elaborates on the challenges that are still to be met in technology transfer to developing countries.

TECHNOLOGY TRANSFER AND CAPACITY-BUILDING

In 1987, five years before the 1992 Rio Declaration, the Parties to the Montreal Protocol debated the practical importance of capacity-building for technology transfer. It was realized up front during the formulation of the Protocol and later during its implementation that the 'software' part of technology transfer is as important as investment components. In fact, in the case of the majority of Parties to the Montreal Protocol – developing countries and, in particular, low volume consuming countries – the information and networking services component forms the major part of the technology transfer.

Many different elements of capacity-building and enabling conditions are essential for successful technology transfer.

**BOX 13.1 THE 1992 RIO DECLARATION ON
ENVIRONMENT AND DEVELOPMENT**

*States should cooperate to strengthen endogenous capacity-building for
sustainable development by improving scientific understanding through
exchanges of scientific and technological knowledge, and by enhancing
the development, adaptation, diffusion and transfer of technologies,
including new and innovative technologies.* (Principle 9, Rio Declaration)

Enabling conditions for technology transfer are as important as the process of transfer itself. Enabling conditions for technology transfer derive from both the building of the capacity of the recipient and the understanding of the needs of the technology by the supplier. Along with the other implementing agencies and bilateral agencies of the MLF, UNEP is working to create enabling conditions and ensure that they are sustained for the long term.

An important context for the UNEP OzonAction Programme is its relation to the Bali Strategic Plan.[3] The Bali Strategic Plan for Technology Support and Capacity-building, which was adopted by UNEP's Governing Council in 2005, is an intergovernmental framework for strengthening the capacity of governments in developing countries and countries with economies in transition (CEITs) to coherently address their needs, priorities and obligations in the field of the environment. It responds to the United Nations Millennium Development Goals. OzonAction directly contributes to achieving the Bali Strategic Plan objectives and has collected valuable experience in this regard over the last 14 years.

Genesis of capacity-building programmes under the MLF

The Montreal Protocol has four implementing agencies: the United Nations Development Programme (UNDP), UNEP, the United Nations Industrial Development Organization (UNIDO) and the World Bank.[4] The UNDP, UNIDO and the World Bank are primarily focused on investment projects, the 'hardware', though they also provide information and networking, or 'non-investment', services such as technical assistance, training and institutional strengthening. UNEP is focused entirely on providing these services, the 'software', which provide an information clearinghouse function as well as assistance to developing countries for their expeditious phaseout of controlled ODSs through the OzonAction programme.

Information and networking activities, as provided by UNEP's OzonAction and other organizations, proved to be important for the success of investment projects implemented by UNIDO, UNDP and the World Bank. As Richard Benedick has put it:

> *of the four implementation agencies responsible for the MLF's activities,
> UNEP received the smallest portion of the budget (less than five per cent).
> Nevertheless, OzonAction generated a vast range of information on the evolv-*

BOX 13.2 PARTNERSHIP CHANGING THE WORLD THROUGH
WELL-DEFINED TECHNOLOGY TRANSFER MECHANISMS:
THE ASIA–PACIFIC EXPERIENCE

*Atul Bagai, Chief, UNEP OzonAction South Asian Network,
Bangkok; N. Balaji, UNEP South Asian Network; and
R. Gopichandran, Centre for Environment Education,
Ahmedabad, India*

Roles played by implementing agencies

The United Nations Development Programme (UNDP), the United Nations Environment Programme (UNEP), the United Nations Industrial Development Organization (UNIDO) and the World Bank are doing much to help countries phase out ODSs. UNIDO uses a preventive approach that begins with the building of awareness of non-ODS technologies, and also helps implement MLF investment projects. UNEP helps countries develop their country programmes for the phaseout of ODS, supports global and regional information exchange, facilitates sharing of experiences through networking, provides technical assistance based on reality checks, and provides policy training to enable implementation of phaseout tasks. UNEP's actions are intended to enable countries to successfully fulfil their commitments as members of multilateral environmental agreements. The UNDP and the World Bank are assisting with the phaseout of ODSs in small and micro-size enterprises in the refrigeration and air-conditioning sector in India, Indonesia, Thailand, Vietnam and Malaysia. They coordinate with local-level groups and governments to provide training, technical assistance, and policy or management support. The UNDP is also helping to eliminate CFCs in the refrigeration manufacturing sector. In the solvent sector in China, the UNDP works with the State Environmental Protection Agency (SEPA) and the Ministry of Industry to develop innovative solutions to phase out ozone depleting solvents used by more than 2200 small and medium-sized consuming enterprises.

As mentioned above, the UNDP assists with the CFC phaseout in the refrigeration and air-conditioning sector in India. One of the components of India's National CFC Consumption Phase-out Plan (NCCOPP) was the Equipment Support Scheme for Refrigeration Servicing Enterprises (ESS). The ESS was created to provide firms that service refrigeration appliances and equipment with assistance to upgrade their tools and equipment and to help them adopt and implement good servicing practices. The Government of India decided who was eligible for assistance, and the UNDP was responsible for procurement and distribution of equipment kits.

Phasing out ODSs in SMEs

One of the greatest challenges addressed through ODS phaseout projects in developing countries was transfer of technologies to small and medium-sized enterprises (SMEs). Due to their size, these enterprises often have limited awareness of alternative technologies and lack the resources to understand and adopt technologies. To ensure that the issues concerning SMEs were adequately addressed, the multilateral agencies and bilateral agencies worked closely with local institutions to understand the needs of SMEs and address their technology transfer requirements. The experience showed that awareness was a significant component of technology transfer, and that increased awareness did facilitate technology adoption.

Awareness for CTC phaseout in SMEs in India

Carbon tetrachloride (CTC) was consumed in significant proportions in solvent applications by SMEs in India. The consumption was predominantly in a few areas close to CTC suppliers and which had industries needing the chemical in solvent applications. To address this consumption, since 2004 the Government of India has followed a strategic approach towards awareness among industry users through an application-level evaluation of ODS use, identification of non-ODS technologies to suit the applications and promotion of these non-ODSs in identified applications. Sometimes a door-to-door approach was adopted in understanding the CTC-consuming applications and promoting non-ODS substitutes.

Over a two-year time frame, this project has contributed to industries using CTC in solvent applications switching over to non-ODS technologies. Awareness activities coupled with price increases in CTC have resulted in a significant drop in CTC consumption in these applications. The project forms a part of the National CTC Phase-out Plan implemented by the Government of India with the World Bank as the lead agency and UNIDO, the Government of Japan (through the UNDP), and the Governments of Germany and France as cooperating agencies. UNEP also engaged in technology transfer to SMEs in certain CTC-consuming areas on a pilot basis during the CTC phaseout strategy development process.

While implementing agencies have attempted to provide project-financing support for SMEs, support for smaller enterprises was more often provided through technology options awareness programmes and demonstration of alternatives. Investment projects for these small enterprises would not have been within the cost-effectiveness thresholds of the MLF. For example, in the case of refrigeration and air-conditioning servicing equipment using non-ODS-based alternatives, multilateral and bilateral agencies sponsored training programmes for service technicians.

Constraints on technology transfer under the Montreal Protocol

While the Protocol facilitated technology transfer for phasing out ODSs, there have been some cases where constraints created challenges. Some key areas that acted as constraints on technology transfer included:

- limited capacity-building on technical know-how (for example non-availability of all design specifications or non-availability of some key components) resulting in inability of recipient enterprise to adapt technology to local conditions;
- absence of technical support for product research, development and upgrading. As a result, the companies that received assistance from the MLF had to depend on suppliers for product development; they could not use money from the MLF to research and develop technologies themselves;
- supply shortages of critical components for non-ODS alternative technology; and
- limited training for staff of the beneficiary enterprise. Often, training was restricted to production personnel only and research and product development personnel were not included in the training activity. This affected development of the capabilities of the organization to take up technology upgradation, as required, on their own.

Limited financing of modernization costs also delayed implementation. For example, in the refrigeration and air-conditioning sector, any enterprise that wanted to adopt advanced technologies like HFC-134a or HC-600a first had to modernize and upgrade their equipment. This directly influenced their ability to adopt the alternative technology. When funds from the MLF were not available for equipment modernization, industry faced a higher cost burden to transfer technology. In some cases, assessments were carried out and enterprises were given approval to use MLF money to modernize and upgrade.

Phaseout of production of CFC-12-based domestic refrigerators in China

The Huari Group, a large refrigerator manufacturer, was helped by UNIDO to pioneer, for the first time outside Europe, the use of cyclopentane insulation foam blowing and isobutane refrigerant. As a result, 338 ODP tons of CFCs were eliminated from use. Prior to the start of the project, CFC-12 was used as the refrigerant and CFC-11 as the foaming agent.

The group invested US$0.7 million of its own funds to complement the US$2.8 million provided by the MLF. Based on the substitution, it also increased annual production of refrigerators by about 5 per cent and started exporting the ODS-free refrigerators to other countries.

The first batch of appliances was redesigned and Huari's staff received training to help them redesign and finalize conversion of their remaining models on their own. All new equipment and the complete manufacturing line were inspected for compliance with prescribed safety standards.

ing treaty, the science, government policies, technologies, industry news, success stories, available publications, meetings and workshops, and training programmes. It utilized an interactive online computerized system and diskettes, as well as traditional publications, including the quarterly newsletter OzonAction, *published in six languages. It undertook policy and technical research and produced sectoral training manuals, technical source books, and multilingual brochures. UNEP also provided databases on technologies, products, enterprises and specialized experts around the world, and it organized technical and training workshops that promoted essential networking between local personnel and experienced specialists from foreign enterprises and governments. In addition, UNEP provided support to developing countries' coordinating offices and helped small, low-consuming parties to prepare their national phaseout programmes. This extensive underpinning work at relatively modest cost was essential in preparing the way for implementation of the MLF investment programmes.*[5]

Capacity-building: Taking account of country-specific needs and barriers

Capacity-building develops and strengthens scientific and technical skills and institutions in developing countries, and enables them to assess, access and adapt environmentally sound technologies needed for compliance with the treaty.[6]

Countries' capacities are very different, however, and it is clear that this has to be taken into account.[7] There are no abstract textbook solutions for capacity-building in the technology transfer process. Solutions have to be evolved, taking into account each country's specific situation.

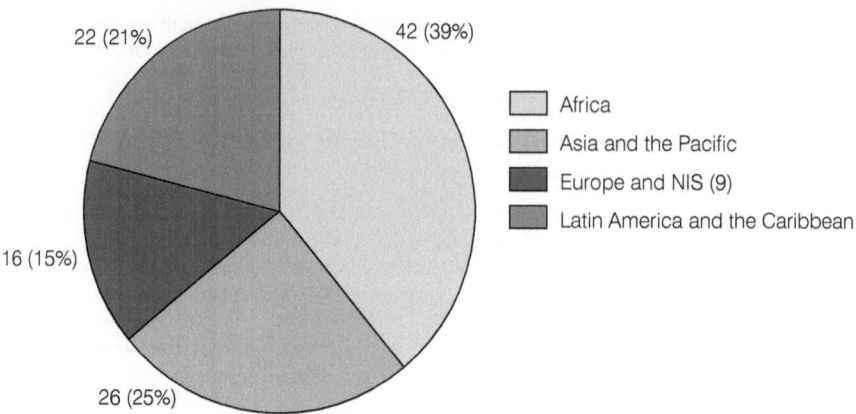

Note: Figure 13.1 shows the number of countries per region and the percentage of the total number of countries assisted by UNEP to prepare their country programmes. NIS stands for 'newly independent states'. 'Asia and the Pacific' includes West Asia.

Figure 13.1 *Regional breakdown of country programmes prepared with assistance from UNEP*

Developing countries differ in their patterns and level of ODS use, character of manufacturing facilities, prices of ODSs and equipment, state of development of the industry, and policies set and enforced by government. The assistance to enable countries to comply with the Montreal Protocol has to take into account these differences.

The OzonAction programme and other implementing agencies have worked with CEITs and developing countries to develop a 'country programme'[8] to identify a specific plan of action. OzonAction helped to prepare the country programmes in 106 countries. It also analysed the needs of the remaining countries based on the country programmes developed by the World Bank, the UNDP and UNIDO. UNEP's OzonAction utilizes regional networks[9] of Ozone Officers to monitor and identify the evolving needs of countries. UNEP was accordingly able to dovetail its capacity-building programme to their needs. A number of industrialized countries participate in the meetings of the regional networks of the National Ozone Units (NOUs) to gain an understanding of the changing scenarios in the developing countries and CEITs.

OzonAction leveraged the processes of:

- awareness raising;
- information exchange;
- training and networking; and
- institutional strengthening.

These are described in detail in the following sections.[10]

Box 13.3 UNEP's instrumental role
IN RAISING AWARENESS ABOUT THE IMPORTANCE
OF OZONE LAYER PROTECTION

UNEP has been instrumental in the campaign to raise public awareness about the hole in the ozone layer. Apart from its many technical reports, UNEP has published a myriad of booklets, brochures and posters aimed at the public, such as 'The Ozone Layer', 'Action on Ozone', and 'The Ozone Story'. In 1994, on the initiative of Venezuela, the General Assembly of the United Nations designated 16 September as the International Day for the Preservation of the Ozone Layer (Ozone Day). From 1995 the Ozone Secretariat, with the help of the Communications and Public Information Division of UNEP, orchestrated the observance of the day by all governments and international organizations as an occasion to spread awareness. Each year a theme is chosen, such as the adverse effects of ozone depletion or the availability of ozone-safe products, and wide publicity is given through posters, radio and TV programmes, special editions of UNEP's flagship magazine, *Our Planet*, workshops, seminars, product exhibitions, essay competitions, painting competitions, and rallies by school children. In some countries such as China, top officials, including ministers of the government, have joined public rallies urging the public to prefer ozone-safe products. UNEP's OzonAction programme has helped organize such events throughout the world. It has prepared and distributed many publications on how every human being can contribute to the protection of the ozone layer and on how every office can be made ozone-safe.

Source: Andersen and Sarma (2002)[11]

AWARENESS RAISING

Key need for awareness raising: **action-oriented awareness about the ozone depletion issue**.

There is a widespread recognition in developing countries that ozone depletion is a global problem, and that everyone is required to do their part to make the world safe for future generations. Other reasons for the ozone-friendly technologies include scarce supply of ODSs resulting from the closure of ODS-manufacturing facilities as mandated by the Protocol, and new marketing opportunities for ozone-friendly products.

Enabling tools for awareness raising

OzonAction uses many tools in its public awareness campaigns. A regular *OzonAction Newsletter* is sent to many of the stakeholders, with sector-specific supplements. An 'information kit' – including videos, posters, radio clips and a national awareness manual – helps NOUs devise their own national awareness campaigns. Regional workshops for policymakers raise awareness of the control measures of the Montreal Protocol and the status of the alternative technologies. The participation of national and international bodies in these workshops, as well as industry representatives and representatives of the Technology and Economic Assessment Panel (TEAP) and the Technical Options Committees (TOC), is strongly encouraged. In addition to general information videos and brochures, OzonAction has a video on CFC-free aerosols, and many posters and stickers on methyl bromide alternatives and mobile air-conditioning alternatives.

INFORMATION EXCHANGE

Key need for information exchange: **awareness of alternatives to ODSs, including cost implications and market preferences.**[12]

Enabling tools for information exchange

OzonAction's main task is to inform countries about the options available for a technology transition to non-ODSs in neutral, up-to-date technical terms.[13] This is essential to build trust in the information disseminated among the relevant stakeholders and to prepare the necessary technology transfer projects on an informed basis. OzonAction has published sectoral technical brochures on legislation, patent information, standards and codes of practice. It has also published papers on retrofitting and emerging technologies, and a technical directory of recovery and recycling machines. Sectoral technical sourcebooks by OzonAction identify worldwide suppliers of alternative technologies, suppliers of products and services, and experts and consultants who provide additional assistance. OzonAction also provides the NOUs with relevant case studies on practical examples of replacement, conversion and successful policy implementation. The OzonAction website (www.unep.fr/ozonaction) is always up to date with the latest information regarding technologies in general and their pros and cons. Finally, with various demonstration projects and activities, OzonAction aims to show that the new technologies not only are environmentally sound but also rest on a viable business model.

BOX 13.4 WEBSITES FOR OZONE PROTECTION PARTNERSHIPS AND PROGRAMMES

- www.halontrader.org
- www.unep.fr/ozonaction
- www.uneptie.org/Ozonaction/topics/capacity.htm
- www.uneptie.org/ozonaction/information/mmc/list.asp?x=a
- www.refrigerantsnaturally.com
- www.solarchill.org/

Halon banks[14]

OzonAction provides the tools for the judicious use of already-produced halon (banked halon) that avoids the demand for new production. The OzonAction 'Business to Business' halon trader on its website is an important tool in this regard.

TRAINING AND NETWORKING

Key need for training and networking: **the skill to assess the technologies and experiences of other countries is important; often developing countries have insufficient expertise or availability of local craftsmanship and skills, or a significant lack of experience in the acquisition and monitoring of technology transfer projects.**

Enabling tools for training and networking

Training

OzonAction organizes training and networking workshops to provide government and industry decision-makers with information on ODS control policies, strategies, and replacement technologies and products. Furthermore, OzonAction organizes 'train the trainer' courses which include information and practical sessions on servicing, maintenance, recovery and recycling. There have, for instance, been safety workshops in China, hydrocarbon workshops in Uruguay, technology transfer workshops in Thailand and Malawi, workshops for small and medium enterprises in South Africa and India, aerosol workshops in Jordan and other countries, and chiller workshops in collaboration with the US Environmental Protection Agency (EPA) in Thailand, the Philippines and Mexico. Training manuals in recovery and recycling, leak reduction, best practices and other topics have been prepared. These efforts complement the work done not only by other implementing agencies, but also by alliances of private stakeholders, such as the Earth Technology Forum.

Regional networking

There are nine regional networks supported by the MLF:

1   the Southeast Asia and Pacific Network (supported by the Government of Sweden);
2   the South Asia Network;
3   the Network for English-speaking Africa;
4   the Network for French-speaking Africa;
5   the Network for the Caribbean;
6   Central America;
7   South America;
8   Europe and Central Asia; and
9   the Network for West Asia.

These networks are coordinated by the four UNEP Regional Offices: Asia and Pacific Region, Africa Region, Latin America and the Caribbean Region, and West Asia Region. The Regional Offices are managed by OzonAction, which is located in Paris. Developed countries also participate in the networks, not only to provide support and advice but also to receive ideas. The Europe and Central Asia Network is managed from Paris.

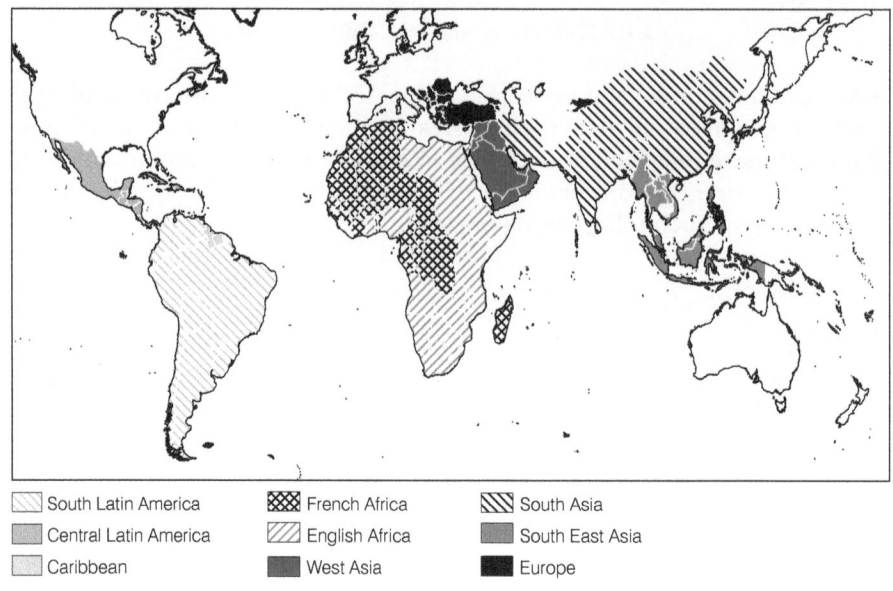

South Latin America
Central Latin America
Caribbean
French Africa
English Africa
West Asia
South Asia
South East Asia
Europe

Figure 13.2 *Regional networking map*

Regional networking is a regularly held, interactive forum and helps the NOUs to develop skills and share experiences and knowledge with counterparts from other countries. The basic principle of regional networking is collective learning by sharing while doing. The networks provide a vital link between policymaking at the international level and actions needed to implement the policies at the national level. Learning about other countries' experiences in facing and overcoming challenges is essential for successful technology transfer. Networks provide a vital element in answering such questions as what technology to adopt, where to get it, how to best implement it, what has to be done on the institutional level to accommodate the technology, and which stakeholders have to be involved in the process of technology transfer. Due to the often similar situations developing countries face, it has proven to be extremely efficient for key actors in developing countries to directly assist each other.

INSTITUTIONAL STRENGTHENING

Key need for institutional strengthening: **to determine which policies need to be set and when, and how to enforce them for the effective transformation to ozone-friendly technologies**.

Institutional strengthening is important because existing legal and regulatory structures are inadequate for effective transfer of technology. The inadequacy of legal and regulatory structures may be due to lack of resources or capabilities to regulate and promote new technologies, organizational rigidities, ineffective and inefficient R&D, or insufficient and poor coordination among different stakeholders at the international or national levels.[15]

Enabling tools for institutional strengthening

The ability of a country to monitor the use of regulated substances as required by the Protocol at the national level must be increased. This requires interaction between relevant partners and government bureaus in order to create the framework and understanding for the implementation of new, technology-related procedures and guidelines. OzonAction provides countries with policy guidelines for industry management of ODS phaseout, for instance. Partnerships with relevant stakeholders or NGOs can further facilitate this process. OzonAction works closely with the NOUs in order to strengthen the capacities of technology transfer and to ensure the development of laws, standards and regulations to encourage the use of new technologies.

STRATEGIC REORIENTATION OF OZONACTION TO FACILITATE THE TECHNOLOGY TRANSFER: CAP (COMPLIANCE ASSISTANCE PROGRAMME)

Some policy tools are aimed at technology 'forcing', while others are aimed more at technology 'fostering'. Technology 'forcing' is generally viewed as a process in which a regulator specifies a standard which can be met only with the new environmentally superior technology.[16] Technology transfer (or extensive R&D) is therefore imposed upon the stakeholders in the concerned country. Technology 'fostering', on the other hand, is a more subtle approach. It is generally viewed as achieving technology transfer by building on partnerships and providing the relevant stakeholders with the appropriate incentives so that they freely adopt new technology.

Whether to adopt a forcing or a fostering approach or a hybrid of the two has proven to be a difficult question.

Since 2002 UNEP's OzonAction has used the Compliance Assistance Programme (CAP) approach for compliance in general, and technology transfer in particular. The CAP is a new approach based on the recognition of the following:[17] first, that the type of assistance needed by developing countries was shifting toward direct assistance for specific compliance issues as the investment projects advanced or neared completion; second, that not enough attention was initially given to institutional strengthening projects; and third, that a constant and direct interaction could increase self-confidence and motivation of the national stakeholders.

The CAP approach has proven very successful as a framework for implementing capacity-building tools, thereby providing effective technology transfer. The CAP moves from a project-by-project approach and assists the countries continuously in Protocol implementation, using a team of professionals with appropriate skills and expertise. This is done by placing the CAP teams in regional offices, where they can work closely with the countries on project implementation, whereas before most activities were centralized in Paris. This regionalization singles out technology needs and requests, and the subsequent implementation of the technologies, through a participatory approach, and

Box 13.5 EXAMPLES OF ENABLING TOOLS FOR
THE AEROSOL SECTOR

Awareness raising

Protecting the Ozone Layer Volume 5: Aerosols, Sterilants, Carbon Tetrachloride and Miscellaneous Uses. Knowing the technology options: Aerosols summarizes the current uses of ODSs in various sectors, the availability of substitutes, and the technological and economic implications of converting to ODS-free technology. The report, first published in 1992, has been updated based on the 1998 reports from UNEP's Technical Options Committees (TOCs).

Information exchange

Sourcebook of Technologies for Protecting the Ozone Layer, Aerosols, Sterilants, Miscellaneous Uses and Carbon Tetrachloride is a guide to sources of technologies, equipment and products that reduce or eliminate ODSs in aerosol, sterilants, miscellaneous uses and carbon tetrachloride applications.

Aerosol Conversion Technology: Handbook provides more information for the aerosol industry to further assess, select and implement alternative technologies for the aerosol sector.

Training and networking

Safety First: Technical Guide for the Safe Handing of Hydrocarbon Propellants provides plant managers and NOUs with critical information related to the safe conversion of aerosol plants to non-CFC propellants, specifically hydrocarbons. The guide presents detailed information about safety, formula adaptations and effective destenching methods for aerosol production using hydrocarbon aerosol propellants.

Institutional strengthening

Elements for Establishing Policies, Strategies and Institutional Framework for Ozone Layer Protection provides practical guidance to developing countries seeking to establish and implement national policies and programmes for phasing out the use of ODSs.

Aerosol Sector Conversion in Action gives 12 technical case studies (from China, Ecuador, India, Indonesia, Jordan, Mauritius, Mexico, the Russian Federation, and Venezuela) that address plant conversions to alternative technologies, including hydrocarbon aerosol propellants, manual pumps and dimethyl ether propellants.

Aerosol products workshops

- Regional Training Course on Aerosol Conversion, 1994, Amman, Jordan
- Training Workshop on Safety Aspects of CFC Substitution, Hangzhou, China, 22–24 May 1996
- Regional Workshop for Aerosol Conversion in the Southeast Asia and Pacific Region, Jakarta, Indonesia, 29 May–1 June 1995

Related workshops

- Transfer of Technology Workshop for the French African Countries, Douala, Cameroon, 27–29 February 1996

- Technology Cooperation Workshop, Hanoi and Ho Chi Minh City, Vietnam, 4–5 July 1996
- Regional Workshop on Lessons Learned from Technology Transfer under the Multilateral Fund of the Montreal Protocol, Asia Pacific Centre on Technology Transfer, New Delhi, India, 4–5 March 1999
- South African Regional Workshop on Lessons Learned from Technology Transfer under the Multilateral Fund of the Montreal Protocol, Mangochi, Malawi, 27–29 May 2002

Source: UNEP[18]

allows OzonAction to be responsive to the needs and capacities of the country in question, according to its country programmes and refrigerant management plans. Important decisions, such as whether an enterprise should keep a certain technology and improve its maintenance through a code of good practice, or whether a new technology should be adopted, are made mindful of the particular situation of the developing country. All the specific tools and activities for general awareness, information clearinghouse activities, training activities and institutional strengthening projects are designed and put to use by specialists of the regional UNEP offices.

Some achievements from 2006 of the CAP approach to technology transfer,[19] and examples of what still needs to be done

The CAP paid special attention to countries that recently ratified the ozone treaties – such as Afghanistan, Bhutan and Eritrea – in order to assess their technological needs and capacities. It utilized the expertise now available in developing countries to assist these countries.

A 'Guide for National Ozone Officers' in simple language was prepared for the benefit of newly appointed Ozone Officers. Fourteen new officers from seven countries were trained with substantive support from the Ozone Secretariat, the World Bank, the UNDP, UNIDO, Japan, Germany, Sweden and Australia. The CAP organized information sessions for capacity-building of Ozone Officers from Paraguay, Guatemala, Honduras and Nicaragua, using experts from UNEP, the Dominican Republic, Costa Rica and Mexico.

In terms of institutional strengthening and policy development, the CAP assisted Afghanistan, Bhutan and Indonesia with the implementation of their commitments for the establishment of ODS management policy instruments in order to assure a smooth technology transfer. Similarly, it assisted Chile, Haiti, the Dominican Republic, Barbados and Guyana by reviewing ODS legislation and providing advice; Botswana, Lesotho, Kenya, Mozambique and Tanzania by providing technical guidance on establishing ODS regulations; and Kyrgyzstan by revising its national ODS legislation.

CASE STUDIES OF TECHNOLOGY TRANSFER ENABLING TOOLS IN ACTION

Croatia's use of technology forcing to phase out ODSs in the refrigeration servicing sector

As a precondition of joining the EU, Croatia was required to implement the EC directives and guidelines regarding the refrigeration servicing sector.[20] The Croatian refrigeration servicing industry was therefore in a position where it had to adapt quickly to the new technology because technology transfer was forced by the EU regulation making compliance with the Montreal Protocol a condition of EC membership.

OzonAction, together with UNIDO, the MLF and Sweden, had undertaken a number of actions to create an enabling environment for technology transfer.[21] One of the first measures under the refrigerant management plan (RMP) was a 'train-the-trainers' programme on proper methods for leak reduction and recovery designed to improve the standard of servicing. Four training centres were established and have so far trained 724 technicians in 48 training sessions.[22] OzonAction was involved in establishing and advising a national workgroup with representatives from different sectors in the industry that established an industry code of practice. OzonAction, together with its partners, is advising Croatia on whether to establish legal requirements to document the sale and consumption of refrigerants as a way to enforce sales restriction. This may further 'force' the industry to adopt the new technology.

Refrigerants, Naturally! and SolarChill Partnerships[23]

At the end of this chapter, Janos Maté of Greenpeace International describes the Refrigerants, Naturally! and SolarChill Partnerships. Refrigerants, Naturally! is a partnership of the Coca-Cola Company, McDonald's Corporation and Unilever with UNEP and Greenpeace to promote a shift in point-of-sale cooling technology in the food and drink, food service and retail sectors towards alternative refrigeration technology that protects the Earth's climate and ozone layer. SolarChill is a partnership of the Danish Technological Institute (DTI), the German Government Development Agency (GTZ–Proklima), Greenpeace International, Programmes for Appropriate Technologies in Health (PATH), UNEP, the United Nations Children's Fund (UNICEF) and the World Health Organization (WHO) that is commercializing a solar-powered climate- and ozone-friendly vaccine cooler.

CFC-113 in solvents for electronics cleaning in Asia[24]

In some Asian manufacturing centres, such as Hong Kong and Singapore, the electronics industry was forced to adopt alternative technologies because their governments engaged in an ambitious phaseout plan and, along with it, imposed restrictions on the quantities of CFCs available to industries. Procuring CFC-113 for electronics cleaning thus became more and more difficult, with increasing prices and a shortage of supply.

OzonAction has engaged in extensive public awareness campaigns which brought 'no-clean' technology (which could be considered as an existing but dormant technology) to life in the solvents sector of CEITs and developing countries. The no-clean technology requires some changes in the manufacturing process; however, this process change eliminates the need for cleaning chemicals – a pollution prevention approach that can lower chemical use and waste production. OzonAction, among other agencies, provides training and technical assistance in this regard.

CFC-free refrigerators in Thailand through corporate leadership with regulation[25]

The phaseout of CFCs from refrigerators in Thailand is a good example of how technology transfer was achieved by industry leadership in forcing technology with national trade barriers against import of ODS products to protect manufacturers of ODS-free products. The Government of Thailand committed to an aggressive CFC phaseout plan, and concerned stakeholders had to cope with this decision. Initially there were doubts concerning the feasibility of the phaseout plan and technology transfer. But the Thai Government took measures to prohibit the manufacture or import of foreign CFC-based refrigerators; hence, those affected were forced to act. The industry leadership and trade barriers came as a result of awareness and capacity-building.

Technology fostering for the CFC aerosol product phaseout through voluntary agreements[26]

Mexico was one of the first signatories to the Montreal Protocol and the first to ratify. The Mexican sectoral example is also one of the first 'technology fostering' examples. The Mexican Government did not have to enforce ODS regulations because:

> the Mexican Aerosol Industry Association (Instituto Mexicano del Aerosol A.C. IMAAC) organized the first voluntary sectoral phaseout of CFCs by any developing country, achieving the phaseout in cosmetic and pesticide products at least five years faster than the European Union.[27]

An appropriate information clearinghouse as well as intensive training and networking proved the key for effective technology transfer. Notably, the president of the Mexican Aerosol Industry Association at that time, Jorge Corona, was also a member of the Montreal Protocol Solvents Technical Options Committee. This gave the Mexican industry associations, in collaboration with international government and business partners, full confidence in the technical alternatives to CFC solvents, allowing the phaseout of existing CFC solvents and the expansion of production in modernized factories that avoided ODSs. In addition to the information clearinghouse activity, extensive use was made of training and networking activities: Mexican factory owners and managers and solvent cleaning experts were brought together through workshops, training and technology study tours. OzonAction, alongside other partners, was actively involved in coordination, organization or facilitation of each of these activities.

Fiji – Good practices in refrigeration

In Fiji, most alternative technology support for industry has been attributed to awareness campaigns at the industry level and government support. In 2000 and 2001, key international companies signed a pledge to reduce CFC emissions and consider environmental issues at the forefront of their activities. Over 60 participants were involved in this technology fostering effort. One of the keys to success in Fiji maintaining its zero consumption of CFCs was running the long-term 'Train the Trainers Program for Good Practices in Refrigeration'. The training programme was initially introduced in 2001 by UNEP, but since then, Fiji has been running it itself, way beyond its expected duration. Over 500 technicians have been trained, and the programme has been successfully incorporated into the curriculum of the Training and Productivity Authority of Fiji. The National Ozone Office was successful in its enterprise as a result of inviting onto the board a former refrigeration technician who had over ten years of experience in the field. For the last five years, the Ozone Office has encouraged industries to work with alternative technologies in the refrigeration and air-conditioning sector, with the assistance of importers and the UNDP Recovery and Recycling Project. Key features in promoting ozone awareness were the numerous awareness workshops and media campaigns targeting the local community, including mobile exhibitions in 2001 and 2003.

LESSONS LEARNED

What can be learned from the many years of experience so far?

Collateral benefits are important for the actual technology transfer. For instance, new refrigeration technologies should both phase out ODSs and at the same time achieve very high energy efficiency, while being consistent with the national programmes of the country in question.

Providing countries with neutral and up-to-date technical information and showing that the technologies rest on viable business models (in other words providing the right incentives) have proven to be very helpful for the actual transfer of technology.

Furthermore, strong partnerships have proven to be critical for technology transfer, giving credibility to the technology as well as a wider scope for information distribution and technology dissemination. For instance, SolarChill and Refrigerants, Naturally! are good examples of the power of partnerships in developing technology that would be difficult for one organization to do alone. Partnerships highlight another important point: that technology cooperation and development lowered the cost of the transformation to an ozone-friendly world. This is significant because during international negotiations, technology innovations and the power of creative development of the industry are never factored in. Often, as the partnerships demonstrate, industries should be viewed as the solution to the problem they themselves created.

Regulation and policy enforcement on behalf of government are also very important. These provide the relevant stakeholders with an obligation to comply with national laws implementing the Montreal Protocol and stimulate them to get involved in R&D. The EU, for instance, having had a successful experience in phasing out CFCs through strong and ambitious regulations, is embarking on a similar path for reducing the use and emissions of HCFCs and HFCs. The approach of the Montreal Protocol is therefore being adopted for other domains, such as climate change.

However, there is no one blueprint for determining a priori whether enabling tools aimed at technology forcing, technology fostering, or a combination of the two is the right approach. The OzonAction CAP approach proved to be responsive to the changing needs and priorities of developing countries. The networking ensures that small countries are not left behind in the process of technology transfer. Not only do small countries get vertical assistance through the CAP and other implementing agencies, they also get horizontal assistance through networking.

OzonAction Pointers for other Multilateral Environmental Agreements

'Carrots' are as important in technology transfer as 'sticks'. This point has been well made by Richard Benedick:

> *Another important lesson from the Montreal history was that not all countries need to agree in order to take a substantial step forwards. In contrast to Kyoto, developing nations did accept limitations on their CFC consumption, but only when they were assured of equitable access to new technologies. Technology development is the missing guest at the Kyoto feast.*[28]

The emphasis should lie on the notion of technology development. In other words, when the actual transfer is concerned, it may not be enough simply to designate a focal point and leave it up to each country to undertake the necessary actions for capacity-building. Active local assistance, as provided by the CAP, may be extremely helpful in terms of public awareness, information exchange, training and networking, and institutional strengthening. Furthermore, the CAP approach can provide essential coordinating activity – it can provide the link between the actors at the international and the national levels (partners, industry, government agencies, technology suppliers, developers, technology buyers, financiers and so forth) which are affected in the process of technology transfer.

Successful technology transfer is possible only because of the many committed people around the world. These people provide a human face, and this is essential. It is not just the big enterprises that are the key ODS players needing assistance, SMEs which use ODSs are much more difficult to reach, and have required considerable effort and ingenuity by OzonAction, industry associations, NGOs and others. Luckily, in the framework of the Montreal

Protocol, information exchange and training had a snowballing effect. The transferred information got further value added from the recipients and beneficiaries, and the entire society benefited from such snowballing. The human part of technology transfer was able to utilize and leverage an optimum combination of awareness, policy and legislation that built the capacities to support the technology transfer.[29]

There is, however, still room to improve this process.

First, a set of indicators is needed to assess capacity-building for technology transfer. Developing indicators would enhance the effectiveness of the actions taken in the long term, because the necessary actions may be identified reliably and have baselines from which to gauge progress.

Second, technology transfer should not be viewed as an isolated process. There is certainly a need to harmonize technology transfer and the issues related to it, such as capacity-building, with the requirements of other MEAs. The examples of SolarChill and Refrigerants, Naturally! demonstrate this.

CASE STUDY: GREENPEACE'S PIONEERING ADVOCACY OF NATURAL REFRIGERANTS

Janos Maté, Political and Business Unit Consultant and Coordinator of the SolarChill Project, Greenpeace International

NGOs are no longer seen only as disseminators of information, but as shapers of policy and indispensable bridges between the general public and the intergovernmental processes. (Kofi Annan, UN Secretary-General, 1998)

Greenpeace

Greenpeace is an international campaigning organization with an uncompromising voice for the environment. It is a catalyst for ethical, political, commercial and technical change in society. Greenpeace campaigns put the spotlight on the causes of environmental degradation. They also identify viable solutions.

Greenpeace began its activities to protect the ozone layer in 1986, with high profile protests against the use of CFCs in aerosol cans. In 1992 Greenpeace made protection of the ozone layer the top priority campaign worldwide and during the following ten years the organization launched over 100 global initiatives to pressure governments and industry to act with a sense of urgency.

The campaign pursued four broad streams of activities:

1 public outreach to generate demand on governments and corporations to take effective and immediate action to protect the ozone layer;
2 policy advocacy to exert continuous pressure on governments to accelerate the ODS phaseout regime of the Montreal Protocol;

3   direct actions to confront multinational chemical corporations with the demand to halt the production of ozone-depleting and global warming substances that threaten life on Earth; and

4   market intervention to persuade companies to manufacture refrigeration and cooling products that use environmentally safer not-in-kind substances and to encourage consumers and developing countries to welcome these products.

This essay elaborates on the motivation, method and success of pioneering market interventions by Greenpeace in the cooling industry sector that is transforming consumer markets as well as corporate practices and demonstrates how citizens' organizations can pick, commercialize and promote environmentally superior technologies. It focuses on three initiatives that arose out of Greenpeace's engagement with the challenges of ozone layer and global climate protection: Greenfreeze, Refrigerants, Naturally! and SolarChill.

Confrontation and solutions

From the late 1980s, when it became apparent that CFC use had to be eliminated, Greenpeace was unequivocal in its opposition to the use of HCFCs and HFCs as CFC replacements. Greenpeace highlighted the fact that HCFCs also deplete the ozone layer, and that both HCFCs and HFCs are potent global warming substances. Greenpeace urged the international community to support the uptake of environmentally sustainable alternatives, promoted the use of 'natural refrigerants' and cautioned against the uptake of fluorocarbon alternatives that 'sweep the ozone crisis under the carpet of global warming'.

In its opposition to the continued production of CFCs and the uptake of fluorocarbon alternatives, Greenpeace engaged in a series of high profile public confrontations with the chemical industry around the world.

The chemical corporations, the producers of CFCs, HCFCs and HFCs, responded by attacking Greenpeace:

* A 1990 DuPont newsletter circulated as late as 1992 ridiculed Greenpeace's position:

    We certainly have the option of no longer refrigerating our food supply, 75 per cent of which is refrigerated as it is harvested, processed, stored, distributed or served. Are we prepared to be totally dependent on food that is consumed as soon as it is harvested? Or food that is dry, canned, smoked, salted or pickled to prevent spoilage?

* In 1991 Hoechst distributed millions of leaflets in Germany claiming that Greenpeace was endangering the lives of children all over the world by opposing refrigerant substances necessary for refrigeration and for the food chain.

* In 1991 ICI wrote to UK Greenpeace supporters:

    Can we all go back to the laboratory and spend the next ten years working on Greenpeace ideas to see if they can be made to work in practice? Greenpeace have refused to join in any discussions of what might actually be done about the problem in a practical way. After all it is so much easier to stand on the sidelines and criticize.

- In 1992 a high-level ICI representative is quoted as saying:
  Greenpeace lacks a sense of urgency. Most alternatives it talks about are not available. They're either pie in the sky or will only be feasible next century. Our alternatives are available now.

Greenpeace picked up the industry's gauntlet and in 1992 published a catalogue of not-in-kind alternatives to CFCs, which was frequently updated with lists of producers and users of non-fluorocarbon technologies in the refrigeration, air-conditioning, transport cooling, foam and solvent sectors. Later, Greenpeace organized conferences, published books and produced videos to promote successful examples of fluorocarbon-free cooling technologies. This information was distributed worldwide as a counterpoint to aggressively marketed HCFCs and HFCs.

The development of Greenfreeze

The most profound market intervention initiative of the Greenpeace Ozone Campaign was the development and the successful commercialization of the 'Greenfreeze' hydrocarbon technology in domestic refrigeration. In 1992 Greenpeace Germany initiated and funded the development of a hydrocarbon domestic refrigerator by bringing together scientists and engineers from Germany's Dortmund Research Institute with the fridge manufacturer DKK Scharfenstein. The team produced a prototype Greenfreeze refrigerator using isobutane for the refrigerant and cyclopentane for blowing the thermal insulation foam – a fluorocarbons-free product (no CFCs/HCFCs or HFCs).

In fact, the use of hydrocarbons in domestic refrigeration pre-dates the invention of the 'miracle' CFCs in the early 1930s. For example, in the 1930s there were 60 refrigerant brands in the US and 11 used isobutane as the refrigerant. Isobutane refrigerators had a poor safety reputation, however, because the refrigerant charge was very large (1.5 kg), because the available technology could not contain the refrigerant, and because the manufacturers of iceboxes and CFC refrigerators did their best to sensationalize any accident. Today's hydrocarbon refrigerators are charged with 30–70 grams of isobutane refrigerant contained in leak-free hermetically sealed compressor systems.

The major German refrigerator manufacturers at first rejected the new technology on the grounds that consumers would be reluctant to purchase refrigerators containing flammable refrigerants. Greenpeace responded by collecting 70,000 pre-sales commitments from the German public, confirming that consumers had a strong appetite for environmentally friendly refrigerators. In addition, the Greenfreeze technology earned the GS (Gepruefte Sicherheit) mark of safety by the German standards authority (Technischer Ueberwachungsverein – TUEV). Greenfreeze also received the German Blue Angel eco-label, a DEM1 million environmental prize from the German Finance Minister and the EU Eco-label for all Greenfreeze domestic fridges.

Within one year, this initiative resulted in the highly successful commercialization of hydrocarbon refrigeration in Germany. The technology then rapidly penetrated the markets of other European countries and made impressive inroads in Japan and many developing countries.

Greenpeace played a proactive role in technology transfer from Europe to China, Japan, Southeast Asia and Latin America.

- April 1993 marked Greenpeace's first visit to China. Greenpeace talked to industry and government officials managing the rapidly growing refrigerator market. After a second visit in the same year and intensive contacts with Chinese industry, the German Government announced its support of the hydrocarbon technology by allocating bilateral MLF moneys to the conversion to hydrocarbon technology. Greenpeace cooperated with China's biggest fridge producer, Kelon, in the production of Greenfreeze refrigerators. Greenpeace also facilitated the cooperation between Western appliance manufactures and Chinese refrigerator manufacturers. Swiss and German governmental agencies began to promote hydrocarbon refrigeration technology in India and Pakistan.
- In 1994 Greenpeace held technical seminars, private briefings and press conferences on hydrocarbon refrigeration in Mexico, Argentina and Brazil. The same year, in cooperation with the Centre for Science and Environment in Delhi, Greenpeace International conducted a lobbying and press tour of India (5–9 September) to promote Greenfreeze and to warn decision-makers against the World Bank's promotion of HCFCs and HFCs at a time when European manufacturers were embracing hydrocarbons.
- In 1997 Greenpeace initiated a three-way collaboration between Cuba, Germany and Greenpeace International to convert Cuba's one and only refrigerator factory (INPUD) from CFCs to hydrocarbons. The project was completed with GEF support in 1999.
- In 1998 Greenpeace Japan hosted a business seminar entitled 'HFC-free trend, the choice of German industry'. Sixty enterprises, representing all major refrigerator manufacturers and top convenience and food chain stores attended. In 1999 Greenpeace Japan submitted more than 10,000 consumer signatures to the President of Matsushita, and the Executive Director of Greenpeace International met the President of the Matsushita Refrigeration Company to demand an early launch of Matsushita's Greenfreeze models in Japan. At the same time Greenpeace held meetings with Matsushita representatives in several European countries to reiterate Greenpeace's expectations from the enterprise. On 28 December 1999 Matsushtia announced plans to launch Greenfreeze in Japan by the end of 2002. Since then all the major Japanese manufacturers have converted their production to Greenfreeze.

Today, in 2007, there are over 200 million Greenfreeze refrigerators in the world. Greenfreeze refrigerators are available in many sizes and a wide variety of models, including units with no-frost freezer compartments. There are over 100 different Greenfreeze models on the market. Greenfreeze technology now dominates the European market and most major Japanese and Chinese enterprises produce Greenfreeze refrigerators.

In 1997 Greenpeace received UNEP's 'Ozone Award' for making Greenfreeze technology freely available to the world. The success of Greenfreeze also changed how Greenpeace perceived its role in society. Providing practical solutions to environmental problems henceforth became an integral part of many other Greenpeace campaigns.

Refrigerants, Naturally!: Bringing 'Greenfreeze' to the corporate world of Coca-Cola, McDonald's and Unilever

The second phase of the Greenfreeze initiative, launched in 1993, was to convince the large commercial users of refrigeration and air-conditioning equipment to shift away from fluorocarbons. The campaign initially focused on European supermarkets. A number of supermarkets responded to Greenpeace's campaign by deciding to shift from fluorocarbon refrigeration and cooling to 'natural refrigerants' such as hydrocarbons, CO_2 and ammonia.

In 1998 the campaign turned its attention to the major corporate sponsors of the 2000 Sydney Olympics, the giant food and beverage enterprises such as Coca-Cola, McDonald's and Unilever, whose daily operations utilize millions of refrigeration units worldwide.

The campaign escalated in 2000 as the opening date of the Sydney Olympics rapidly approached. Greenpeace published a stinging report entitled *Green Olympics, Dirty Sponsors*.[30] The report criticized the major sponsors for violating the spirit and the Environmental Guidelines of the 'Green Olympics' through their use of HFCs and HCFCs in cooling equipment. The report was followed up with a high-profile internet campaign, and public protests aimed at Coca-Cola.

The campaign achieved its goals much sooner than expected. On 28 June Coca-Cola announced plans to globally phase out the use of hydrofluorocarbons (HFCs) and other fluorocarbons in refrigeration, wherever commercially and technically feasible, by the Athens Olympic Games in 2004. Coca-Cola also announced its intention to expand research into refrigeration alternatives. The Coca-Cola announcement was soon followed by similar commitments from McDonald's and Unilever.

These developments sent shock waves through the global refrigeration industry, rattled the confidence of the chemical enterprises in the future of their fluorocarbon refrigerants and blowing agents, and were acknowledged as being very significant by delegates in the corridors of the Montreal Protocol.

Greenpeace views these corporate commitments as the most enduring achievement of the Sydney 2000 Olympics. The organization now points to the commitments of these enterprises as positive examples of corporate environmental responsibility.

Since their public commitments in 2000 to phase out their reliance on fluorocarbons, the three enterprises have committed extensive resources to the development, testing and deployment of HFC-free cooling equipment. In 2004 Coca-Cola, McDonald's and Unilever, supported by UNEP and Greenpeace, launched the Refrigerants, Naturally! partnership. In 2006 the initiative was formally joined by Pepsi Cola, Carlsberg Beer and IKEA. Refrigerants, Naturally! promotes the development HFC/HCFC-free technologies for point of sale commercial refrigeration. Refrigerants, Naturally! is recognized by the UN Commission on Sustainable Development as a voluntary, multi-stakeholder 'Partnership for Sustainable Development' that contributes to the implementation of Agenda 21.

Parallel to Refrigerants, Naturally!, at the Refrigeration and Air-Conditioning Exhibition 2007 (Birmingham, UK, March 2007) six major supermarket chains (ASDA, Marks & Spencer, Sainsbury's, Somerfield, Tesco and Waitrose) announced their intention to phase out HFC-based cooling equipment and to invest in refrigeration systems based on natural refrigerants such as carbon dioxide or hydrocarbons. Refrigerants, Naturally! welcomed this announcement.

Today, McDonald's is focusing on eliminating HFCs in their air-conditioning, walk-in freezers and ice-cream coolers. Unilever has begun a large-scale roll out of hydrocarbon-based ice-cream freezers – by 2006 it had installed 100,000 units in Europe and currently it has plans to switch to the same technology in Brazil, Mexico and Asia.

Coca-Cola has developed a new technology for bottle coolers and vending machines, using carbon dioxide as the refrigerant. All of Coca-Cola's cooling equipment at showcase events, such as the Olympics and the World Cup are now HFC-free. HCFCs and HFCs have been eliminated from the insulation foam of all new Coca-Cola equipment, and the enterprise expects to enter into large-scale commercialization of its CO_2 equipment within the next three years.

SolarChill: Harnessing the power of the sun for human health

In 2000 the Greenfreeze campaign extended into the realm of human health technologies through the SolarChill Project. The aim of the SolarChill Project is to deliver environmentally sustainable, solar-powered and lead battery-free vaccine and food refrigeration to locations in the world without reliable electrical supply. Over 2.5 billion people live in such regions, where the delivery of immunization and public health programmes, as well as the preservation of food supplies, is challenging due to the absence of adequate cooling technologies.

SolarChill incorporates Greenfreeze technology with the added feature of harnessing solar energy to power the refrigerators. Solar energy powers a direct current compressor, which then runs the refrigerant cycle and creates an ice bank. The energy of the sun is thus stored in ice, instead of in lead batteries, and the ice, with the aid of a thermostat, then maintains the required temperatures in the cabinet. Thick insulation ensures that the unit can maintain the required temperatures for up to four days even without any electric power source. There are now two SolarChill models: a 50 litre vaccine cooler and a 100 litre food refrigerator.

Co-initiated by Greenpeace and UNEP, the SolarChill Project also includes the Danish Technological Institute (DTI), the German Government

Development Agency (GTZ ProKlima), Programs for Appropriate Technologies in Health (PATH), the United Nations Children's Fund (UNICEF) and the World Health Organization (WHO). During the six-year R&D phase, the SolarChill Project also involved the participation of industry, primarily the Danfoss and Vestfrost enterprises.

Greenpeace has coordinated the project since its inception. Funding for research and development was provided by Greenpeace Netherlands (through a sizeable grant from The Netherlands National Lottery), Greenpeace Germany, Greenpeace International, PATH and GTZ ProKlima. The development of the technology was overseen by the DTI and guided by the collective expertise of the Project Partners. Extensive field tests were conducted in Senegal, Indonesia and Cuba.

On 4 October 2006 at the Cooling Industry Awards Night at the Royal Lancaster Hotel in London, the SolarChill Vaccine Cooler and Refrigerator Project was the winner of the prestigious Cooling Industry Award under the Environmental Pioneer-Refrigeration category. On 1 November 2006 Dr A. P. J. Abdul Kalam, President of India, inaugurated the installation of two SolarChill Vaccine Coolers at Rashtrapati Bhavan (the President's residence). Dr Kalam thus became the very first purchaser of this breakthrough technology.

Today, SolarChill is receiving keen interest from around the world, and at least one major manufacturer is ready to begin commercialized production in 2007.

CONCLUSION

Greenfreeze, Refrigerants, Naturally! and SolarChill are practical responses to the duel atmospheric crises of ozone layer depletion and global warming. They are vivid examples of how it is possible to meet human needs with environmentally responsible technologies and of the old dictum 'where there is a will there is a way'.

Lessons[1]

INTRODUCTION

Technological innovation, market transformation and technology transfer are critical to the success of many global environmental agreements, including the Montreal Protocol on Substances that Deplete the Ozone Layer and the Kyoto Protocol to the United Nations Framework Convention on Climate Change. The implementation of the Montreal Protocol shows that regulatory regimes that set strict standards but allow flexibility in implementation can spur competition, promote innovation, and speed the commercialization of environmentally superior technologies. This speeds the phasing out of old chemicals and technologies that are harmful to the environment.[2] The Montreal Protocol experience also demonstrates that when regulatory regimes are designed to reflect the perspectives and experiences of both developed and developing countries and are tailored to local conditions, the transition to the new technologies is easier, more cost-effective, and more environmentally and economically beneficial than maintaining the status quo.

At first glance, and given that it regulates 96 different ozone-depleting substances (ODSs), it is easy to regard the Montreal Protocol simply as a chemical treaty. But it is more accurate and perceptive to regard the Montreal Protocol as an environmental technology treaty. The ODSs addressed in the treaty were used in thousands of products across nearly 250 sectors. A process of continuous technological innovation was required to develop both chemical substitutes for ODSs and non-chemical and not-in-kind substitutes, such as roll-on instead of aerosol deodorants. This process worked: 99 per cent of ODS production and consumption has been eliminated in developed countries, and by 2010 developing countries will have eliminated more than 95 per cent. One extraordinary measure of success is that enterprises and their customers are generally more satisfied with the replacements than with the old ODS-based products and technologies. This was a no-compromise market transformation: in addition to being ozone-safe, alternatives and substitutes are equally safe or safer for the Earth's climate, more energy-efficient, lower in toxicity, superior in safety, and more reliable and durable. More often than not, alternatives have reduced costs to businesses and increased employment. Corporate, military, environmental and citizen stakeholders are proud of what they have accomplished through consensus, cooperation and regulation.

This continuous process for technological innovation and diffusion has placed the ozone layer on the path to recovery. It also has done more to mitigate climate change than the Kyoto Protocol will do even if it is fully implemented, removing by 2010 between five and six times the CO_2 equivalent that the Kyoto Protocol will remove during its initial commitment period from 2008 to 2012.[3] Remarkably, the Montreal Protocol can deliver additional climate protection by accelerating the hydrochlorofluorocarbon (HCFC) phaseout while insisting alternatives and substitutes for HCFCs have the highest life-cycle climate performance[4] and by collecting and destroying ODSs contained in products.

How did the Montreal Protocol do all this? What are the secrets to its success? The answer starts with leadership of an astonishing variety of organizations, people from governments, international organizations, non-governmental organizations and industry associations, scientists, engineers, and many others who took early action to tackle the problem of ozone depletion and inspired others to follow their example. Institutional arrangements also were important, including the Ozone Units – the focal points for action on the Montreal Protocol in each of its now 191 Parties. The Ozone Unit focal points are dedicated groups of professionals who understand their mission and who have formed both formal and informal networks with the broader ozone community to carry it out. Financial assistance to developing countries was another critical part of the institutional arrangements, as was the Montreal Protocol's visionary technology assessment process, including the Technology and Economic Assessment Panel (TEAP). Another factor in the Protocol's success was the ability of the regime to identify and remove barriers to technology transfer, including through changes in national laws and voluntary codes.

The factors that contributed to the success of the Montreal Protocol thus offer lessons and insights that might be useful to the future operation of the climate change treaties.

The Montreal Protocol: Then and now

When the Montreal Protocol was negotiated in the mid-1980s, the Parties faced an unprecedented environmental problem that was global and plagued with scientific, technical and economic uncertainty, especially with regard to ODS substitutes, alternatives and other needed technological innovations.

Making the transition from ODSs was expected to be very costly. Some CFC industry stakeholders and even some scientists argued that there was no need for an immediate response, saying that the threat CFCs posed to the ozone layer was only a hypothesis based on chemical reactions in a laboratory. They claimed that more time was needed to iron out the science and that any action risked the hundreds of thousands of jobs and billions of dollars of income that depended on CFCs. Fortunately, negotiations on ODS controls continued.

Negotiations on the Montreal Protocol concluded on 16 September 1987, and it entered into force on 1 January 1989 after the requisite 20 Parties representing a majority of production ratified the treaty. It imposed its first control measures on developed country Parties that year, but would not impose control measures on developing country Parties until ten years later in recognition of the principle of 'common but differentiated responsibilities'. The justification for allowing developing countries to phase out later was that developed countries had invented, profited from, and excessively used and emitted ODSs,[5] and that the costs of reductions in developing countries would be lower once new technology was widely available and benefiting from economies of scale. There was also an appreciation that many developing countries lacked the financial and technical capacity to make the transition to non-ODS products and technologies.

Today, there are 191 Parties to the Montreal Protocol preparing to eliminate the remaining production of ODSs. With few exceptions, Parties to the Montreal Protocol have managed to comply with its control measures for ODS production and consumption and have helped place the ozone layer on a path to recovery.

The 2006 United Nations Environment Programme (UNEP)/World Meteorological Organization (WMO) 'Scientific Assessment Panel report', which synthesizes all the scientific observations of the ozone layer from satellites, balloons and terrestrial instruments, described the Montreal Protocol's success:[6]

> *The total combined abundances of anthropogenic ozone-depleting gases in the troposphere continue to decline from the peak values reached in the 1992–1994 time period. [...]*

> *The combined stratospheric abundances of the ozone-depleting gases show a downward trend from their peak values of the late 1990s, which is consistent with surface observations of these gases and a time lag for transport to the stratosphere. [...]*

> *The Montreal Protocol is working: there is clear evidence of a decrease in the atmospheric burden of ozone-depleting substances and some early signs of stratospheric ozone recovery.*

Economic analyses of the Montreal Protocol's control measures have found that the speed of the phaseout has been faster, the costs have been lower, and the alternatives and substitutes have been more environmentally acceptable than the Parties anticipated during the Protocol's negotiations.[7]

Overall, the success of the Montreal Protocol has been due to the process of continuous technological innovation and diffusion. A significant part of the technology change achieved by the Protocol came through voluntary transfer of technologies, processes and best practices through multinational enterprises, industrial and professional associations, and NGOs in both developed and developing countries. This process is documented in Chapters 4–11. The other part of the technology change came through the Protocol's financial mechanism, which covers the incremental costs of technology transfer for developing country Parties.

The implementation of the Protocol and the ozone layer recovery are by no means complete. There is much left to be done. However, the Parties to the Protocol have the system in place to meet the remaining challenges confronting the Montreal Protocol and the ozone layer. Meanwhile, the experience of the Montreal Protocol to date offers many important lessons on how to facilitate technology transfer and change that are applicable to the United Nations Framework Convention on Climate Change and its Kyoto Protocol.

Technology transfer under the UNFCCC and the Kyoto Protocol: What can be learned from the ozone experience?

The ultimate success of the UN Framework Convention on Climate Change (UNFCCC) and the Kyoto Protocol in mitigating climate change will depend on continuous technological innovation and the rapid and widespread transfer and diffusion of technologies, including 'know-how'. This technological innovation and diffusion must occur rapidly and continue until the objectives of the climate change treaties are achieved. Both evolutionary and revolutionary technology changes must occur worldwide.

At their first meeting in 1995, the Parties to the UNFCCC requested the Intergovernmental Panel on Climate Change (IPCC) to include in its assessments an elaboration of the terms under which transfer of environmentally sound technologies and know-how could take place.[8] The result was the IPCC special report *Methodological and Technological Issues in Technology Transfer*, completed in 2000.

Not surprisingly, the barriers to technology transfer identified in the IPCC report were similar to the barriers encountered under the Montreal Protocol. These barriers include poor regulatory motivation, lack of appropriate policies, slow and ineffective systems of governance, financial sector weakness and inhibited competition, business uncertainty, lack of consumer awareness and environmental market demand, imperfect information, inhibited access to technology, and inadequate incremental financing for environmental goals. (See Chapter 12 for a detailed discussion on barriers to technology transfer encountered under the Montreal Protocol.)

Unfortunately, in its first 15 years, from 1992 to 2007, the UNFCCC has not achieved the same level of success as the Montreal Protocol. One of the latest reports of the IPCC shows how little has been achieved:

- *the global atmospheric concentration of carbon dioxide has increased from the pre-industrial value of about 280 parts per million (ppm) to about 379ppm in 2005;*
- *the atmospheric concentration of carbon dioxide in 2005 exceeds by far the natural range over the last 650,000 years (180 to 300 ppm), as determined from ice cores;*
- *the annual carbon dioxide concentration growth rate was greater during the last ten years (1995–2005 average increase of 1.9ppm) than ever before since measurements began (1960–2005 average increase of 1.4ppm); and*
- *the annual fossil carbon dioxide emissions increased from an average of 6.4 gigatonnes of carbon (Gt C) per year in the 1990s to 7.2 Gt C in 2000–2005.*[9]

Clearly, the Parties to the UNFCCC had not made sufficient progress over the last 15 years to remedy, to any extent, the causes of climate change.

At the Delhi Sustainable Development Summit in 2007, UNFCCC Executive Secretary Yvo de Boer reported that the root cause of inaction was an unwarranted fear of economic hardship:

> *Industrialized countries fear unwillingness on the part of their developing country competitors to act and are therefore reluctant to take the first step themselves. Developing countries fear that a new round of climate negotiations would impose on them obligations that would hurt their economic goals. [...T]he key to the climate change problem is to provide incentives for economies to grow along a greener path and to put in place mechanisms to ensure that resources required to green the growth are available.*

The experience of the Montreal Protocol in protecting the ozone layer has been quite successful in addressing this type of problem by including developing countries in the control measures and deeply integrating them into the treaty's administrative processes at all levels. Thus the climate change treaties might experience similar success through voluntary country programmes as described under Article 10 of the Kyoto Protocol (and explained below).

Since nearly all of the Parties to the Montreal Protocol are also Parties to the UNFCCC, and since most are Parties to the Kyoto Protocol, there is the potential for fruitful dialogue both within and among governments to break down barriers to technology transfer and clear the way for successful multilateral actions on the important issue of climate change. This can start by looking at the key lessons from the success of the Montreal Protocol and how they might be applied to the climate change treaties.

Leadership

Success is not accidental, and leadership is always important. This is especially true in the early phases of problem-solving, when there is uncertainty about the pace and cost of the technological innovation and diffusion required.

Leadership under the Montreal Protocol was exercised by UNEP and key Parties, including their military organizations and enterprises, to create awareness regarding the depletion of the ozone layer, to take early action in phasing out ODSs, to develop new, ozone-friendly technologies, and to diffuse the environmentally superior technologies throughout the world.

LESSON 1: ACT NOW

> *The world can only be grasped by action, not by contemplation – the hand is the cutting edge of the mind.* (Jacob Bronowski)[10]

Taking early action to protect the environment generally results in lower costs and greater benefits compared with postponing action to some future period.

One of the best ways to lead is through market actions that set the example and set the pace.

Scientists Mario J. Molina and F. Sherwood Rowland hypothesized the link between CFCs and ozone depletion in their 1974 paper published in *Nature*. The Governing Council of UNEP responded early in 1976, and, on its instructions, UNEP organized a meeting of experts from many countries in Washington, DC, in 1977. In 1978 UNEP initiated the diplomatic process of intergovernmental meetings to negotiate a solution to the problem of ozone depletion, even though the chemical industry and some countries were sceptical of the link with CFCs. UNEP took seriously the warnings of scientists and invoked the 'precautionary principle' to avoid irreversible damage to the environment.

Many citizens and governments, particularly of the US and some Scandinavian countries, also immediately responded with product boycotts and eventual bans on use of CFCs in cosmetic aerosol products. These actions alone virtually halted the growth in ODS production and emissions for the decade that was required to organize an international response.

Despite continued scientific uncertainty, the international community negotiated first the Vienna Convention for the Protection of the Ozone Layer in 1985 and two years later the Montreal Protocol, with its agreed specific control schedules for the production and consumption of CFCs and halons. The Parties have since strengthened the Montreal Protocol several times by adding additional ODSs and accelerating ODS phaseout schedules, based on continuous technological innovation and continuous assessment of technical and economic feasibility.

The initial control measures under the Montreal Protocol involved a 50 per cent reduction in consumption of five CFCs and a freeze in consumption of three halons by 2000, as most countries felt that more would not be technically or economically feasible.

Despite the fact that only 25 Parties, including just four developing countries, had ratified the Montreal Protocol by 1988, UNEP believed from the beginning that every country was committed to protecting the ozone layer and to taking immediate action to reduce consumption of ODSs. Even before the Protocol entered into force on 1 January 1989, UNEP invited all the countries in the world, along with scientific and technical experts and industry representatives, to a 1988 workshop on substitutes and alternatives to CFCs and halons in The Hague. Countries and experts gave presentations on ODS substitutes and alternatives that were either already available or under development, demonstrating that immediate action for reducing ODS consumption was feasible. This meeting, anticipating the Montreal Protocol's ratification at the first Meeting of the Parties in 1989, resulted in the informal establishment of scientific, environmental, technical and economic assessment panels and Technical Options Committees (TOCs) for all sectors of ODS use, which would later be formally organized under Article 6 of the Protocol. The purpose of the assessment panels and the TOCs is to advise the Parties on strengthening the Protocol. In addition, the establishment of the TOCs resulted in the added benefit of starting all countries on the path to implementing the ODS alternatives readily available, such as non-ODS replacements for aerosols.

Before the Montreal Protocol entered into force, many countries and enterprises took early action to eliminate their ODS consumption, eagerly embracing the business opportunity to develop and introduce new products and technologies to their markets. Consider, for example, the following actions by countries and enterprises in 1988, a full year before the Montreal Protocol entered into force:

- AT&T announced it had developed an alternative to CFC-113, used as a solvent, at the first international conference on alternatives to CFCs and halons in January 1988, which drew more than 1000 participants;
- Brazil banned CFCs in household and cosmetic products;
- Norway introduced an aggressive plan to phase out 50 per cent of its CFC consumption by 1991 and 90–100 per cent by 1995;
- Sweden set timetables for product bans and an escalating fee schedule for the use of CFCs;
- DuPont and Pennwalt announced that CFC production should halt as soon as practical;
- the US EPA, with industry groups and NGOs, announced a sector-wide phaseout of CFCs in food packaging; and
- Nortel and Seiko Epson committed to an accelerated phaseout of CFC-113.

Furthermore, the US had already achieved dramatic reductions in ODS consumption in the 1970s, more than a decade before the Montreal Protocol, by prohibiting cosmetic and convenience aerosol products.

The Montreal Protocol of 1987 was but a first small step, requiring only a 50 per cent reduction in consumption of five CFCs and a freeze on consumption of three halons by 2000. It was the early steps for reduction mentioned above, coupled with the warnings of scientists that the steps were not adequate and the assurance of the TEAP that much more was feasible, that gave confidence to the Parties to strengthen the Protocol through adjustments and amendments and take steps to fully phase out 96 ODSs much earlier (see Appendix 1 for the control measures).

Throughout, the Parties found that the costs of phaseout were lower than anticipated and often provided unexpected benefits in the form of superior products, greater productivity and higher profits for both developed and developing countries.

The bold actions by many countries and enterprises were based on a desire to protect the ozone layer. The Montreal Protocol was but an expression of this desire. Looking back, the success of the Montreal Protocol now makes it seem predestined. But the boldness of the ozone pioneers offers a key lesson for climate campaigns, which might be illustrated in the passage attributed to Goethe (as elaborated by William H. Murray):

> *Until one is committed, there is hesitancy, the chance to draw back, always ineffectiveness [...] the moment one definitely commits oneself, then Providence moves too. All sorts of things occur to help one that would never otherwise have occurred. A whole stream of events issues from the decision, raising in*

one's favour all manner of unforeseen incidents and meetings and material assistance, which no man could have dreamed would have come his way.[11]

Pointers for the climate change treaties

Action to reduce greenhouse gases (GHGs) could begin immediately, even while waiting for the perfect legal solution to the climate change issues. Like action under the Montreal Protocol, countries would see many benefits, would get over their imaginary fears of difficulties and would create the momentum for faster action. Many countries and enterprises have demonstrated commitments to early action on climate change, pledging to reduce GHG emissions voluntarily and to prioritize the use of more energy-efficient products and equipment.[12] The climate change treaties could make these activities the focal point of their efforts to protect the climate.

LESSON 2: DEVELOP VISIONARY TECHNOLOGY ASSESSMENT

Advice by objective, independent and visionary bodies of experts on a regular basis on the latest developments can ensure that alternatives and technologies that meet a treaty's goals are identified, and capacity to develop, adapt and transfer technology for such alternatives and technologies is enhanced.

The Parties to the Montreal Protocol benefit from annual, up-to-date technical assessments from its Technology and Economic Assessment Panel (TEAP) and its Technical Options Committees (TOCs) for the six sectors of ODS use. The TOC reports are consolidated by the TEAP, and the results synthesized with findings of the Scientific Assessment Panel and the Environmental Effects Assessment Panel. The Scientific Assessment Panel assesses the state of the ozone layer and the levels of ODSs in the atmosphere, and the Environmental Effects Assessment Panel assesses the environmental impacts of ozone depletion. They issue reports every four years, which are then condensed into a synthesis report, along with the findings of the TEAP. The latest synthesis report is made available on the website of the Ozone Secretariat.[13]

The TEAP consists of the co-chairs of the TOCs and a few other experts. It has three co-chairs. Each TOC has co-chairs from both developing and developed countries and 20–35 members from all parts of the world. Members of the TEAP are appointed by Meetings of the Parties (MOPs). Governments may propose members for TOCs. TOC memberships consist of representatives of government environment ministries, industry and academia and a few professional consultants. The co-chairs of the TOCs have full freedom to choose whom they want as members in consultation with the TEAP; their choices will be based on the expertise needed, which may vary over from time.

Because of the industry presence on the TOCs and the TEAP, there was concern that industry participants would understate technical feasibility of alternatives or try to use the reports to promote specific products. But industry representatives generally serve on TOCs and in the TEAP alongside representatives from competitors and knowledgeable academics, which creates a critical check against undue influence. Most important, in the many years of practice since 1988, there has never been a significant misleading or misstated finding on technology. The point has always been to provide the Parties with up-to-date information on what is working.

In addition, a code of conduct incorporated in the terms of reference, approved by the MOPs in decision VIII/19 in 1996, seeks to avoid conflicts of interest and resolve any that occur. This code of conduct was further elaborated by the 18th MOP in 2006. Each member serves in his or her individual capacity and pledges not to promote the interests of his or her company, government or employer. All members of the TEAP and TOCs declare their source of funding for participation and any financial interest they may have in companies connected with ODSs or alternatives. The travel costs of members from developing countries or countries with economies in transition (CEITs) are met by the Ozone Secretariat. Some members give their time for free while some are paid by their employers or others.

Indeed, the presence of industry on the TOCs and the TEAP is a distinguishing feature that provides access to cutting-edge data often not yet published in scientific or technical journals, since industry rarely publishes on emerging technologies it has developed for commercial purposes. As a result, reports from the TOCs and the TEAP often provide the Parties with the first public disclosure of the latest developments in new environmentally friendly technologies. This greatly contrasts with the approach of the IPCC, which relies on what has already been published.

While the TEAP and TOCs, along with the Scientific and Environmental Effects Assessment Panels, were originally constituted under Article 6 of the Protocol to advise the Parties at least once every four years on strengthening the Protocol, the MOPs have actually used the TEAP and TOCs to spearhead more aggressive phaseout and to solve the many problems faced by the Parties. For example, every three years the TEAP and the TOCs are asked to review the replenishment requirements of the Multilateral Fund (MLF). They have also established and defined process agent, essential use and critical use exemptions for ODS use as well as set limits on ODS emissions to allow some continued use while keeping adverse impacts on the ozone layer to a minimum.

The reports of the TEAP or TOCs are presented to the Parties as they are written, without any editing by policymakers. Parties are free to disagree with the reports, but cannot amend them. The Panels can present information that is relevant for policymaking, but do not issue specific policy recommendations.

Technical assessment under the climate change treaties

The UNFCCC established a Subsidiary Body for Scientific and Technological Advice (SUBSTA) as its key technical advisory body. SUBSTA is open to participation by all Parties and generally consists of government representatives with expertise in relevant fields. It is guided by the Parties to the UNFCCC and reports regularly to the Conferences of the Parties. Overall, its objectives are similar to those of the Montreal Protocol's assessment panels.

Table 14.1 *Comparison of the technical assessment bodies for the Montreal Protocol and the UNFCCC*

	TEAP/TOCs	UNFCCC/SUBSTA	IPCC
Members	TOC members selected or rejected by TEAP and TOC co-chairs, even if proposed by governments; TEAP selected by governments	Selected by governments	Selected by governments
Report approval	By consensus of TEAP and TOC members, right of signed minority report rarely exercised; no government review	Governments review, edit and approve	Governments review, edit and approve
Frequency	Yearly progress and topical reports; frequent special assignments; full technical assessment every four years; MLF replenishment study every three years	Variable	Once in about five years

SUBSTA relies heavily on the reports by the expert panels organized by the IPCC. It bases its assessments mainly on published and peer-reviewed scientific and technical literature and operates out of three Working Groups and a Task Force.

The IPCC reports have played a key role in providing technical data for negotiations of the climate change treaties. The first IPCC report was published in 1990 and was relied on during the negotiation of the UNFCCC in 1992. The second IPCC report was published in 1995 and helped spur the negotiation and adoption of the Kyoto Protocol in 1997. The third report was published in 2001 and the fourth will be published in 2007.

Unlike the reports from the TOCs and the TEAP under the Montreal Protocol, the IPCC and SUBSTA reports are limited to information that has already been published and peer-reviewed. This may be acceptable and even desirable for assessments of atmospheric or environmental effects science, but it is entirely unacceptable and counterproductive in the assessment of technology. The Parties to the UNFCCC do not have access to cutting-edge technological developments that have not yet been reported in the scientific or technical literature, which stifles their ability to understand and promote technological development and transfer. Worse still, industry participants in the IPCC assessments are rewarded only with the satisfaction of accurately describing the published literature while participants in the TEAP assessments are rewarded with having early access to, and influence over, the cutting-edge technology that they will implement in their own enterprises.

Pointers for the climate change treaties

The UNFCCC and Kyoto Protocol could create committees of experts like the TEAP/TOCs to identify and report, on an annual basis, feasible technologies and best practices from throughout the world that can be implemented by the Parties. TOCs could be established for each of the sectors identified by SUBSTA, with a mandate to submit reports directly to the MOPs every year. The selection of experts, the process of their functioning and their reports could be made independent of approval by governments to ensure objectivity, as in the Montreal Protocol. The IPCC could continue its five-year report cycle on science and impacts, but the reports would be more credible and useful if governments were not allowed to edit them prior to their publication.

This objective and forward-looking approach would provide enterprises in developed and particularly developing countries with up-to-date information on technologies that are available, or soon to be available, and objective assessments of their technical and economic feasibility.

LESSON 3: ENCOURAGE LEADERSHIP BY MULTINATIONAL AND DOMESTIC ENTERPRISES

Successful multilateral environmental agreements benefit from voluntary commitments by multinational enterprises to meet or exceed regulatory standards in all the countries they operate in ahead of phaseout dates or other deadlines.

Both before and after the Montreal Protocol was negotiated, many multinational enterprises took leadership roles in phasing out CFCs or other ODSs and in introducing alternative technologies that were ozone-friendly. Although many of these multinational enterprises were based in developed countries, they carried out their leadership commitments in the developing countries in which they operated, which played an important role in technology transfer (see Chapters 4, 7, 9 and 10 for details).

UNEP, NGOs and other organizations could organize pledges by industry to go beyond their commitments under the Kyoto Protocol and to provide advice and assistance to other enterprises (see Chapter 13).

Pointers for the climate change treaties

Many major corporations are already convinced of the inevitability of action on climate change,[14] and a growing number of companies believe that inaction is no longer a viable option. All companies will be affected to some degree, and all have an obligation to examine the issue and take prudent action.

There are many enterprises in the developed countries that have already achieved considerable gains in reducing GHGs. These enterprises could be persuaded to introduce the climate-friendly options in all countries where they operate and to mentor enterprises in developing countries on the best practices to reduce GHGs.

Human and institutional capital

Protecting the ozone layer drew together individuals and organizations, creating an epistemic community that has transcended the bare requirements of the Montreal Protocol, vigorously exchanged information and experiences, and promoted the enthusiastic participation of many Parties, technical experts, NGOs and enterprises. The success of the Montreal Protocol reflects the efforts of the many individuals and organizations working at all levels to empower the Protocol's institutions and implement its control measures.

LESSON 4: IDENTIFY AND INVOLVE ALL STAKEHOLDERS, AND DEVELOP LOCAL AND INTERNATIONAL PARTNERSHIPS

Many small drops make a mighty flood. (Tamil proverb)

Partnerships, synergies and other forms of cooperation among all stakeholders can be instrumental in ensuring effective implementation and compliance. Stakeholders include officials at international and regional UN institutions and implementing agencies, government officials at the national and local levels, scientific and technical experts and consultants, non-governmental organizations, multinational and small and medium-sized enterprises, consumer groups, and industry groups and associations.

National partners

Nearly every Party to the Montreal Protocol has engaged in some form of stakeholder dialogue and collaboration, and most operate national steering committees comprising representatives of government ministries (for example agriculture, defence, environment, finance and industry) and industry associations, technical experts, NGOs, and others such as international implementing organizations or bilateral donor agencies. These efforts extend far beyond working within the TOCs and the TEAP and include, for example, partnerships with doctors and other health-care professionals and associations on the need for a transition from CFC-based metered-dose inhalers, and with farming associations on alternatives to methyl bromide (see Chapters 11 and 13 for further details).

Military leadership

The Montreal Protocol is unique in having harnessed the cooperation of the military in many countries to take the lead in phasing out ODSs. Military organizations realized early that their preparedness to serve their countries could be affected by the phaseout of ODSs and rushed to develop and implement alternatives. The military itself discovered and developed many of these themselves. In some cases, the military changed to ozone-friendly procurement specifications for supplies and forced suppliers to develop alternatives. In addition, they rendered great public service by sharing their techniques with other countries (see Chapter 5 for further details).

International-level partners

Stakeholder groups at the international level involved in technology transfer include the Alternative Fluorocarbons Environmental Acceptability Study (AFEAS), the American Refrigeration Institute (ARI), the American Society of Heating, Refrigerating and Air Conditioning Engineers (ASHRAE), the Halons Alternative Research Corporation (HARC), the Halon Users National Consortium (HUNC), the Industry Cooperative for Ozone Layer Protection (ICOLP), the Japan Industrial Conference for Ozone Layer Protection (JICOP), the Mobile Air Conditioning Society (MACS), the Programme for Alternative Fluorocarbon Toxicity Testing (PAFT), the International Institute of Refrigeration (IIR), and military organizations in many countries, including Australia, Canada, Sweden, the UK and the US (see Chapters 4, 5, 7, 9 and 10 for further details).

The MLF, the Global Environmental Facility (GEF), and the implementing agencies of the United Nations Development Programme (UNDP), UNEP, the United Nations Industrial Development Organization (UNIDO) and the World Bank were fully involved in implementing the Protocol. In addition, the WMO has assumed the leadership since the 1970s for finalizing science assessments.

Pointers for the climate change treaties

Climate change has already attracted many stakeholders and their organizations, as evidenced by the thousands of individuals and organizations that turn out as observers at the Conferences of the Parties to the UNFCCC and Kyoto Protocol. These include representatives from industries involved in energy supply, energy conservation, transportation, buildings, agriculture, forestry, waste management and chemicals, as well as consumer associations, academics and NGOs from all parts of the world that deal not only with climate change but also with many other environmental issues affected by climate change. Most of these stakeholders are already convinced of the great dangers of climate change and knowledgeable about how to mitigate the climate change to the extent possible and adapt to any inevitable climate change. Their involvement, however, has so far been at the level of exchange of ideas and has not been translated to action and leadership at the local level.

Military organizations are aware of the strategic dangers of climate change, and in some developed countries they are already partners with their governments in combating it. But military organizations have yet to take a leadership role on climate change on a global scale comparable with their contribution to the phaseout of ODSs (see Chapter 5).

Spurring these groups into action, both at the domestic level and at the international level, would lead to many 'water drops' of mitigating the causes of climate change or adapting to its impacts that combined could become a mighty flood of achievement, as was demonstrated by the Montreal Protocol.

Lesson 5: Raise Awareness

Human history becomes more and more a race between education and catastrophe. (H. G. Wells)[15]

One of the most important first steps in successfully implementing an environmental treaty is to inform all stakeholders about the causes and impacts of the environmental problem; the international and domestic regulatory efforts to address it; and the environmentally friendly products, processes and practices that are either available as alternatives or are or should be under development. Stakeholders also benefit from information about the feasibility of switching to alternatives, the mandated time frame and the accessibility of institutions capable of providing financial and technical assistance.

Under the Montreal Protocol, education and outreach programmes, such as those of UNEP's OzonAction Unit and the US EPA's Stratospheric Protection Division, motivated industry and consumer groups to promote ozone solutions through education programmes, voluntary agreements and other initiatives.

For example, in 1986 and 1987 children and grass-roots organizations in the US successfully campaigned for McDonald's to stop using CFC packaging for its fast food products. The EPA's Stratospheric Protection Division, Friends of the Earth-USA, the Center for Global Change, the Natural Resources Defense Council (NRDC), the Foodservice and Packaging Institute (FPI) and the Environmental Defense Fund (EDF, now Environmental Defense) reached agreement with the food-packaging industry to voluntarily phase out CFCs in the manufacture of foam packaging. The foam manufacturers agreed to end the use of CFCs within one year (by December 1988) and agreed to use HCFC-22 only as an interim alternative.

UNEP's OzonAction and numerous countries and enterprises took other steps to raise awareness, including:

- convening seminars and workshops, including exhibitions showcasing alternative technologies;
- recognizing and promoting environmentally friendly government and private enterprise efforts;
- encouraging industry pledges to adopt alternatives and substitutes;
- giving awards to recognize outstanding contributions from both public and private sectors;
- distributing environment-friendly messages via banners, posters, bags, key fobs, clocks, stationery, car stickers and so on;
- holding competitions and field activities;
- celebrating the International Day for the Protection of the Ozone Layer (16 September, the day in 1987 when the Montreal Protocol was signed); and
- disseminating information through all media outlets, including the internet, television, radio, books, magazines, newspapers and newsletters, and using public figures to promote key messages.

OzonAction also prepared many publications and technical resource materials based on TEAP reports to familiarize the industries of developing countries with the advantages and disadvantages of the different technologies. In the case of foams, where there are many alternatives, the MLF funded projects demonstrating different technologies (see Chapter 13 for further details).

Pointers for the climate change treaties

All of humanity is a stakeholder when it comes to climate change. Thus everyone needs to be aware of their power to change the dismal climate scenarios for the better. Climate change will impact on every aspect of human existence – weather patterns; food and natural resource supplies; the spread of diseases among humans, animals and plants; and even the very geography of the planet, with the potential for rising sea levels that will flood coastal areas and redraw the world's maps. Adequate awareness of the issue and of the means to mitigate or adapt should spur all stakeholders to take action in their own interests.

LESSON 6: REQUIRE COUNTRY PROGRAMMES FROM EACH DEVELOPING COUNTRY, WITH SPECIFIC VOLUNTARY GOALS TOWARDS GREEN GROWTH

Country programmes are implementation road maps: they set short- and long-term goals based on a country-specific assessment of regulatory, institutional, technological and financial capacity. Country programmes also develop an approach to ensure compliance that is consistent with each country's domestic policy priorities, including economic development. Setting specific voluntary or mandatory goals tailored to individual countries or industry sectors is an essential ingredient for success.

Under the Montreal Protocol, country programmes for developing countries and CEITs were developed and implemented with technical and financial assistance from the MLF and the GEF and based on some of the experiences and strategies deployed by developed countries in their country programmes. The programmes were periodically evaluated and updated by countries to reflect current conditions. Many countries often needed as much assistance in developing country plans as they did in implementing them.

One benefit from country programmes is that many countries have found it in their interests to adopt ODS phaseout schedules that are more aggressive than those mandated by the Montreal Protocol. Mexican enterprises, for example, profited from their accelerated phaseout of CFC aerosol products; Colombian flower growers took pride in being first to phase out methyl bromide for soil fumigation for flowers marketed worldwide; and India demonstrated leadership in halting the use of halons in portable fire extinguishers. Overall, the actual levels of ODS production and consumption in developing countries have been lower than permitted levels in every year since the Montreal Protocol entered into force, in part due to the country programmes (see Chapters 8, 9 and 11).

The UN's Millennium Development Goals provide another, broader example of a process that sets ambitious goals and provides financial and technical assistance to help achieve them. The goals are:

- reducing by half the proportion of people living on less than one US dollar a day;
- eliminating gender disparity in primary and secondary education preferably by 2005 and at all levels by 2015;
- reducing the mortality rate among children under five by two-thirds; and
- achieving significant improvement in the lives of at least 100 million slum dwellers by 2020.[16]

Possible action by UNFCCC and the Kyoto Protocol

Although the US is currently the largest emitter of GHGs worldwide, it has not ratified the Kyoto Protocol, choosing instead to use voluntary programmes and research incentives to reduce its GHG emissions. One reason given for this strategy is that any reduction would be wiped out by the growing emissions of the fast-growing economies of developing countries such as Brazil, India and China. On the other hand, some developing countries argue that they are still mired in poverty and that large-scale GHG emissions reductions are not feasible during rapid economic growth. In the last 15 years, the arguments have endlessly and unproductively gone back and forth while the emissions continue to grow. The real scope for a universal push for a greener growth path has not been exploited by the UNFCCC since the Conferences of the Parties to UNFCCC have been busy debating the legal issues.

Article 10 of the Kyoto Protocol mandates every Party to:

> *formulate, implement, publish and regularly update national and, where appropriate, regional programmes containing measures to mitigate climate change and measures to facilitate adequate adaptation to climate change.*

The developed countries that have ratified the Kyoto Protocol have mandatory goals and have announced their national or regional programmes to attain the goals. The developed countries that have not ratified the Kyoto Protocol, such as the US and Australia, and the developing countries, which do not have mandatory goals, do not have national programmes that spell out how they will mitigate climate change, as implied by their ratification of the UNFCCC. That does not mean that they are doing nothing. They have simply not committed themselves publicly to a programme to reduce their GHG. The developing countries that ratified the Kyoto Protocol are obliged to prepare such a programme. Given the success of country programmes under the Montreal Protocol to focus on reduction of consumption of ODS, it stands to reason that climate-oriented country programmes specifically suited to each developing country's needs would reduce the growth rate of the GHG emissions. Setting voluntary goals through country programmes would enable donor countries and GEF to focus their programmes to assist the developing countries and such assistance would enthuse the developing countries to adopt the goals.

This is especially true because reducing GHG emissions offers even greater collateral benefits than ozone layer protection. For example, implementing the Montreal Protocol sometimes required replacing chemicals and applications that were quite profitable for their manufacturers with more costly substitutes. Indeed, the transition from CFCs to ozone-friendly substitutes in developing countries may not have been possible without financial assistance to cover the incremental costs of the phaseout. On the other hand, protecting the climate is profitable when it involves the transition to more energy-efficient equipment, and in many cases manufacturers and consumers both benefit. Thanks to the rising oil prices, almost all countries have realized that better transport systems, energy conservation and alternative sources of energy to fossil fuels could lead to less dependence on oil imports – with their uncertain and growing costs – and result in greener and more steady economic development. Other collateral benefits include less pollution, less congestion on roads and improved quality of life.

But despite these benefits, when it comes to addressing climate change the world has not seen the kind of leadership asserted by countries to protect the ozone layer. This failure to recognize and capture opportunities with such potentially large economic, environmental and other benefits is puzzling.

Technical options to mitigate climate change and their costs

Parties to the UNFCCC could base goals and possibly country plans on the numerous options put forward by the IPCC, the International Energy Agency, the Stern Report and others that have identified low-cost and no-cost alternatives. Consider the following excerpt from the IPCC's 2001 report on mitigating climate change:

> *The technological and economic potential to reduce GHG emissions is large enough to hold annual global GHG emissions to levels close to or even below those of 2000 by 2010 and even lower by 2020. Realization of these reductions requires combined actions in all sectors of the economy, including adoption of energy-efficient technologies and practices, increased fuel switching toward lower carbon fuels, continued growth in the use of efficient gas turbines and combined heat and power systems, greater reliance on renewable energy sources, reduced methane emissions through improved farm management practices and ruminant methane reduction strategies, diversification of land use to provide sinks and offsets, increased recovery of landfill methane for electricity production and increased recycling, reduction in the release of industrial gases, more efficient vehicles, physical sequestration of CO_2, and improving end-use efficiency.*

> *Some of the costs associated with sector-specific options for reducing GHG emissions may appear high (for example US\$300/tCeq). However, we estimate that there is technological potential for reductions of between 1900 and 2600 MtCeq/yr by 2010 and 3600 to 5050 MtCeq/yr by 2020. Half of these reductions are achievable at net negative costs (value of energy saved is greater than capital, operating and maintenance costs), and most of the remainder are available at a cost of less than US\$100tCeq/yr. The*

continued development and adoption of a wide range of GHG mitigation technologies and practices will result not only in a large technical and economic potential for reducing GHG emissions but will also provide continued means for pursuing sustainable development goals.[17] *[...]*

Certain energy-efficient, renewable and distributed energy options offer non-energy benefits. One class of such benefits accrues at the national level, for example via improved competitiveness, energy security, job creation and environmental protection, while another relates to consumers and their decision-making processes. From a consumer perspective, it is often the non-energy benefits that motivate decisions to adopt such technologies. Consumer benefits from energy-efficient technologies can be grouped into the following categories: 1) improved indoor environment, comfort, health, safety and productivity; 2) reduced noise; 3) labour and time savings; 4) improved process control; 5) increased reliability, amenity or convenience; 6) water savings and waste minimization; and 7) direct and indirect economic benefits from downsizing or elimination of equipment. Such benefits have been observed in all end-use sectors. For renewable and distributed energy technologies, the non-energy benefits stem primarily from reduced risk of business interruption during and after natural disasters, grid system failures or other adverse events in the electric power grid.[18]

Pointers for the climate change treaties

Setting goals and then providing appropriate levels of technical and financial assistance to developing countries can signal to both governments and industry that complying with environmental standards is in their best interests and that the benefits of compliance outweigh the costs.

It is significant that many of the industrial enterprises already working to avoid climate change have a large presence in the US: they have arrived at their conclusions in spite of the fact that countries like the US have not ratified the Kyoto Protocol.

All the IPCC reports make it clear that developing countries will suffer most from climate change since they have the least capacity to adapt. Where backed by an assurance of technical and financial assistance and when convinced of the usefulness to them of specific goals, developing countries have in the past never hesitated to adopt specific goals voluntarily.

The UNFCCC could go beyond the legalities of the Kyoto Protocol and harness the energies of the leaders of enterprises and NGOs throughout the world to create a wave of voluntary action that could compel all countries to join the mainstream.

LESSON 7: EMPOWER THE FINANCIAL MECHANISM TO BE A PROACTIVE INSTRUMENT FOR TECHNOLOGY TRANSFER

The financial mechanism of the Montreal Protocol, the MLF, is based on the recognition that many developing countries lack the capacity to comply with treaty obligations, and that developed countries, which are often disproportionately responsible for causing the problems the treaty is designed to address, should provide financial assistance to developing countries to ensure compliance.

The Parties to the Montreal Protocol created the MLF to provide financial assistance to developing countries to implement and comply with the Protocol's control measures, including by facilitating the transfer of technology. The contributions of the MLF have been critical to the success of the Montreal Protocol (see Chapter 3).

The GEF, though not a financial mechanism of the Protocol, provided financial assistance to CEITs, which were not eligible to receive financial assistance from the MLF. The GEF had the difficult task of assisting CEITs to return to compliance by phasing out ODSs in line with developed countries.

CEIT Parties have nearly completed a full phaseout of CFCs and halons, with consumption decreasing to less than 350 ODP Mt in 2003 from about 296,000 ODP Mt in the late 1980s – a 99.8 per cent reduction.[19] At the individual country level, the GEF has funded projects that have phased out 20–60 per cent of a country's ODS consumption. Overall, the GEF has helped phase out 28 per cent of total global ODS consumption.

The success of both the MLF and GEF in the Montreal Protocol is largely a result of the freedom and flexibility granted to financial mechanisms by the Protocol's Parties. The indicative list of incremental costs (Appendix 2) gave good guidance to the MLF, but the MLF had the right of interpreting each entry in the list to suit effective achievement of its goals. These had to be to the satisfaction of both the Article 5 and non-Article 5 Parties, who were equally represented on the Executive Committee and among whom there were many opposed goals. There were extensive discussions in the Executive Committee, elected every year by the MOPs, on all issues; these were resolved by arriving at compromises that allowed projects to proceed. Rarely have such disputes been taken to the MOPs. Sometimes the Executive Committee even went beyond the indicative list – institutional strengthening and networks in developing countries were funded by the MLF, for example, even though they were not mentioned in the list (see Lesson 8).

The Executive Committee also had the freedom to experiment with new techniques. It gradually progressed from projects for each enterprise through sector-wide projects to National Terminal Phaseout Plans. The GEF had similar freedom to improvise solutions in CEITs for their ozone projects, even though GEF Council Members are not elected by the MOPs.

The MLF is also the focus of all the activities to assist developing countries. Donor countries could have their own bilateral Montreal Protocol programmes (up to 20 per cent of their contribution due to the MLF) in developing countries, but such programmes had to be approved by the Executive Committee. This avoided confusion and duplication of activities. Many of the donors chose to use only a small part of this allowance and left most of the activities to the MLF.

The MLF does not distinguish between Parties on any political basis. All members of the Executive Committee have one vote (unlike in the World Bank, where the voting strength is proportional to contributions). Developing countries and developed countries are equal in number, and the chairmanship rotates between the groups.

Another reason for the success of the MLF is the replenishment process, which occurs every three years. The TEAP estimates the funding required for each replenishment period, taking into account the obligations of the developing countries, the projects already approved and the lead time for completion of projects. The TEAP report is reviewed and decided upon by the Parties at the MOP. It is remarkable that even after extensive negotiations, the Parties generally approve a replenishment figure very near the one recommended by the TEAP. In fact, there have been several occasions where the Parties approved funding for Article 5 Parties that would reduce their consumption of ODSs by more than that required by the schedules of the Protocol. This has the added bonus of allowing developing countries to plan country programmes and other implementation projects with a high degree of confidence that the necessary funding will be available to execute their plans.

Pointers for the climate change treaties

The GEF can incorporate the innovative techniques and approaches it used in financing the CEIT ODS phaseout when it functions as the principal financial mechanism for the UNFCCC.[20] The GEF's 2005 Performance Study concludes that many lessons from the ozone experience could assist in developing and refining regulatory regimes for other chemicals.

While the GEF is the financial mechanism of the UNFCCC and the Kyoto Protocol, it is not the sole focal point of developing country programme approval and implementation. The UNFCCC and the Kyoto Protocol have created,[21] or are in the process of creating, several other funds, such as the Kyoto Protocol Adaptation Fund, the Least Developed Countries Fund and the Special Climate Change Fund. This is highly unproductive. The MLF functioned well because the developing countries knew exactly where and how to access funds for their projects, and the donors knew periodically how much replenishment the Fund needed. All developing country programmes and projects, including those financed by bilateral aid programmes, were approved by the Executive Committee to avoid duplication and confusion.

The GEF receives annual guidance from the UNFCCC Conference of the Parties. It will be productive if the guidance from the UNFCCC is on general lines, such as time limits for the GEF to assist the developing countries to prepare their time-bound voluntary programmes and goals under Article 10 of the Kyoto Protocol. The UNFCCC and the Kyoto Protocol can set up a system for:

- preparing country programmes for all the developing countries, with voluntary goals for a greener growth path;
- obtaining commitments from the developed countries to fund the preparation and implementation of these country programmes;
- providing periodical replenishment for the GEF for the purpose of achieving the goals;

- ensuring that the needs of the Parties are reflected in the programmes of the GEF by setting up a coordination mechanism; and
- giving the necessary freedom to the GEF to improvise, while setting up a system of monitoring and evaluation.

LESSON 8: CREATE FOCAL POINTS AND NETWORKS

Almost from the start, developed countries each had a focal point within their governments to deal with the ozone issues. The creation of an office, or focal point, within each developing country's government, with appropriate financial assistance to ensure adequate resources and prevent excessive staff turnover, can help compliance and implementation efforts. International and regional networks of focal points and others are important in sharing experiences, exchanging knowledge and furthering skill development.

The MLF financed such an office, or 'ozone cell', at relatively modest cost and consistent with local needs in each developing country Party. This made it possible for each Party to have a consistent presence on ozone issues, with a single focal point that was versed in the technical details of the ozone regime and in close communication with the Ozone Secretariat and ozone counterparts in other Parties. Focal points and ozone cells have helped governments to:

- coordinate country activities for phaseout;
- consult with industry and other interested organizations on the steps to be taken for phaseout;
- prepare a country programme setting out a strategy and plan of action;
- coordinate the technical and financial support of the implementing agencies, the bilateral agencies and the Fund Secretariat;
- design and implement national law and the financial measures to facilitate phaseout;
- organize awareness and training programmes for industry and the public; and
- create an effective national system for monitoring and reporting on national production and consumption of ODSs.

Once established, focal points in each country were organized into nine regional and global networks to facilitate the exchange of information, best practices and technology transfer. These networks played an extremely important role in providing feedback to the MLF and other Parties on which projects within country plans worked or did not work, and in facilitating transfer of expertise and technology from developed to developing countries.

Sweden established a network as early as 1990 for countries in Southeast Asia, and the concept was quickly adopted by the MLF and implemented through UNEP's OzonAction Programme. There are now nine networks, comprising every developing country and 14 developed countries, which are administered by Regional Network Coordinators working out of UNEP's regional offices. The global hub of these networks operates at the Paris offices of the UNEP/DTIE, which has organized a variety of network activities that have led to improvements in data reporting and refrigerant management plans, and helped exert professional 'peer pressure' on Ozone Officers to ensure aggressive implementation of the Montreal Protocol. The networks also have fostered steady south–south and south–north cooperation on technology transfer and, along with NGOs such as the Environmental Investigation Agency, on curbing illegal trade.

Pointers for the climate change treaties

This system of focal points and networks could be followed by the climate change treaties. Currently, larger developing countries tend to have separate officers dealing with ozone and climate issues, while smaller countries have a single officer to deal with many issues, including ozone and climate. The GEF could include establishing 'climate cells' as part of its efforts to develop and implement country programmes in developing countries, and provide appropriate training and network-building. The mandate of the climate cell in each developing country could assist the government to:

- coordinate the country activities for green growth;
- consult with industry and other interested organizations on the steps to be taken for the delineation of the growth path and implementation;
- prepare a country programme setting out a strategy and sectoral goals and a plan of action to achieve these goals;
- coordinate the technical and financial support of the implementing agencies, the bilateral agencies and the GEF;
- organize awareness and training programmes for industry and the public; and
- create an effective national system for monitoring and reporting on national progress towards the goals the country set itself.

Developed countries could be invited to join networks that would promote exchange of knowledge and experience, mutual assistance, and competitive spirit to achieve the goals.

LESSON 9: DEVELOP AND IMPLEMENT TRAINING PROGRAMMES

No training, no technology transfer

Adequate human and institutional capacity is essential at every stage of every transfer process. The transfer of many environmentally sustainable technologies demands a wide range of technical, business, management and regulatory skills. Technology transfer should be accompanied by training for the recipient country to the extent that it can not only use the technology, but also possibly improve upon it.

The MLF and GEF have developed and implemented training programmes for policymakers, industry executives and technicians to promote technology transfer. In addition, UNEP is implementing regional and sub-regional customs training as a cost-effective substitute for national customs training, making use of existing regional customs training facilities. National customs training takes advantage of a 'train-the-trainers' approach to maximize its reach, and some countries make use of a certification system to recognize trained personnel and ensure quality control. Training programmes also benefit from monitoring systems to report on current levels of training activity (see Chapter 13 for further details).

Pointers for the climate change treaties

The UNFCCC and GEF could expand their training programmes to spread best practices for green growth, possibly based on practices from leading enterprises that could be collected and compiled by the proposed Technical Options Committees for climate change. The awareness, education and information, and training programmes on different global environmental issues could be organized by the GEF in a coordinated way to take advantage of economies of scale and ensure quality control. UNEP's Green Customs initiative is a useful model, as it coordinates the aspects of awareness, information and training of several multilateral environmental agreements, including the Montreal Protocol, the Basel Convention and the Convention on International Trade in Endangered Species.

LESSON 10: USE REGULATIONS AND POLICIES TO PROMOTE TECHNOLOGY TRANSFER

Domestic law and technology transfer

Multilateral environmental agreements (MEAs) must be implemented through domestic laws. Law at the domestic level can greatly influence the transfer of technologies as part of implementing an international environmental agreement. Depending on how regulatory regimes are structured, the law can be a great enabler, or occasionally a great barrier. These policies can combine a mix of specific standards, financial incentives and information-based mechanisms to encourage innovation and dissemination of new technologies.

All the Parties to the Montreal Protocol deployed restrictions on the producers and users of ODSs, which included outright bans on production and imports. Many Parties took action before the Montreal Protocol was negotiated. A recent report (March 2007) by the World Business Council for Sustainable Development[22] asserts that the only way to combat climate change is through decisive, concerted and sustained actions between governments, businesses and consumers.

For example, in 1978 the US became the first country to ban the manufacture and import of most cosmetic and convenience CFC aerosol products.

Several other countries enacted similar bans by the end of 1990 (Sweden in 1979, Canada in 1980, Norway in 1981, Taiwan in 1983, Australia in 1989, Austria in 1989, Brazil in 1989, Indonesia in 1990 and Thailand in 1990).

In Thailand, Japanese and Thai manufacturers and the US EPA negotiated the first national phaseout of the production and import of CFCs in refrigerators. Thailand was also the first developing country to enact an environmental trade restriction prohibiting the manufacture and import of refrigerators containing ODSs. All Parties to the Montreal Protocol have since followed suit.

Other policies involved the use of financial incentives to spur technology transfer. Many countries have taxes or fees on ODSs that are intended to discourage use and raise revenue. In some cases, the revenue was spent on programmes to encourage ozone layer protection, while in others the revenue was not used to fund environmental activities.

For example, the US levied one of the most substantial taxes on ODSs, imported products containing ODSs and 'floor stocks' of ODSs held for future use. The imposition of the tax drove up prices for ODSs, which helped signal to both manufacturers and consumers that the time had come for the transition to superior technologies. The tax also made such a transition more cost-effective.

Revenues from these taxes were often applied specifically to other efforts to address ozone depletion, such as research, education and technical assistance to enterprises. The Republic of Korea was among the first to use ODS tax revenues in this manner.

Other kinds of financial mechanism include Singapore's efforts to reduce ODS supplies through permitting processes designed to capture the 'monopoly rent' that chemical suppliers would have charged customers as ODS supplies became more scarce under national phaseout plans. The permit price was set at a level that kept the total ODS use within the Protocol limits. Revenue from the permits financed research and technology assistance to companies seeking alternatives and substitutes.

Another kind of measure countries have used with success is labelling programmes. Labelling programmes help inform consumers which products and processes are 'ozone-safe', for example. This educates consumers about the extent of ODSs in products and the environmental consequences of ozone depletion, and empowers them to avoid and/or boycott ODS products. Labelling also encourages product manufacturers to halt ODS use to satisfy customers and avoid administrative expenses and penalties. Many companies marketing ODS-free products apply labels proclaiming them to be 'CFC-free' or 'ozone-safe'. Additionally, labelling programmes provided an opportunity for the MLF and GEF to develop product-labelling schemes (see Chapter 4).

Pointers for the climate change treaties

Governments could induce technology change through regulation of energy markets, environmental regulations, energy efficiency standards, and energy and emission taxes. Indeed many countries, including developing countries such as Brazil, China and India, are already implementing many of these policies, resulting in benefits that extend beyond mitigating climate change. For example, many GHG reduction policies also help reduce air pollution, provide energy independence through less reliance on imported fossil fuels, improve the efficiency and operation of transportation systems, and protect biodiversity.[23] The policies aim to:

- encourage use of natural gas instead of coal or petroleum;
- improve the efficiency of energy production;
- promote non-fossil-fuel energy sources such as nuclear and renewable energy;
- improve end-use efficiency in buildings and industry; and
- improve the efficiency of automobiles, reducing transport demand.

Governments as prods and guides

The following are some excerpts from the background paper of the UNFCCC Secretariat:[24]

- Laws and regulations can have a major impact on GHG emissions because they affect business behaviour and public habits. Some governments encourage the use of mass transit; some – through tax arrangements, road-building programmes and even subsidies – encourage the burning of fossil fuels. One way (admittedly not always popular) of changing behaviour is to make it illegal. Another is to make it more expensive through taxes or penalties.
- Some governments have cut GHG emissions with a mixture of carrots and sticks – with inducements, subsidies, voluntary programmes, regulations and fines. Several have attacked the problem directly by imposing 'taxes' on carbon use. Others have established 'carbon markets', where units of energy use may be bought and sold.
- Minimum standards for energy efficiency in new buildings were updated recently in a number of countries, including Austria, France, Japan, New Zealand and the UK. Such measures can include requirements for walls and roofs that limit heat loss and can require a minimum level of thermal efficiency for furnaces and water heaters.
- Standards for energy efficiency in electrical appliances have been established by some governments. A programme begun by Japan in 1998, for example, is expected to cut the energy requirements of home video recorders by 59 per cent, refrigerators by 30 per cent and computers by 83 per cent.
- Economic and fiscal instruments have been used to spur shifts in freight transport from roads to rail and ships, which use less fuel per tonne of goods transported. Examples are the introduction of road tolls in Austria and mileage-based tolls for lorries in Austria, Germany and Slovenia; increased investments in rail systems in Austria and Belgium; and promotion of ship and rail use for freight in Belgium, Switzerland and Japan.
- Among voluntary arrangements is an agreement reached between the EU and European and Asian automobile manufacturer organizations setting reduced targets for carbon dioxide emissions from passenger cars and light commercial vehicles.

- 'Green tariffs' have been used by Belgium, Germany, Hungary and Switzerland to spur the use of renewable energy. The 'tariffs' guarantee electrical-generating companies a higher price for a unit of renewable energy than the prevailing market price.
- Steps have been taken by a number of governments to require waste firms and landfills to prevent the escape of GHGs such as methane. Landfill taxes per tonne of waste in Switzerland and Norway are higher for facilities that are not sealed, while in Austria, taxes are higher for landfills that do not recover the methane emitted.

The challenge for the UNFCCC and the Kyoto Protocol is to get every country to adopt the best and most suitable regulations and policies. The UNFCCC and GEF could campaign to raise awareness and educate developing countries in this regard.

LESSON 11: REMOVE LEGAL AND INSTITUTIONAL BARRIERS, AND IMPROVE SYSTEMS OF GOVERNANCE

Parties must review their regulatory regimes to remove barriers and streamline permitting and other processes to allow for swift and effective technology transfer. Barriers to technology transfer include lack of financial institutions to help finance the associated costs; high or uncertain import duties; high or uncertain inflation, interest rates and tax policies; institutional corruption; transaction costs; and intellectual property protection.

There are numerous examples where even a small glitch in a country's regulations prevented or significantly delayed technology transfer. For example, one country had received a shipment of new equipment that had been purchased with MLF funds. The MLF had required that none of its funds be used to pay customs or other import duties on the new equipment. The country had to request that its customs office waive the duties on the new equipment; this was eventually done, but only after the equipment had languished in storage at the airport for two years.

In a number of countries, the delays in banking systems held up technology transfer, and in others many projects were postponed due to delays in the issue of necessary permits and licences. These experiences and others indicate that regulatory processes may need to be adjusted to accommodate the swift introduction of new technologies by streamlining and coordinating different agency approval processes.

Most of the technologies needed were in the public domain, and what was needed was awareness of these technologies and training to adopt them. In most instances where the technology was not in the public domain, it could be obtained at a reasonable price and on reasonable terms. There were a few instances, however, where owners of a new technology refused to accommodate a transfer on reasonable terms, making intellectual property protection a barrier. In such instances, compulsory licences were available, but in practice were rarely used. Rather, the terms for technology transfer became more attractive over time as competing alternatives were developed (see Chapter 12 for details).

Pointers for the climate change treaties

The improvement of systems of governance will assist not only the technology change for GHG reduction but also all other aspects of economic development. The UNFCCC, Kyoto Protocol and GEF must make the improvement of systems of governance, to an extent relevant for the implementation of the country programmes, a part of their financial assistance.

LESSON 12: USE PUBLIC PROCUREMENT TO PROMOTE ALTERNATIVES

Government agencies, and particularly military organizations, are among the largest purchasers of goods and services in almost every country, whether developed or developing. When governments set high environmental standards and criteria for the goods and services they purchase, private sector suppliers will quickly change their production practices to meet them.

Many governments prohibited the purchase of ODS-based products and services, which made suppliers confident that transitioning out of CFCs and other ODSs would not adversely impact their businesses and helped jump-start large-scale production of new, environmentally friendly technologies.

Military organizations in both developed and developing countries played a prominent part in developing substitutes and alternatives across a wide range of applications, catalysing the widespread use of these substitutes and alternatives in civilian applications (see Chapter 5). The need to address climate change poses similar challenges to military readiness, particularly in terms of energy supply and efficiency and potential disruptions in the availability of natural resources and food supply lines. Some military organizations have already begun to address this issue by investing in the development of alternative and renewable forms of energy, such as solar and wind power.

Governments and their military services could give preference to climate-friendly products and services, and thereby set an example for others to follow.

CONCLUSION

Continuous technological innovation and diffusion has been the key to success for the Montreal Protocol, and it will prove to be the key to success for climate protection too. The Montreal Protocol demonstrated many ways both to promote the innovation and diffusion of technology and to remove legal and institutional barriers. The strong commitment by many Parties, military organizations, and multinational and domestic enterprises to protect the ozone layer made the legal obligations of the Montreal Protocol more an expression of the will of the citizens of the world than an imposition.

Box 14.1 The lessons

Lesson 1: Act now with the best available technologies – do not wait for an ideal legal solution.

Lesson 2: Develop visionary technology assessment – with cutting-edge know-how – free from political and commercial pressures.

Lesson 3: Encourage leadership by multinational enterprises and major enterprises in each country.

Lesson 4: Identify and involve all stakeholders, and develop local and international partnerships.

Lesson 5: Raise awareness among stakeholders, including decisionmakers, senior industry executives, technicians, associations and the general public.

Lesson 6: Require country programmes from each developing country, as mandated by Article 10 of the Kyoto Protocol, with specific voluntary goals set by the countries themselves.

Lesson 7: Empower the financial mechanism to be a proactive instrument for technology transfer – make it the sole focal point of assistance and give it freedom to innovate.

Lesson 8: Assist developing countries to institute focal points for action on climate change, and form networks of countries to exchange knowledge and experience.

Lesson 9: Develop and implement training programmes for stakeholders on technologies and techniques for a green growth path.

Lesson 10: Use regulations and policies to promote technology transfer. Many countries have excellent regulations and policies on some aspects of the activities needed to reduce the emissions of GHGs – ensure the spread of the best of these technologies to all countries.

Lesson 11: Remove legal and institutional barriers, and improve systems of governance.

Lesson 12: Use public procurement to promote climate-safe technologies.

The specific lessons that combined to produce this extraordinary result should be studied and applied to the Kyoto Protocol as well. Mechanisms like the Clean Development Mechanism and Joint Implementation are useful tools, but more important is the harnessing of the will of enterprises, governments and citizens to protect the climate through implementation of the many practical ideas available. The path to success outlined by the Montreal Protocol (see Box 14.1) is well worth following.

APPENDICES

Control Measures of the Montreal Protocol

The continuous strengthening of the Montreal Protocol from 1990 through the adjustments in 1990, 1992, 1995, 1997 and 1999, and amendments in 1990, 1992, 1997 and 1999, ended in specific schedules of phase-outs for all the ozone-depleting substances (ODSs) identified.

The lists of controlled substances are annexed to the Protocol as:

*   Annex A (Group I – CFCs, Group II – halons);
*   Annex B (Group I – other CFCs, Group II – carbon tetrachloride, Group III – methyl chloroform);
*   Annex C (Group I – HCFCs, Group II – HBFCs, Group III – bromo-chloromethane);
*   Annex E (methyl bromide).

The following is a summary of the phase-out schedules as of 1 July 2001.

SCHEDULES FOR NON-ARTICLE 5 PARTIES (PARTIES OTHER THAN DEVELOPING COUNTRIES OPERATING UNDER ARTICLE 5)

Annex A, Group I; Annex B, Groups I, II, III; Annex C, Group II

Production and consumption to be phased out by the end of 1995, but for possible essential-use exemptions granted from year to year by Meetings of the Parties.

For meeting the basic domestic needs (BDN) of Article 5 Parties, the following quantities were permitted.

Annex A, Group I (CFCs)

*   Until the end of 2002: annual average of its production to meet the BDN for the period of 1995 to 1997 inclusive (base).
*   Until the end of 2004: 80 per cent of the base.
*   Until the end of 2006: 50 per cent of the base.
*   Until the end of 2009: 15 per cent of the base.
*   From 1 January 2010: zero.

Annex B, Group I (other CFCs)

- Until the end of 2002: 15 per cent of production in 1989.
- Until the end of 2006: 80 per cent of the (base) production for meeting the BDN during 1998–2000.
- Until the end of 2009: 15 per cent of the base.
- From 1 January 2010: zero.

Annex B, Groups II and III (carbon tetrachloride and methyl chloroform)
15 per cent of the production in 1989.

Annex C, Group II (HBFCs)
None.

Annex A, Group II (halons)

Production and consumption to be phased out by the end of 1993 but for possible essential-use exemptions. The additional production permitted to meet the basic domestic needs (BDN) was:

- Until the end of 2001: 15 per cent of the production in 1986.
- Until the end of 2004: annual average of production to meet the BDN in 1995–1997 (base).
- Until the end of 2009: 50 per cent of the base.
- From 1 January 2010: zero.

Annex C, Group I (HCFCs)

Consumption frozen at the base level (1989 HCFC consumption +2.8 per cent of 1989 CFC consumption) in 1996; 35 per cent reduction from 1 January 2004; 65 per cent reduction from 1 January 2010; 90 per cent reduction from 1 January 2015; 99.5 per cent reduction from 1 January 2020 and consumption restricted to servicing; and 100 per cent phase-out from 1 January 2030. Production frozen at the base level (1989 HCFC production +2.8 per cent of the 1989 HCFC production) from 1 January 2004; 15 per cent additional production allowed to meet the BDN.

Annex C, Group III (bromochloromethane)

Production and consumption phase-out from 1 January 2002. No exemptions.

Annex E (methyl bromide)

Production and consumption frozen at the base level of 1991 until the end of 1998; 25 per cent reduction until the end of 2000; 50 per cent until the end of 2002; 70 per cent until the end of 2004; complete phase-out from 1 January 2005 with possible critical-use exemptions.
    Production to meet the BDN is as follows:

- Until the end of 2001: 15 per cent of the base level production.
- Until the end of 2004: 80 per cent of production in 1995–1998 to meet the BDN.
- From 1 January 2005: zero.

SCHEDULES FOR ARTICLE 5 PARTIES
(DEVELOPING COUNTRIES)

Annex A, Group I (CFCs)

Production and consumption frozen at the level of average during 1995–1997 (base) from 1 July 1999; 50 per cent reduction from 1 January 2005; 85 per cent reduction from 1 January 2007; and 100 per cent phase-out from 1 January 2010 with possible essential-use exemptions; 10 per cent base level production permitted to meet BDN until the end of 2009.

Annex A, Group II (halons)

Production and consumption frozen at the average 1995–1997 level (base) from 1 January 2002; 50 per cent reduction from 1 January 2005; 100 per cent phase-out from 2010 with possible essential-use exemptions; 10 per cent of base production allowed to meet BDN until the end of 2009.

Annex B, Group I (other CFCs)

Production and consumption reduction of 20 per cent from the level of 1998–2000 (base) from 1 January 2003; 85 per cent reduction from 1 January 2007; 100 per cent phase-out from 1 January 2010 with possible essential-use exemptions; 10 per cent of base level production allowed to meet the BDN until the end of 2009.

Annex B, Group II (carbon tetrachloride)

Production and consumption reduction of 85 per cent from 1 January 2005 from the level of 1998–2000 (base); 100 per cent phase-out by 2010 with possible essential-use exemptions; 10 per cent additional production allowed to meet the BDN until the end of 2009.

Annex B, Group III (methyl chloroform)

Freeze of production and consumption at the 1998–2000 level (base) from 1 January 2003; 30 per cent reduction from 1 January 2005; 70 per cent reduction from 1 January 2010; 100 per cent phase-out from 1 January 2015 with possible essential-use exemptions; 10 per cent additional production allowed to meet the BDN until the end of 2009.

Annex C, Group I (HCFCs)

Freeze of production and consumption from 1 January 2016 at 2015 level; phase-out of consumption from 1 January 1940; 15 per cent of base level allowed until the end of 2039.

Annex C, Group II (HBFCs)

Phase-out of production and consumption from 1 January 1996 with possible essential-use exemptions.

Annex C, Group III *(bromochloromethane)*

Phase-out of production and consumption from 1 January 2002 with possible essential-use exemptions.

Annex E *(methyl bromide)*

Freeze of production and consumption at 1995–1998 level (base) from 1 January 2002; 20 per cent reduction from 1 January 2005; 100 per cent phase-out from 1 January 2015 with possible essential-use exemptions; amounts used for quarantine and pre-shipment applications exempted at all stages.

Indicative List of Categories of Incremental Costs: Annex VIII of the Fourth Meeting of the Parties to the Montreal Protocol, 1992

The evaluation of requests for financing incremental costs of a given project shall take into account the following general principles:

(a) The most cost-effective and efficient option should be chosen, taking into account the national industrial strategy of the recipient party. It should be considered carefully to what extent the infrastructure at present used for production of the controlled substances could be put to alternative uses, thus resulting in decreased capital abandonment, and how to avoid de-industrialization and loss of export revenues;

(b) Consideration of project proposals for funding should involve the careful scrutiny of cost items listed in an effort to ensure that there is no double-counting;

(c) Savings or benefits that will be gained at both the strategic and project levels during the transition process should be taken into account on a case-by-case basis, according to criteria decided by the Parties and as elaborated in the guidelines of the Executive Committee;

(d) The funding of incremental costs is intended as an incentive for early adoption of ozone protecting technologies. In this respect the Executive Committee shall agree which time scales for payment of incremental costs are appropriate in each sector.

Incremental costs that once agreed are to be met by the financial mechanism include those listed below. If incremental costs other than those mentioned below are identified and quantified, a decision as to whether they are to be met by the financial mechanism shall be taken by the Executive Committee consistent with any criteria decided by the Parties and elaborated in the guidelines of the Executive Committee. The incremental recurring costs apply only for a transition period to be defined. The following list is indicated:

(a) Supply of substitutes:

(i) Cost of conversion of existing production facilities; cost of patents and designs and incremental cost of royalties; capital cost of conversion; and cost of retraining of personnel and cost of research to adapt technology to local circumstances;

(ii) Costs arising from premature retirement or enforced idleness, taking into account any guidance of the Executive Committee on appropriate cut-off dates; of productive capacity previously used to produce substances controlled by existing and/or amended or adjusted Protocol provisions; and where such capacity is not replaced by converted or new capacity to produce alternatives;

(iii) Cost of establishing new production facilities for substitutes of capacity equivalent to capacity lost when plants are converted or scrapped, including: cost of patents and designs and incremental cost of royalties; capital cost; and cost of training and cost of research to adapt technology to local circumstances;

(iv) Net operational cost, including the cost of raw materials; and

(v) Cost of import of substitutes.

(b) Use in manufacturing as an intermediate good:

(i)   Cost of conversion of existing equipment and product manufacturing facilities;

(ii) Cost of patents and designs and incremental cost of royalties;

(iii) Capital cost;

(iv) Cost of retraining;

(v)   Cost of research and development; and

(vi) Operational cost, including the cost of raw materials except where otherwise provided for.

(c) End use:

(i)   Cost of premature modification or replacement of user equipment;

(ii) Cost of collection, management, recycling, and, if cost-effective, destruction of ozone-depleting substances; and

(iii) Cost of providing technical assistance to reduce consumption and unintended emission of ozone-depleting substances.

List of Project Completion Reports Studied

MULTILATERAL FUND

UNDP

Aerosols
Burundi: 39-9. Bangladesh: 17-5. India: 22-115, 117, 118, 135, 136, 138; 24-16,171, 172-174, 179; 28-213, 227. Malaysia: 18-64. Mauritius: 12-3. Sri Lanka: 18-7.Thailand: 17-45, 46; 23-83, 84; 25-94, 96. Vietnam: 17-7; 18-10, 11.

Compressors
Colombia: 13-7.

Foams
Argentina: 14-13; 15-14; 18-26, 27 to 33; 20-48; 22-55, 56; 23-61, 63; 25-76; 26-78; 29-94 to 96; 30-10; 31-11; 34-12. Burundi: 3510. Benin: 32-12. Brazil: 1-13; 17-21; 18-27 to 29; 19-34 to 46; 20-53; 22-64 to 72; 23-77 to 80, 82, 84, 87, 89, 92 to 95; 25-101,102, 104, 105, 107 to 109; 26-110, 112 to 114, 116; 27; 117, 120, 125, 129, 130; 28-131, 135 to 138, 140, 143, 144; 29-144 to 152, 154 to 158; 31-167 to 169, 173, 175, 178 to 185, 187 to 198, 201, 211 to 218, 220, 221, 224 to 226, 227, 229 to 235, 237 to 240, 243 to 250, 252, 254 to 259. Colombia: 13-9; 29-36 to 38; 32-47 to 49. China: 6-3, 7, 13,; 10-32; 11-54; 13-73, 74; 15-87 to 96, 98 to 103,; 17-125 to 127, 134; 18-141 to 143; 19-160, 161; 22-202, 206; 23-228, 229; 24-236; 25-250, 254, 258,; 26-260, 263, 266; 27-279; 28-286, 289, 290, 297; 29-305, 312 to 316, 318, 319, 325, 326, 329 to 332, 334. Dominican Republic: 29-21, 22, 24. Egypt: 9-10; 11-20, 12-22a and b, 26, 28, 29; 15-36, 46, 48. Gambia: 22-6. Ghana: 12-7. Indonesia: 17-32; 18-34; 20-44, 46, 47; 22-60; 23-63, 65, 67, 68, 71, 81, 82; 25-86, 89,91; 26-10, 93, 95 to 98, 100, 101, 103, 106; 29-112, 114,; 35-126. India: 12-16 to 18; 13-27, 30; 19-71 to 76, 78 to 83; 20-96 to 102; 22-111, 128, 130, 132, 133; 23-143, 146 to 151, 153 to 158; 28-189, 190, 192, 193; 27-196 to 198, 202, 205, 207; 28-210 to 212, 214, 215, 218, 219, 2224, 226, 228, 229; 28-231 to 237, 239 to 242, 245, 246; 31-258 to 261, 263, 267 to 277; 32-280, 285, 289; 34-304, 305, 312, 321, 322, 324, 330, 331, 33 to 335. Iran: 32-76; 34-88; 35-1. Jamaica: 22-6. Laos: 38-10. Malaysia: 12-22, 23, 25, 33; 13-38, 41, 43; 15-40, 48, 49, 56; 17-61; 18-65 to 68, 70, 71; 19-88, 89; 22-94 to 96, 98; 23-99, 104 to 107; 25-111. 26-115, 116; 28-121 to 123; 29-130, 131; 31-136 to 138. Mexico: 18-41, 42; 19-46, 47; 20-48, 49; 22-54 to 57; 23-66. 69, 71 to 73; 25-82, 84; 26-87; 31-98; 35-103. Malawi: 26-12. Morocco: 20-4; 23-14 to 16; 25-23; 32-43, 44. Nigeria: 20-12 to 15; 23-21 to 29; 26-31 to 39, 42, 43; 28-46, 47, 49, 50; 29-55 to 59; 31-70; 34-83 to 86, 88,

89, 92; 35-94, 95. Panama: 13-5. Paraguay: 23-4. Peru: 17-9. Philippines: 12-26 to 28; 13-31, 32, 34, 37; 17-37; 18-37,62, 91 to 93; 22-48, 50; 23-51, 52; 31-64. Thailand: 12-27, 29, 30 a and b, 31; 13-35, 36; 15-41, 42, 44; 18-47; 19-51; 20-55,56; 22-63, 64, 66, 67; 23-68 to 71, 75; 25-90; 26-98 to 103, 105, 106; 28-117, 118; 29-119, 122, 123. Uruguay: 13-6, 9.

Halons

China; 12-66. India: 24-163,165,168,170, 175 to 177; 28-208, 209, 216, 220, 222, 256. Malaysia: 18-172. Philippines; 18-28. Uruguay: 18-17, 18.

Refrigeration

Argentina; 23-6, 66, 81, 86, 88, 90, 91; 28-92. Brazil: 23-81, 86, 88, 90, 91; 25-111, 115, 118; 28-133, 134; 31-166. Colombia; 13-4, 5, 8; 15-6, 11, 12, 14; 17-15, 16; 19-22. Costa Rica: 18-8; 27-17 to 19. China; 17-123, 124; 19-167, 168; 23-232, 234; 24-242; 32-367. Cuba: 23-6. Dominican Republic: 22-9; 4-11; 25-16. Egypt: 12-30, 31; 15-44. 45; 18-49, 50; 20-58. El Salvador: 22-3. Guatemala: 15-6, 7; 23-16. Indonesia: 23-66; 25-88, 90, 92; 26-99, 102, 104, 105; 35-131 to 136. India: 31-257; 32-282, 286; 34-323; 35-339 to 342. Iran: 28-43, 44, 46, 65; 29-55, 56, 58, 60; 31-66 to 68, 71; 34-82, 84 to 87, 89 to100, 109 to 112; 35-131, 132, 134 to 140. Lebanon; 29-34. Malaysia: 13-45; 18-73, 74; 23-103; 28-126; 29-128; 32-140 141. Mexico: 5-61, 63; 15-32, 33. Peru: 15-5, 6, 8; 19-13, 14, 16, 44. SriLanka: 17-4to 6. Syria: 26-33, 38, 39; 28-46; 29-52; 31-64 to 67; 32-69, 79. Thailand: 13-33.Venezuela; 18-43, 44; 19-46, 48, 49.

Solvents

Brazil: 18-36, 37. China: 7-19,20; 10-36, 37; 12-65; 18-152; 19-169, 171, 172; 20-178, 186; 22-195, 216; 23-224; 28-287. Malaysia: 11020. Mexico: 15, 36, 37. Philippines: 19-46; 25-56.

UNIDO

Albania: 10. Algeria: 13, 28, 34, 37 to 39, 41, 47, 56. Argentina: 47, 117. Botswana: 5. Brazil: 170 to 172, 174, 176, 177, 219, 222, 241. Cameroon: 10, 16. Colombia: 32. China: 248, 302, 306, 369, 376, 389, 26. DER Korea: 18 to 22. Egypt: 79. Honduras: 5 to 7. Indonesia: 119, 141, 143, 152. India: 223, 255, 266, 283, 284, 290, 290, 303, 306, 308, 311, 313, 314, 316, 327, 338. Iran: 2, 26, 29, 52, 73, 98, 101, 103 to 105, 107, 108, 113, 115, 119, 130. Jordan: 47, 66, 68, 71, 72, 74, 75. Kenya: 17. Lebanon: 36, 39, 45. Libya: 3. Malaysia; 143. Macedonia: 16, 19. Mexico: 81, 90, 91, 99. Mali: 12. Morocco: 45 to 47. Nigeria: 10, 11, 30, 40, 44, 48, 51 to 53, 71, 76, 97, 98. Oman: 2, 4. Pakistan: 9, 10, 14, 17, 43, 47, 51, 52. Panama: 16. Qatar: 4, 5. Romania: 15, 20. Senegal: 16. Sudan: 6. 9, 10, 11, 13. Syria: 11, 15, 61, 71 to 74, 76, 88. Thailand: 97. Tunisia: 35. Turkey: 68, 72. Venezuela: 78, 83, 84, 91, 94. Vietnam: 23. Yemen: 8, 10 to 12. Yugoslavia: 12, 16.

World Bank

Argentina: Year 2000- Interclima, Mirgor. 2001- Asisthos, Sistemaire, Whirlpool. 2002- Briket, El Dorado. 2003- Cachan, Mclean. Brazil: 2001- Gelopar, Geltec, Reubli. 2005- Refripar Curitiba, Refripar San Carlos, Esmalte, IBBL, MAgostini, MetaLfri, Recrusul, Reubli, Sao Rafael.

Chile: Teefin. China: 25th Excom- Chengde, Dalian, Henanxinfei, Qingdao haier, Shenyang, Shuangyan, Tonxiang, Wuxifoam, Zhoziang. 26th Excom- Beijing, Changling, Liming, Shanghai LPG, Shanglingfoam, Tianjian, Zheijang Fire. 27th

Excom- Changling, Lanzhou, Shanghai 18,. 29th Excom- Hangzhou 4, Huayi 2, Huayi 1, Yifeng, Yinguang, Zhonghan, Zibo 7. 1999- Shangahi Compro, Dalian. 2000- Ekchor, Jinanshiyan, SAAC, Xinfei, Dongfeng, Penglai. Cangzhou, Wuxi, Yantai, Shanghai 6, Shuanglu, Taizho, Yueyang. 2001-BeiComRef, Gaofeng, Handan, Nanjing, Qinghuangdao, Sshangling. 2002- Changfeng 20.176, Changging 20. 182, Heilong 28.292, Liangzhu 17.137, Qingdao 7.270, Qingyang 9.157, Qingwangdao, Wanbao 13.69, Yangzhai Tongli 29.321, Yantaifoam 23.227, Zhenjiang 5 17.133, Zhejiang 28.299. 2003- Changde 28.195, China MAC srctor 26-255, Haiou 27.280, Hangxing 29.320, Haohua 18.149, Qianjin 29. 317, Shaghai Furong 29.335, ZhenjiangcomRef 27.276, Zhenjiang 20.183. 2004- Anhui 16.111, Ningbo 22.215. 2005-Changzhou 28.293, Chengdu 20.179, Chongging Bingyang 22.214, Fujian Aerosol 24.244, Guangzhou 22.198, Handan 29.322, Hefei 23.218, Huojia 27.275, Jiangsu Haimen 27.277, Jinjiang 27.285, Nantong 31.362, Putuo, 27.281, QUJING 31.258, Shanghai Baoshan 27.284, Shanghai General Machinery 20.180, Shengzhufoam 29.311, SUBEI 22.200, Tianjin 20.175, Wuhan New World 22.208, Wuxian 31.361, Xiashan 27.271, Xinzhuang 27.274, Yueyang 28.303, Zheijiangchunhui 22.210, Zhejiang 28.300, Zhengzhou 15.86.

Colombia: 2003- Daniel Fernandez 26.28, Polares 28.34, Rojas 28.30, Supernordico 28.44. 2004-Friotermica 26.29, Indufrio 26.30.2005- Mac 34.51, Rymco 31.45,

Ecuador: October 1999- Drex 2, Ecasa 2, Indu, Mafrico, Ecasa 26.26, Elasto 26.24, Indurana- 26.25,

Egypt: MCMC. India: 25th Excom- Bharat, Modixerox. 26th Excom- Hindus. 2th Excom- Bluestar, Pfeda, Modixerox, Shriram, Tanquil, Vijay. Mar 99- Polyflex, Seepra. April 99- Bluestar Foam, Duroflex, Industrial Foams. Feb 00- Ishwar Arts, Ishwar Aashis, Milton Plastics, Milton Polyply, Panorma Plastics, Vikram, MachooMadras, Polyp 1, Repoly, SDC. June 00- Alpha Hadirole. Sept 00- Meghdoot, Stella. Nov 00- Aeroind, Aeropres. 2001- Cellop Cellothermo, Freezeking, Friztech Ind 18.64, Godrej, Kiroloskar, Kurlonkan, Manali, Murali, Polar, Polynote, Rockwell, Sandeep, Sethia, Sheetal, Shroff, Sidwal, SPIC, Standard, Subros, V. Krishna, Wimco. 2002- My Fairlady 22.119, Pranav 22.116, Sandan 22.121. 2004- Aarkay 23.144, AccraPak 22.114, Aeropharma 11.10, Excel 28.217, Dabirun 19.89, Saikrupa 19.91, Supercold 20.105, VKrishna 18.63. 2005- Attarwale 22.137, BPL Refrig 25.183, Godrej 30.337, Pranav Vikas 22.116, Rishiroop 34.320, Subros 38.357, Videocon 22.134, Whirlpool 27.204.

Indonesia: 25th Excom- Royal. 26th Excom- ITI. 27th Excom-Multihalon. 29th Excom- INTRI, Sanyo,Topjaya. March 99- National Gobel, Uppindo TA, Whana Drby. April 99- Tulus. 2000- AeroTA, Afi, Dasa, Garuda, Interfoam Dabli. 2001- Cahaya, LGElectronics, Maspion. 2002- Kimura 29.118, Porkka 23.74, Tara 23.73, DuaRoda 35.128, Erlangga 13.16, Foam Indo 11.12, Maspion 21.52, Positive 15.19, Samsung Maspion Indonesia, Sumberlogam 29.111, Anto Indo 31.120, bostino 35.127, Dawaniba 29.117, Multikriya 15.20, Intimas 29.116, Mac 15.29.

Jordan; 26th Excom- HAddad, Household, Jorpet.co, Kolaghassi. March 1999- Insttit, JCP, JorAntsep,. 2005- Abushaka 31.63, Alhussam 22.31, Arabchemical 31.64, FAA 23.37, Fivestar 17.24

Malaysia: 25th Excom- Aetech, Macconserv, Mac. 26th Excom- Bansle, Melcom, Tenco. 27th Excom- Argon, Widetech. Oct 99- Ontrak, Nippondenso. 2000- Engtek, Halon. 2002- APM 18.76, Penang 19.86, Transfreeze 19.87, UCM 19.90. 2003- Starfoam 17.59. 2004- Mashrae 18.77

Mexico: 25th Excom- IMSS, Jinen, Jed 1, Jed 2, Styl, QOMEGA, QWIMO. Oct 99- Gigan. Mar99- MAC. 2000- Climas, Gigante 2, Jed 3. 2003- Airtemp 34.102. 2004-

Aurera 1 5.62, Aurera 2 5.65, Chiller 28.95

Pakistan: 2000- Saleem. 2001- Spel 23.16, United 26.29. 2003- Koldkraft 23.18, Shadman 25.28. 2004- Razi 18.07, Singer 23.21, Jaguar 29.34.

Philippines; 25th excom- IONICS, 27TH Excom- Concept, EATP, SAN. July 1999- TUC. 2000- Foiline. 2001- GroupRef. 2002-TAS 109.17.

Thailand; 25th Excom- CRT, HANA, Kulthok, Kulthorn Phase I, SAHA, Sanyo Phase 1, CIG, Denso 1, GSS, Hita, KKZ, Ky 1, Sancom II, San ref, Toshiba 1, Toshiba 2. 26th Excom-CLC. 27th Excom- APN, Sanyo 2. 29th Excom- Padrie, Viriakat, Getaa-PU foam, Teamtronics. March 00- Siri. June 00- CANASIA, Siam, Technic foam. Thermobond. Threatt. Sept 00- Thai Airways. 2001- Genjit 23.77, Jennings 23.82, PE Containers, Shahakarn, Siamcool 23.82, Somerville 25.95. 2002- Arco 28.115, Makassan 28.116, Plastmate 27.115. 2003- Isotech 32.13, Siamsteel 27.111, Willich 27.112. 2004- Siamcargo 31.132.

Tunisia: 27th- Instrength. Mar 99- Cenafi. April 99- Sotum. 2000- Aero, Tabid

Turkey: 25th- Dolu. 26th- ARCELIK, Profilo, TAS. Feb 99- IFC Assan, IFC Tekiz, Klimasan, Suntas. Feb 2000- Teba. March 2000- Costidas, Iedas. June 00- Malzeme. 2001- Izopuli 24.41, Pimsa, Purplast, Thermaflex. 2002- EPS 24.42. 2003- Ugur 19.21, 2005- Aselsan 25.50, Beta 31.66, Elele 24.40, Elta 28.55, Gumaksan 22.26, Kulahcioglu 22.27, Safas 23.29, SFA Sogutma 22.25, Solvent Umb 25.49, Urosan 32.70.

Uruguay: 2000-Colder, 2002- Etcheparel 15.14. 2005-Nevol 19.21

Venezuela: Jan 00- Aaisa, FAACA. Sept 00- Clinica Atias, Impres

Zimbabwe: 2002- Capri 20.8, Commercial 20.7, Imperiale 20.9, Refair 20.8.

GEF PROJECTS AND SUB-PROJECTS

We analysed 20 Projects (including 144 sub-projects) in 18 Countries with Economies in Transition for the book.

Armenia (GEF project ID: 1226)
6 sub projects in the refrigeration and aerosol sectors and technical assistance. 5 UNDP sub-projects (recovery and recycling; monitoring the RMP; awareness and incentives; commercial refrigeration phase-out; aerosol manufacture phaseout) and 1 UNEP sub-project (institutional strengthening and customs training).

Azerbaijan (GEF project ID: 463)
6 sub-projects in refrigeration, one sub-project in halons, one recovery and recycling sub-project for refrigerants and two technical assistance and training components.

Belarus (GEF project ID: 108)
9 sub-projects to phase out direct annual consumption of at least 806 metric ODP tons or about 77 per cent of Belarus's 1994 consumption, through phaseout of CFCs in three refrigeration manufacturing and servicing sub-projects, and five sub-projects in the solvent industry. In addition, it supported a small programme in technology transfer and training for the fire protection sector, and a national ODS Phaseout unit.

Bulgaria (GEF project ID: 93)
12 sub-projects to phase out sixty-five per cent of Bulgaria's 1993 consumption (468 metric tons ODP) through sub-projects in the refrigeration, foam-blowing and solvents

sectors. A recycling and servicing component was planned to phase out further ODS already in use in refrigerator, and a third component will strengthen the national Phase-out Task Force. The project consisted of technology conversion in 12 sub-projects in seven enterprises and technical assistance and training component involving 4 sub-projects in the Institute of Refrigeration and the Ministry of Environment.

Czech Republic (GEF project ID: 588)
5 sub-projects to eliminate production of CFCs in the Czech Republic (approximately 2000 tons ODP annually in the early 1990s). It established a national refrigerant Recovery/Reclamation/Recycling (3R) programme. It phased out CFCs in certain commercial, industrial and transport refrigeration systems, as well as introduced low and non-ozone-depleting foam technologies. An estimated 390 metric tons ODP was scheduled for phaseout.

Estonia (GEF project ID: 768)
3 sub-projects to implement a comprehensive National Programme for Recovery/Recycling of refrigerants in the refrigeration and air conditioning sub-sectors as part of the Refrigerant Management Plan (RMP).

Hungary (GEF project ID: 94)
13 sub-projects to phase out over half of Hungary's 1993 consumption (approximately 1150 metric tons ODP) through 13 sub-projects in the solvents, foam, aerosol, halon and refrigeration sectors, and through a recovery, recycling and reclamation component.

Kazakhstan (GEF project ID: 769)
5 sub-projects to phase out 612.91 ODP tons, approximately 47 per cent of the total ODS consumption at the1998 level. The rest of the ODS in use, amounting to 691.94 ODP tons, would be eliminated through introduction of a ban on imports of equipment using and containing ODS, through import quotas, technological improvements and economic incentives alongside the support of various public awareness campaigns.

Latvia (GEF project ID: 343)
3 sub-projects to assist phaseout of ODS.

Lithuania (GEF project ID: 344)
6 sub projects: two sub-projects in refrigeration, one sub-project in aerosols, one for recovery and recycling for refrigerants and two technical assistance and training components.

Poland (GEF project ID: 115)
6 sub-projects to phase out at least 1054 metric ODP tons or about 55 per cent of Poland's 1994 weighted consumption, through phaseout of CFCs in three refrigeration sub-projects, two sub-projects in other industries requiring foam-blowing, and one sub-project in the medical aerosol industry.

Russia (GEF project ID: 74, 114, and 655)
31 sub-projects. Investment Component to cover the refrigeration servicing, medical aerosol, non-insulating foam, solvent, and fire protection sectors as well as the original aerosol and refrigeration sectors.

Slovak Republic (GEF project ID: 590)
2 sub-projects eliminate annual consumption of 280 metric ODP tons (23 per cent of 1991 consumption) through phasing out the use of CFCs in two Slovakian manufacturing refrigerators and freezers enterprises.

Slovenia (GEF project ID: 589)
6 sub-projects to phase out approximately 36 per cent (345 metric tons ODP) of Slovenia's 1993 consumption through six sub-projects in the refrigeration, foams, aerosol and solvent sectors.

Tajikistan (GEF project ID: 15)
3 sub-projects. Technical assistance at the institutional level; phaseout priority in the refrigeration sector to enable the phase out of 91 MT of annual ODS consumption required for compliance with MP control provisions.

Turkmenistan (GEF project ID: 593)
5 sub-projects comprising one recovery and recycling sub-project for refrigerants and four technical assistance and training components.

Ukraine (GEF Project ID: 107 and 2147)
17 sub-projects .16 sub-projects in the refrigeration, aerosol, foam and solvent sectors, phasing out annual consumption of approximately 2273 tons of ODS, equivalent to 69 per cent of Ukraine's 1994 consumption. One project targeted methyl bromide phase-out.

Uzbekistan (GEF Project ID: 594)
6 sub-projects: one in refrigeration, one recovery and recycling sub-project for Refrigerants and four technical assistance and training components.

Regional Project (GEF Project ID: 1305)
Initiating early phaseout of methyl bromide through awareness raising, policy development and demonstration/training activities (Bulgaria, Czech Republic, Estonia, Hungary, Latvia, Lithuania, Poland, Slovak Republic).

Regional Project (GEF Project ID: 2118)
Total sector methyl bromide phaseout in countries with economies in transition.

A Technology Transfer Agreement

Beamech Group Limited
CO-2 PROCESS LICENCE (FLC)
MONTREAL PROTOCOL MULTILATERAL FUND
UNIDO / UROSAN

This agreement is made the day of between Beamech Group Limited whose principle office is at Tenax Road, Trafford Park, Manchester, M17 1JT, England (hereinafter referred to as 'the Licensor') of the one part and Urosan Kimya Sanayii A.S. whose principal office is at Cumhuriyet cad No. 361, 80230 Harbiye, Istanbul, Turkey (hereinafter referred to as 'the Licensor') of the other part.

Whereas:

1. The Licensor possess patent rights, technical knowledge, know-how and expertise relating to processes and apparatus for manufacturing polyurethane or other cellular foam using a continuous or discontinuous process utilizing the addition of CO_2.

2. Insofar as the aforesaid patent rights, technical knowledge, know-how and expertise relate to foam manufacturing apparatus the Licensor has granted the exclusive right to manufacture and sell such apparatus to Beamech Group Limited of Tenax Road, Trafford Park, Manchester, England and its sub-licensees (hereinafter individually referred to as a 'Licensed Apparatus Manufacturer').

3. The parties hereto have agreed to enter into this agreement upon the terms and conditions hereinafter contained.

4. The payment for the License will be made by the United Nations Industrial Development Organisation (UNIDO) P.O. Box 300, A-1400, Vienna, Austria under UNIDO Contract No. 97 / 201 with the Licensor.

Now it is hereby agreed as follows:

1. Definitions

In this agreement the following expressions shall have the following meanings respectively:

(a) 'the Location(s)' shall mean Urosan Kimya Sanayii A.S., Cumhuriyet cad No. 361, 80230 Harbiye, Istanbul

(b) 'the Territory' shall mean Turkey

(c) 'the Patents' shall mean the Letters Patent short particulars of which are set out in Part I of the Schedule hereto

(d) 'the Applications' shall mean the application or applications for Letters Patent short particulars of which are set out in Part II of the Schedule hereto

(e) 'the know-how' shall mean the technical knowledge know-how and expertise possessed by the Licensor for carrying out the Processes as hereinafter defined

(f) 'the Products' shall mean processes or methods for producing the products as described in the specification of the Patents or the Applications

(g) 'the Apparatus' shall mean apparatus for producing the products by any of the processes which apparatus is for the time being protected by any patent in force or applied for within the territory and belonging to or licensed exclusively to the Licensor and which apparatus is to be acquired by the Licensee under UNIDO Contract No. 97 / 201 with the Licensor from a Licensed apparatus Manufacturer pursuant to Clause 3 hereof

2. Grant of license

(a) Subject to the terms and conditions hereinafter set forth the Licensor grants to the Licensee a non-exclusive license to use the Processes on the apparatus at the Location(s) and to use and sell within or without the Territory the Products produced by the Processes.

(b) The license granted under sub-clause (a) hereof is personal to the Licensee and excludes any right to assign the license or to grant sub-licenses.

(c) The License granted under Clause 2(a) hereof shall include a license under each of the Patents.

3. Apparatus

(a) The licensee will obtain the apparatus from a Licensed Apparatus Manufacturer under UNIDO Contract No. 97/201 with the Licensor.

(b) The Licensor undertakes to use its best endeavours on behalf of the Licensee to ensure that a Licensed Apparatus Manufacturer supplies installs and commissions the Apparatus at the Location(s) without unreasonable delay.

4. Payment for the License

Payment for the license hereby granted in the total all-inclusive amount of US $50,000 will be made by UNIDO under Contracts No. 97 / 201 with the Licensor.

5. Taxes

Insofar as any payment as provided for in Clause 4 herein is subject to or shall become subject to any tax or other deductions whatsoever in the Territory such tax or deduction shall be for the account of the Licensee absolutely so that the payment received by the Licensor shall be the full amount of the payment as provided for in Clause 4 herein free of all liability to such tax or deduction.

6. Licensor's Obligations

The Licensor during the continuance of this agreement hereby:

(a) Undertakes to communicate the know-how to the Licensee to be used by the Licensee for the purposes of manufacturing the Products as the Location(s) by the application of the processes and for no other purpose.

(b) Undertakes at the expense and at the request of the licensee to send to the Location(s) a qualified technical representative of the Licensor to assist the Licensee in the use of the know-how and the processes for the purposes aforesaid and for no other purpose.

(c) Undertakes to communicate to the Licensee without delay particulars of all inventions, know-how, technical knowledge and expertise relating to the processes [which cannot be used without infringing the patents or any letters patent granted pursuant to the applications] which may be acquired in the future by the licensor for use by the Licensee for the purposes aforesaid and for no other purpose and any patent or application for patent arising from such inventions shall be a patent or application for patent within the scope of this agreement but for the avoidance of doubt no other invention, know-how, technical information knowledge or expertise belonging to or acquired by the Licensor shall be subject to or within the scope of this agreement.

(d) Undertakes to pay all fees and expenses for obtaining and maintaining in force the patents and the applications and other patents or application for patent within the scope of this agreement provided that this undertaking shall not obligate the Licensor to prosecute any patent application which in its discretion it may deem appropriate or to defend any proceedings for the revocation of any patent.

7. Licensee's obligations

The licensee during the continuance of this agreement

(a) Undertakes to communicate to the Licensor without delay all inventions, know-how, technical knowledge and expertise relating to the process acquired by the Licensee and to assign to the Licensor all rights to apply for or hold patents relating to such inventions. Any such patents or applications for patent shall thereafter be within the scope of this agreement.

(b) Undertakes not to raise or cause to be raised or assist others to raise any question concerning the validity of the patents or any objection to the grant of letters patent on the applications or any such question or objection to the grant of letters patent or applications for patent within the scope of this agreement on any ground whatsoever.

(c) Undertakes in terms similar to the undertaking given in sub-clause (b) of this clause in relation to any corresponding patent or application for patent in any country outside the territory.

8. Litigation and Infringement

(a) Any litigation relating to the Patents or the Applications or any letter of patent granted pursuant to the applications or any other letters patent or applications for patent within the scope of this agreement shall be initiated, defended or prosecuted at the sole discretion of the Licensor.

(b) The Licensee shall notify the Licensor of any infringement or threatened infringement of the Patents or the Applications or any letters patent granted pursuant to the applications or any other letters patent or applications for patent within the scope of this agreement which shall come to the notice of the Licensee.

(c) In relation to any litigation for infringement as aforementioned the Licensee at the request of and the expense of the Licensor give all reasonable assistance to the Licensor.

9. Liability

(a) The licensor does not guarantee and the effectiveness of this agreement shall not be affected by the validity of Patents or any other letters patent within the scope of this agreement.

(b) The Licensor shall not be responsible for any loss or damage whether direct or consequential incurred or sustained by the Licensee arising out of or incidental to this agreement or relating to the Apparatus or the use of the processes irrespective of the cause of such loss or damage.

10. Secrecy

(a) The Licensee acknowledges that any information from the Licensor in pursuance of this agreement is given by the Licensor in confidence. In Clause 8 (a) hereof and the Licensee undertakes to take such precautions against disclosure of such information to others a it would take or could be reasonably expected to take in respect of its own trade secrets.

(b) On termination of this agreement whether by effluxion of time or by the provisions of Clause 14 herein the Licensee shall be bound by the undertaking of sub-clause (a) of this clause for a period of 5 years commencing with the date of termination.

11. Duration

(a) Subject to the provisions of Clause 14 herein this agreement shall continue in force for a period of ten years from the Commencement date or until the expiry date of the last of the Patents or last of the letters patent to be granted on any of the applications whichever is greater provided that if the Licensor allow any such Patent or patent to be granted on application to lapse unintentionally and subsequently succeeds in obtaining an order for restoration of the lapsed patent then during the period between said lapse and said order for restoration this agreement shall be deemed to have continued in force.

12. Termination

(a) The licensor shall have the right to terminate this agreement and all rights hereby granted upon the happening one or more of the following events.
   i)   If the Licensee shall make default in the payment or any payments payable hereunder as and when the same shall become due.
   ii) If the Licensee shall make an arrangement with its creditors or shall go into liquidation whether compulsory or voluntary except for the purpose of amalgamation or reconstruction.
   iii) If the Licensee shall commit any breach of the agreements or undertakings on its part contained and shall fail to remedy such breach within thirty days after written notice thereof to the Licensee by the Licensor specifying the nature of the breach.

(b) Any termination under sub-clause (a0 of this clause shall be without any prejudice to the right of the Licensor to recover any payments due to the Licensor under this agreement and to the rights or remedies of either Party in respect of any antecedent breach of this agreement.

13. Assignment

Nether Party may assign its rights or obligations under this agreement to any third party without the consent of the other party but this will not prevent the Licensor from assigning its rights and obligations hereunder to any person, firm or company to whom it assigns the Patents and Applications.

14 Law

This agreement shall be read and construed according to and shall be governed by the Law of England.

15. Arbitration

In the event of a difference arising between the parties hereto on the construction of this agreement or any clause herein contained or any matter in any way connected therewith or the rights, duties and obligations of either Party hereunder it shall failing agreement between Parties hereto within three months of the date of written notice of such difference being given by one party to the other be determined in accordance with the Rules of conciliation and Arbitration of the International Chamber of Commerce by one or more arbitrators appointed in accordance with those Rules. The Arbitration shall be held at ————- in the English language.

In Witness whereof the Licensor and the Licensee have caused these presents to be executed by their duly authorized officers this day and the year first above written.

Schedule

Part I
Patent Number   Date

PART II
Application Number   Filing date

List of Military ODS Management and Phaseout Initiatives in the US

US step-by-step ODS management and phaseout initiatives

Aircraft/aircraft systems

- CFC-12 and CFC-500 replacement in fixed and rotary-wing aircraft air-conditioning.
- Replacement of ODS refrigerants in the environmental control units of Army aerial reconnaissance and airborne battle management systems.
- Replacement of ODS refrigerants, air-transportable galley lavatory.
- Installation of ODS-refrigerant recovery equipment for aircraft systems.
- CFC-113 elimination in the B-2 Aircrew Training Device programme.
- Removal of ODS solvent references from technical manifest (C/KC-135 aircraft).
- Elimination of Class I ODS solvents from aircraft manufacturing, depot maintenance and field maintenance processes.
- Installation of water-based landing-gear cleaning system.
- Implementation of ODS-free solvent alternatives, ground-radar simulators.
- CFC-12 to HFC-134a conversion in ground support equipment.
- Replacement of CFC-12 with HFC-134a (T-1A Jayhawk aircraft).
- Replacement of 1,1,1-trichloroethane in aircraft fuel-tank repair.
- Elimination of the use of ODS solvents through the use of aqueous-based cleaning systems from the repair and maintenance of Army rotary-winged aircraft.
- Elimination of the use of ODS solvents in the repair and remanufacture of aircraft honeycomb composite skins.
- Elimination of the use of ODS solvents in the repair of propeller blades and critical landing-gear components for Army rotary-winged aircraft.
- Replacement of 1,1,1-trichloroethane (conformal stripping with ODS-free alternatives).
- P-20 ramp truck conversion from halon-1211 to aqueous film-forming foam.
- P-23 CFR truck conversion from halon-1211 to dry-chemical system.
- Crash rescue truck air-conditioning conversion.
- Removal of halon-based fire-suppression system (C-17 aircrew training system).
- Replacement of halon-1301 portable extinguishers with CO_2 units.
- Development and implementation of the first non-halon aircraft engine nacelle fire suppression systems in the world.
- Development and implementation of non-halon dry bay fire suppression systems.
- Development and implementation of on-board inert gas generator systems (OBIGGS) for fuel tank explosion inertion.

- Replacement of halon-1211 fire extinguishers on army flightlines (the staging areas of a military airport) with dry chemical, carbon dioxide and aqueous foam systems.

Missile and satellite systems
- ODS solvent elimination, Missile Warning and Space Surveillance System.
- Replacement of CFC-113 with non-Class I ODS alternative in NAVSTAR (satellite) global positioing system (GPS).
- ODS elimination in processing of solid rocket motor unit in Titan IV, Stage IV shipping and storage container environmental control unit in Peacekeeper missile.
- Refrigerant replacement, ballistic-missile early-warning system in Cyber Mainframe Computer.
- Brine chiller modification in Minuteman and Peacekeeper missiles.
- CFC-12 chiller modification in Minuteman guidance and control system (G&C).
- Air-conditioning service certification in aerospace ground equipment.
- On-pad spacecraft ODS-free air-conditioning in Delta-II launch vehicle.
- Replacement of the use of ODS solvents in the manufacture of Stinger, Hellfire and TOW missiles.

Ships, submarines and other watercraft
- CFC-12 replacement using HFC-134a and HCFC-22 retrofit kits in watercraft refrigeration and cooling equipment, US Army.
- Retrofit of halon-1301 fire suppression systems on Army watercraft with HFC-227ea/water spray systems.
- Development and implementation of HFC-227ea, water mist and HFC-227ea/water spray fire suppression systems to replace halon-1301 in new-construction ships.
- CFC-114 replacement using HFC-236fa retrofit kits in Navy surface ship air-conditioning (chillers).
- CFC-12 replacement using HFC-134a retrofit kits in Navy surface ship air-conditioning and refrigeration systems and submarine refrigeration systems.
- Development of shipboard HFC-134a high-efficiency air-conditioning systems (chillers) for use on new-construction ships.
- Retrofit of tactical CFC-12 air-conditioning, refrigeration and water-cooling systems on Army Logistics Support Vessels (LSVs).
- Retrofit of halon-1301 and halon-1211 fire suppression systems in the engine rooms of Army landing craft, Logistics Support Vehicles, tugboats and bridging support vessels.

Field equipment
- CFC-12 replacement in field refrigeration equipment.
- Replacement of CFC-12 in tactical shelter air-conditioning.
- Replacement of thousands of CFC-12 Army Environmental Control Units.

Ground combat vehicles
- ODS removal from Milstar mobile constellation control vehicles and Milstar communication vehicles.
- CFC-12 replacement in field ambulance air-conditioning systems.
- Retrofit of the air-conditioning systems in Army ambulances, replacing CFC-12 with HFC-134a refrigerant.
- Elimination of ODS solvents from the depot repair and maintenance of Army ground combat vehicles.

- Retrofit of the engine compartment fire suppression system of the Abrams Main Battle tank, replacing halon-1301 with dry powder.
- Retrofit of the engine compartment fire suppression system of the Bradley Fighting Vehicle System, replacing halon-1301 with HFC-227ea.
- Retrofit of the engine compartment fire suppression system of the Multiple Launch Rocket System, replacing halon-1301 with HFC-227ea.
- Retrofit of the engine compartment fire suppression system of the M9 Armored Combat Earthmover, replacing halon-1301 with dry powder.
- Retrofit of the engine compartment fire suppression system of the M992 Field Artillery Ammunition Supply Vehicle, replacing halon-1301 with HFC-227ea.
- Retrofit of the engine compartment fire suppression system of the Fox M93A1 Nuclear, Biological and Chemical Reconnaissance System, replacing halon-1301 with nitrogen.
- Replacement of halon-1301 with carbon dioxide in the hand-held fire extinguisher of all Army ground systems except the Abrams Main Battle Tank.
- Replacement of halon-1301 hand-held fire extinguisher of the Abrams Main Battle Tank with one using a 50/50 mixture of water and potassium acetate.
- Development and implementation of the world's first non-halon crew compartment explosion suppression system.

General
- DoD participation in the development of an industry standard for recycled halon-1301.
- CFC-12 replacement in water chillers.
- DoD participation in the Halon Alternatives Research Corporation.
- DoD participation in the International Cooperative for Environmental Leadership (formerly the Industry Cooperative for Ozone Layer Protection (ICOLP)) – an industry group dedicated to identifying and testing ODS-free solvents.
- DoD participation in the Next Generation Program (NGP) and Advanced Agent Working Group (AAWG).

Munitions
- General purpose bomb-and-fuse ODS solvent elimination in conventional weapons.
- Elimination of ODS solvents in the manufacture of small-calibre ammunition for US DoD, NATO and other allies.
- Elimination of the use of ODS solvents in the manufacture of mines, Bangalore torpedoes, flares, smoke-generation systems and many other battlefield munitions.
- Elimination of ODS solvents in the formulation of tactical high-explosive materials.

Useful Websites for Information on Military Phaseout

OzonAction Programme
www.unep.fr/ozonaction
Electronic versions of many publications, as well as an exhaustive list of available publications, can be obtained from the OzonAction Programme.

Enviro$ense/Integrated Solvent Substitution Data System (ISSDS)
www.epa.gov/envirosense/
This website is a gateway to a number of databases on the internet that contain substitution and process alternatives to ODSs. Inquiries can be addressed to more than one database at a time so that information can be retrieved from several with just the one query.

Environment Canada
www.ec.gc.ca/ozone/index.htm
Technical, policy and general background information related to ozone protection, the Montreal Protocol and ODS substitutes.

Halon Alternatives Research Corporation (HARC)
www.harc.org
HARC is a voluntary, non-profit trade association formed by concerned halon users and the fire protection industry. It assists halon users to redeploy existing halon banks from applications where alternatives have replaced halons to those applications still requiring halons. HARC facilitates halon recycling, helps determine critical use, acts as an information clearinghouse and is a focal point for national/international halon recycling.

US Air Force Centre for Environmental Excellence (AFCEE)
www.afcee.brooks.af.mil/pro-act
AFCEE is a field-operating agency of the Civil Engineer of the US Air Force, providing a complete range of environmental, architectural and landscape design, planning and construction management services and products. The site includes success stories, fact sheets and discussion forums that address ozone protection and other environmental topics affecting the US Air Force.

US Army Environmental Support Office (ESO)
www.environmentalsupportoffice.com
Previously known as the US Army Acquisition Pollution Prevention Support Office, the ESO is the central manager of the US Army Ozone Depleting Substances Elimination Program, tasked with eliminating ODSs from Army weapon systems and industrial operations; the manager of the US Army ODS Reserve; and the source for US Army regulations and expertise on ODS elimination policies and projects.

The US EPA's Solvent Alternatives Guide (SAGE)
http://clean.rti.org/
A comprehensive online guide designed to provide pollution prevention information on non-ODS solvent and process alternatives for component cleaning and degreasing. It operates both as an 'expert system', evaluating alternative processes and chemicals for particular problems, and as a hypertext manual on cleaning alternatives.

The US EPA Stratospheric Ozone Protection Homepage
http://earth1.epa.gov/ozone/
Website with information on the science of ozone depletion, US regulations designed to protect the ozone layer, information on methyl bromide, flyers about the UV index, information for the general public and other topics.

Joint Service Pollution Prevention Technical Library
http://p2library.nfesc.navy.mil/
Website with information on environmental issues and technologies, including ODS alternatives for use by DoD installations and operations.

Next-Generation Fire Suppression Technology Program (NGP)
www.bfrl.nist.gov/866/NGP/index.htm

Ozone and Climate Protection Awards Won by Military Organizations

The Montreal Protocol is a historic success story on many fronts – democratic, scientific, diplomatic, regulatory and business – and in the area of technology cooperation between the public and private sectors. All the successes in protecting the stratospheric ozone layer depended on the work of confident and committed individuals and organizations.

Awards presented by the United Nations, the Nobel Committee, national environmental authorities and other organizations are one way to identify the most significant contributions in ozone protection. These awards recognize exceptional leadership, personal dedication and technical achievements in eliminating ODSs. Individuals and organizations from dozens of countries – including many from military organizations – have earned these prestigious awards. They exemplify those who have accomplished the extraordinary, the brilliant and the merely successful, serving as a source of inspiration and encouragement and reminding us that it is often necessary to stand up and act based on what science says is necessary and take risks to protect the global environment. A list of the awards given to leaders of the military ODS phaseout is given below.

US STRATOSPHERIC OZONE PROTECTION AWARDS FOR MILITARY LEADERSHIP

1997 US EPA Best-of-the-Best Awards

Military organizations and military contractors
- Lockheed Martin Corporation
- Raytheon/Texas Instruments Systems
- Thiokol/NASA
- US Air Force Space Launch Programs
- US Army Acquisition Pollution Prevention Support Office
- US Department of Defense (DoD)
- US Naval Research Laboratory
- US Naval Surface Warfare Center

Associations supporting military leadership
- Halons Alternatives Research Corporation (HARC)
- Institute for Interconnecting and Packaging Electronics Circuits (IPC)
- CFC Benchmarking Team
- International Cooperative for Environmental Leadership (ICEL)

Individuals for military contribution
- Philip J. DiNenno, Hughes Associates
- Stephen P. Evanoff, Lockheed Martin Corporation
- Joe Felty, Raytheon/Texas Instruments Systems
- Joel Krinsky, US Navy
- E. Thomas Morehouse, Jr, US Department of Defense
- Ronald W. Sibley, US Defense Logistics Agency
- Daniel Verdonik, US Army
- Gary D. Vest, US Department of Defense

Annual US EPA Stratospheric Ozone Protection Awards

1990 to individuals for military contribution
- Joe Felty, Texas Instruments
- Kathi Johnson, China Lake Navy Weapons Center Electronics Manufacturing Facility
- Robin Sellers, Naval Avionics Center, US Navy

1991 to individuals for military contribution
- Thomas E. Daum, US Defense Reutilization and Marketing Service, Defense Logistics Agency
- E. Thomas Morehouse, Jr, US Air Force

1992 to military organizations and military contractors
- The Boeing Company
- British Aerospace Airbus
- General Dynamics, Space Systems Division
- General Dynamics, Fort Worth Division
- Naval Air Warfare Center, Lakehurst, US Navy
- York International

1992 to associations
- Halon Alternative Research Corporation
- Halon Essential Use Panel, EPA, Victoria, Australia
- Royal Norwegian Navy Materiel Command
- US Army Acquisition Pollution Prevention Support Office

1992 to individuals for military contribution
- Bryan H. Baxter, British Aerospace
- Philip J. DiNenno, Hughes Associates
- Stephen Peter Evanoff III, General Dynamics
- Colin Lewis, UK Ministry of Defence
- Tony L. Phillips, General Dynamics
- Henry J. Weltman, General Dynamics

1993 to military organizations and military contractors
- Defense Electronics Supply Center, Columbus
- Defense Logistics Agency
- Department of the Navy, US Chief of Naval Operations
- Hill Air Force Base

- Hughes Aircraft
- Kelly Air Force Base, Texas
- Martin Marietta Astronautics
- Naval Aviation Depot, Cherry Point
- Naval Aviation Depot, Norfolk
- Rockwell International/US Army Air-to-Ground Missile Systems Project Office
- Texas Instruments, Missile Systems Division
- Thiokol, Space Operations
- US Air Force, Air Base Fire Protection and Crash Rescue Systems Branch

1993 to associations
- Industry Cooperative for Ozone Layer Protection (ICOLP/ICEL)

1993 to individuals for military contribution
- Nicholas T. Castellucci, Northrop Grumman
- Timothy Crawford, Electronic Manufacturing Production Facility
- John Fischer, Naval Air Warfare Center
- Terry Schaumberg, San Antonio Air Logistics Center
- Angie Criser Schurig, Texas Instruments
- John R. Stemniski, The Charles Stark Draper Laboratory
- Robert E. Tapscott, New Mexico Engineering Research Institute
- Gary D. Vest, Principal Assistant Deputy Under Secretary of Defense (Environmental Security)
- Clare Vinton, National Center for Manufacturing Sciences

1994 to military organizations and military contractors
- Aeronautical Systems Center, Wright Laboratory, Aircraft Halon Replacement Team
- The Aerospace Guidance and Metrology Center, Newark Air Force Base
- US Army Communications-Electronics Command/Tobyhanna Depot
- Falcon Halon Team, Wright-Patterson Air Force Base
- Honeywell
- Lockheed
- Martin Marietta
- Norsk Forsvarsteknologi (Norway)
- Northrop Grumman
- Saab-Scania (Sweden)

1994 to individuals for military contribution
- James A. Fain, Jr, Aeronautical Systems Center, Wright-Patterson Air Force Base
- Mary Beth Fennell, Naval Aviation Depot, Cherry Point
- Joel Krinsky, US Navy
- Barbara Kucnerowicz-Polak, State Fire Service (Poland)
- Marion McQuaide, UK Ministry of Defence
- Steven Rasmussen, Hill Air Force Base
- Franklin Sheppard, Jr, Office of the Chief of Naval Operations, US Navy
- Ronald Sibley, Defense Logistics Agency
- Jack Swindle, Texas Instruments
- James Vincent, US Army Aviation and Troop Command

1995 to military organizations and military contractors
- Aberdeen Test Center, US Army
- Advanced Cruise Missile DSO, US Air Force
- AGM-130 Systems Program Office, US Air Force
- Annapolis Detachment, Carderock Division, Naval Surface Warfare Center, US Navy
- Defence Institute of Fire Research (India)
- GEO-CENTERS
- Low-Residue Soldering Task Force
- Navy Technology Center for Safety and Survivability, US Naval Research Laboratory
- Texas Instruments
- Titan IV Program ODS Reduction Team

1995 to individuals for military contribution
- Neil Antin, US Naval Sea Systems Command
- David Breslin, US Naval Sea Systems Command
- Robert Gay, US Defense Logistics Agency
- Casey Grant, National Fire Protection Association
- Michael C. Grieco, ICBM Systems Program Office, US Air Force
- Michael J. Leake, Texas Instruments
- Cynthia Lingg, US Air Force
- C. K. Marfatia, Real Value Appliances (India)
- Daniel P. Verdonik, Hughes Associates
- Hans U. Wäckerlig, Swiss Institute for the Promotion of Safety and Security

1996 to military organizations and military contractors
- Advanced Amphibious Assault Vehicle, US Marine Corps
- Center for Technical Excellence for ODS Solvents, Corpus Christi Army Depot
- Draper Laboratory
- F/A-18 Program Office and V-22 Program Office, US Navy
- Tank-Automotive Research, Development and Engineering Center, US Army
- ICBM System Program Office, US Air Force
- Lockheed Martin Aeronautical Systems
- Lockheed Martin Skunk Works
- Philadelphia Detachment of the Carderock Division of the Naval Warfare Center, US Navy

1996 to individuals for military contribution
- Thomas A. Bush, US Army
- John King, US Air Force
- Peter Mullenhard, US Navy Clearinghouse
- Larry Novak, Texas Instruments
- Ronald S. Sheinson, US Naval Research Laboratory

1997 to military organizations and military contractors
- Lockheed Martin Michoud Space Systems
- Naval Sea Systems Command
- Navy Strategic Systems Programs Fire Control and Guidance Branch
- RAH-66 Comanche Program Manager's Office

- US Army, Pacific
- HQ US Air Force Weapon System ODS Management Team

1997 to individuals for military contribution
- Robert L. Darwin, Naval Sea Systems Command
- Richard L. Helmick, Naval Surface Warfare Center
- John A. Manzione, US Army CECOM
- Leroy E. Sanderson, US Marine Corps
- Bruce G. Unkel, Naval Sea Systems Command

1998 to military organizations and military contractors
- General Headquarters of the State Fire Service of Poland
- Lockheed Martin Missiles and Space
- Lockheed Martin Tactical Aircraft Systems
- US Marine Corps, Advanced Amphibious Assault
- US Navy LPD 17 Amphibious Transport Dock Ship Team
- US Navy New Attack Submarine Program Office

1998 to associations supporting military leadership
- Halon Alternative Options Committee (India)

1998 to individuals for military contribution
- Stephen O. Andersen, US EPA
- Michelle Maynard Collins, NASA
- David A. Koehler, Ocean City Research
- Thomas M. Landy, US Army Tank-Automotive and Armaments Command
- Gregory Toms, US Naval Sea Systems Command

1999 to military organizations and military contractors
- Canadian Forces Fire Marshal, Department of National Defense
- Idaho Army National Guard, Combined Support Maintenance Shop
- Project Management Office for Bradley Fighting Vehicle Systems

1999 to individuals for military contribution
- Robert T. Wickham, Wickham Associates

2000 to military organizations and military contractors
- US Air Force Research Laboratory

2000 to individuals for military contribution
- David Ball, Kidde (UK)
- David J. Liddy, Ministry of Defence (UK)
- Steve McCormick, US Army Tank-Automotive Command

2001 – EPA Stratospheric Awards not presented due to scheduling change.

2002 to individuals for military contribution
- Nikolai Kopylov, All-Russian Research Institute for Fire Protection, Russia
- James Frederick O'Bryon, US Department of Defense

- Reva Rubenstein, US Environmental Protection Agency
- Darrel A Staley, The Boeing Company
- Howard L. Wesoky, NASA

2003 – No EPA Stratospheric Awards Presented to military entities.

2004 to military organizations and contractors
- NASA White Sands Test Facility
- Netherlands–US Military and Environmental Halon Leadership Team

2005 to individuals for military contribution
- Mark L. Robin, Hughes Associates

2006 to organizations for military contribution
- Environmental Investigation Agency

Other international awards for military ozone protection

1994 International Cooperative for Ozone Layer Protection Global Achievement Award
- Stephen O. Andersen, US EPA

1997 Vietnam Ozone Award
- People's Army Newspaper

2001 US DoD Award for Excellence
- Stephen O. Andersen, US EPA

Notes

CHAPTER 2

1   United Nations Conference on Trade and Development (2001a) *Environment*, UNCTAD/ITE/IIT/23, United Nations, New York and Geneva.
2   United Nations Conference on Trade and Development (2001a), op cit (Note 1).
3   United Nations Conference on Trade and Development (2001b) *Achieving Objectives of Multilateral Environmental Agreements: A Package of Trade Measures and Positive Measures*, Jeena Vha and Ulrich Hoffman (eds), United Nations, New York and Geneva.
4   Intergovernmental Panel on Climate Change (2001) *Methodological and Technological Issues in Technology Transfer: Special Report of the Intergovernmental Panel on Climate Change*, Cambridge University Press, Cambridge, UK.
5   United Nations Conference on Trade and Development (2001c) *Transfer of Technology*, United Nations, Geneva, available at www.unctad.org/en/docs/psiteiitd28.en.pdf.
6   Maskus, Keith E. (2004) *Encouraging International Technology Transfer*, International Centre for Trade and Sustainable Development (ICTSD) and United Nations Conference on Trade and Development (UNCTAD), Geneva.
7   Kranzberg, M. (1986) 'The technical elements in international technology transfer: Historical perspectives', in J. R. McIntyre and D. S. Papp (eds) *The Political Economy of International Technology Transfer*, Quorum Books, New York, pp31–46.
8   Intergovernmental Panel on Climate Change (2001), op cit (Note 4).
9   Andersen, S. and Sarma, K. M. (2002) *Protecting The Ozone Layer: The United Nations History*, Earthscan, London.
10 Intergovernmental Panel on Climate Change (2001), op cit (Note 4).
11 Intergovernmental Panel on Climate Change (2001), op cit (Note 4).
12 Science and Development Network, excerpt from 'Technology transfer', www.scidev.net/dossiers/index.cfm?fuseaction=dossierfulltext&Dossier=12, accessed 6 March 2007.
13 International Environmental Technology Centre and United Nations Environment Programme (2003) *The Seven 'C's for the Successful Transfer and Uptake of Environmentally Sound Technologies*, International Environmental Technology Centre and United Nations Environment Programme, Osaka, Japan.
14 United Nations Conference on Trade and Development (2001a), op cit (Note 1).
15 United Nations Conference on Trade and Development (2001c), op cit (Note 5), p63.
16 United Nations Conference on Trade and Development (2001c), op cit (Note 5).
17 Organisation for Economic Co-operation and Development (2005) 'Achieving the successful transfer of environmentally sound technologies: Trade related aspects', Working Paper No 2005-02, OECD, Paris, p8.
18 Organisation for Economic Co-operation and Development (2005), op cit (Note 17).
19 Intergovernmental Panel on Climate Change (2001), op cit (Note 4).
20 Stephen O. Andersen, Ajay Mathur, Sukumar Devotta, Daniel M. Kammen, Jeanne Townend and Laura Van Wie McGrory (1998), 'Conceptual framework for organizing

chapters', developed for Intergovernmental Panel on Climate Change (2001), op cit (Note 4).

21   Intergovernmental Panel on Climate Change (2001), op cit (Note 4).

22   Maskus (2004), op cit (Note 6).

23   United Nations Conference on Trade and Development (2000) 'The role of publicly funded research and publicly owned technologies in the transfer and diffusion of environmentally sound technologies', available at www.unctad.org/en/docs/psiteiipd9.en.pdf, p25.

24   Zaelke, D., Kaniaru, D. and Kcuzikova, E. (eds) (2005) *Making Law Work: Environmental Compliance and Sustainable Development*, Cameron May, London; Hunter, D., Sulzman, J. and Zaelke, D. (2007) *International Environmental Law and Policy*, 3rd Edition, Foundation Press, New York

25   Andersen, S. O. and Zaelke, D. (2003) *Industry Genius: People and Inventions Protecting the Climate and the Fragile Ozone Layer*, Greenleaf Press, London. This book documents eight companies and two government programmes whose visionary leadership in technology innovation helped both the environment and the companies' bottom line.

26   This should not be interpreted to mean that licensing only applies to products. Anything that is covered by intellectual property right laws (such as processes, trademarks and copyrighted materials) can be licensed.

27   Maskus (2004), op cit (Note 6).

28   Intergovernmental Panel on Climate Change (2001), op cit (Note 4); Maskus (2004), op cit (Note 6).

29   Maskus (2004), op cit (Note 4).

30   Intergovernmental Panel on Climate Change (2001), op cit (Note 4).

31   Corral, C. M. (2002) *Environmental Policy and Technological Innovation: Why Do Firms Adopt or Reject New Technologies?* Edward Elgar, Cheltenham, UK.

32   Intergovernmental Panel on Climate Change (2001), op cit (Note 4), Section 1.5: 'Stakeholders, decisions and policies'.

33   Czech Republic Project Completion Report, Document No 18384, World Bank, August 1998.

34   Strelneck, D. and Linquiti, P. (1999) 'Environmental technology transfer to developing countries: Practical lessons learned during implementation of the Montreal Protocol', ICF Consulting, presented at the 17th Annual Research Conference of the Association for Public Policy and Management, Washington, DC, November 1995.

35   Intergovernmental Panel on Climate Change (2001), op cit (Note 4).

36   Organisation for Economic Co-operation and Development (1997) 'National innovation systems', OECD, Paris, available at www.oecd.org/dataoecd/35/56/2101733.pdf.

37   Organisation for Economic Cooperation and Development (1997), op cit (Note 36).

38   Intergovernmental Panel on Climate Change (2001), op cit (Note 4); UN Millennium Project (2005) *Innovation: Applying Knowledge in Development*, Earthscan, London; Organisation for Economic Co-operation and Development (1997), op cit (Note 36).

39   Organisation for Economic Co operation and Development (2005), op cit (Note 17).

40   Jaffe, A. B., Newell, R. G. and Stavins, R. (2001) 'Technological change and the environment', Discussion Paper 00-47REV, Resources for the Future, Washington, DC.

41   Robalino, David A. (2000) *Social Capital, Technology Diffusion and Sustainable Growth in the Developing World*, RAND, Washington, DC, Publication Number RGSD-151; Blackman, Allen (1999) 'The economics of technology diffusion: Implications for climate policy in developing countries', Discussion Paper 99-42, Resources for the Future, Washington, DC; Jaffe et al (2001), op cit (Note 40); IPCC (2001), op cit (Note 4).

42   Organisation for Economic Co-operation and Development and the International Energy Agency (2003) *Technology Innovation, Development and Diffusion*, OECD Publications, Paris.

43 Blackman (1999), op cit (Note 41); United Nations Conference on Trade and Development (2001c), op cit (Note 5); Intergovernmental Panel on Climate Change (2001), op cit (Note 4).

44 Blackman (1999), op cit (Note 41).

45 Intergovernmental Panel on Climate Change (2001), op cit (Note 4); United Nations Conference on Trade and Development (2001c), op cit (Note 5).

46 Intergovernmental Panel on Climate Change (2001), op cit (Note 4).

47 Robalino (2000), op cit (Note 41); Blackman (1999), op cit (Note 41); Intergovernmental Panel on Climate Change (2001), op cit (Note 4); United Nations Conference on Trade and Development (2001c), op cit (Note 5).

48 Robalino (2000), op cit (Note 41).

49 Maskus (2004), op cit (Note 6); Intergovernmental Panel on Climate Change (2001), op cit (Note 4); United Nations Conference on Trade and Development (2001c), op cit (Note 5).

50 Blackman (1999), op cit (Note 41).

51 Some authors suggest that the availability of leasing and finance for leasing is a key factor in the diffusion of expensive and risky technologies: by leasing a technology rather than purchasing it outright, the firm reduces the risk of its investment.

52 Blackman (1999), op cit (Note 41); Robalino (2000), op cit (Note 41); Stoneman, Paul (2002) *The Economics of Technological Diffusion*, Blackwell Publishers, Oxford.

53 International Energy Agency and Organisation for Economic Co-operation and Development (2003) *Technology Innovation, Development and Diffusion*, OECD Publications, Paris.

54 Intergovernmental Panel on Climate Change (2001), op cit (Note 4).

55 Stoneman, Paul (2002), op cit (Note 52).

56 Organisation for Economic Co-operation and Development (2005), op cit (Note 17).

57 'What is GEF', www.gefweb.org/What_is_the_GEF/what_is_the_gef.html.

58 'Projects', www.gefweb.org/Projects/projects-projects/projects-projects.html.

59 'Operational policies', http://thegef.org/Operational_Policies/Eligibility_Criteria/eligibility_criteria.html.

60 World Bank 'Multilateral development banks', http://web.worldbank.org/WBSITE/EXTERNAL/EXTABOUTUS/0,,contentMDK:20040614~menuPK:41699~pagePK:43912~piPK:44037~theSitePK:29708,00.html.

61 World Bank 'Multilateral development banks', op cit (Note 60).

CHAPTER 3

1   The science portion of this chapter is condensed from Lani Sinclair (2002) 'The science of ozone depletion: From theory to certainty', Chapter 1 in *Protecting the Ozone Layer: The United Nations History*, Earthscan, London. For an elaboration of science beyond that original chapter see also: Appendix 1, 'Ozone layer timelines: 4500 million years ago to present,' in *Protecting the Ozone Layer*; 'Ozone Milestones', US National Aeronautics and Space Administration/Jet Propulsion Laboratory. http://atmos.jpl.nasa.gov/milestones.htm.; Stolarski, R. (1999) 'History of the study of atmospheric ozone'. http://hyperion.gsfc.nasa.gov/Personnel/people/Stolarski_Richard_S./history.html; and Leeds, A. R. (1880) 'Lines of Discovery in the History of Ozone,' Annals of the New York Academy of Sciences, vol I, no 3, pp363–391 (as cited in Stolarski, 1999).

2   Because ozone-depleting substances were already scheduled for phaseout under the Montreal Protocol, they are not included in the Kyoto Protocol. However, SF_6, PFCs

and HFCs, which are substitutes for ozone-depleting substances in some applications, are controlled by the Kyoto Protocol. The HFCs controlled under the Kyoto Protocol have global warming potentials (GWPs) that are generally lower than the CFCs they replaced. For example, the GWP of HFC-134a is about one-sixth the GWP of the CFC-12 it replaces as a refrigerant. See 'The implications to the Montreal Protocol of the inclusion of HFCs and PFCs in the Kyoto Protocol', Report of the HFC and PFC Task Force of the Technology and Economic Assessment Panel (Ozone Secretariat, United Nations Environment Programme, October 1999).

3   Crutzen, P. (1970) 'The influence of nitrogen oxides on the atmospheric ozone content', *Quarterly Journal of the Royal Meteorological Society*, vol 96, no 408, pp320–325.

4   Lovelock, J. E., Maggi, R. J. and Wade, R. J. (1973) 'Halogenated hydrocarbons in and over the Atlantic', *Nature*, vol 241, p195 (cited in Kowalok, M. (1993) 'Common threads: Research lessons from acid rain, ozone depletion, and global warming', *Environment*, vol 35, no 6, pp12–20 and 35–38, www.ciesin.org/docs/011-464/011-464.html).

5   Harrison, H. (1970) 'Stratospheric ozone with added water vapour: Influence of high-altitude aircraft', *Science*, November 13. At the time, Boeing may not have appreciated the value of its Scientific Research Laboratories in developing science to guide product development. Boeing abandoned SST development, while British Aerospace and its customers continued – under criticism from environmental NGOs – without financial success.

6   Johnston, H. S. (1971) 'Reduction of stratospheric ozone by nitrogen oxide catalysts from supersonic transport exhaust', *Science*, vol 173, pp517–522 (cited in Stolarski, 1999, op cit, Note 1).

7   Crutzen, P. (1972) 'SSTs – A threat to the Earth's ozone shield', *Ambio*, vol I, no 2, pp41–51.

8   Glas, J. P. (1989) 'Protecting the ozone layer: A perspective from industry', National Academy of Sciences, www.nap.edu/openbook/030900426X/html/137.html. By 1975 DuPont had financed studies by J. E. Lovelock (Reading University, UK), J. N. Pitts and J. A. Taylor (University of California), C. Sandorfy (University of Montreal) and R. A. Rasmussen (Washington State University) that had confirmed that CFCs were rapidly accumulating in the atmosphere. J. W. Swinnerton (US Naval Research Laboratory) reported similar results. See McCarthy, Ray and Schuyler, Roy L. (for DuPont) (1975) 'Testimony to the Subcommittee on the Upper Atmosphere', US Senate, October.

9   Molina, M. and Rowland, F. S. (1974) 'Stratospheric sink for chlorofluoromethanes: Chlorine atom-catalyzed destruction of ozone', *Nature*, vol 249, pp810–812.

10   Brodeur, P. (1986) 'Annals of chemistry', *The New Yorker*, 9 June.

11   The Dobson instrument was so important to science that its inventor was able to publish many ground-breaking papers based on 'first-look' measurements; see, for example, Dobson, G. M. B. and Harrison, D. N. (1926) 'Measurements of the amount of ozone in the Earth's atmosphere and its relation to other geophysical conditions', *Proceedings of the Royal Society of London*, Series A, vol 110, pp660–693 (cited in Stolarski, 1999, op cit, Note 1).

12   Cagin, S. and Dray, P. (1993) *Between Earth and Sky: How CFCs Changed our World and Endangered the Ozone Layer*, Pantheon, New York.

13   Chubachi, S. (1984) 'Preliminary result of ozone observations at Syoma Station from February 1982 to January 1983', *Memoirs of National Institute of Polar Research*, Special Issue No 34: 'Proceedings of the Sixth Symposium on Polar Meteorology and Glaciology'.

14   Farman, J. S., Gardiner, B. G. and Shanklin, J. D. (1985) 'Large losses of total ozone in Antarctic reveal seasonal ClO_x/NO_x interaction', *Nature*, vol 315, pp207–210.

15   Cagin and Dray (1993), op cit (Note 12).

16   Ibid.

17   In the early 1900s methyl bromide was used as a medicine and anaesthetic and later in organic synthesis as a methylating agent (Torkelson, T. R. and Rowe, V. K. (1994) 'Methyl bromide', in R. L. Harris, L. J. Cralley and L. V. Cralley (eds) *Patty's Industrial Hygiene and Toxicology*, 3rd revised edition, March, John Wiley and Sons, New York, pp3442–3446) and as a low boiling-point solvent to extract oils from nuts, seeds and flowers (Matheson Gas Products (1980) 'Matheson Gas data book', Matheson Gas Products, East Rutherford, New Jersey, vol 6, pp361–363).

18   Carbon tetrachloride is also used in the production of plastics, semiconductors and petrol additives, and in recovery of tin from tin-plating waste.

19   Methyl chloroform is used as a solvent ingredient in aerosol products and in miscellaneous solvent, coating and adhesive applications.

20   Methyl bromide is used in organic synthesis as a methylating agent (Torkelson and Rowe (1994), op cit, Note 17) and as a low boiling-point solvent to extract oils from nuts, seeds and flowers (Matheson Gas Products (1980), op cit, Note 17).

21   Methyl bromide was used as a medicine and anaesthetic in the early 1900s.

22   CFC-11 is used as a solvent for flushing CFC refrigerating systems during service, as a solvent ingredient in aerosol products using CFC-12 propellants, and in miscellaneous solvent and adhesives applications.

23   The cumulative use of CFC-11 in miscellaneous uses was significant. Uses included tobacco expansion, wind tunnel test gases and product ingredient.

24   Minor quantities of CFC-12 were used as a propellant in hand-held aerosol fire extinguishers.

25   The cumulative use of CFC-12 in miscellaneous uses was significant. Uses included contact food freezing, explosion inerting of ethylene oxide mixtures used in pest fumigation and sterilization, vessel leak testing, nuclear fuel processing, aluminium processing, insulating between double glazing, and as a dielectric medium in scientific and medical equipment, thermostats and solar window controls, and tyre inflators.

26   The Chief Officer is nominated by the Executive Committee for appointment by the Executive Director of UNEP.

27   Annex V of UNEP/Ozl.Pro/9/12, dated 25 September 1997.

28   Up to 20 per cent of a contribution due to the Fund can be spent by a contributor on regional and bilateral cooperation if it is related to compliance with the provisions of the Protocol and meets agreed incremental costs. In practice, the Executive Committee has to approve each of the bilateral activities in advance.

29   Information from www.multilateralfund.org/implementing_agencies.htm.

30   Solomon, S., Portmann, R. W. and Thompson, D. W. J. (2007) 'Contrasts between Antarctic and Arctic ozone depletion', *Proceedings of the National Academy of Science*, Washington, DC, January.

31   UNEP (2006) *The Scientific Assessment of Ozone Depletion*, Ozone Secretariat, UNEP, Nairobi

32   Velders, G. J. M., Andersen, S. O., Daniel, J. S. Fahey, D. W. and McFarland, M. (2007) 'The importance of the Montreal Protocol in protecting climate', *Proceedings of the National Academy of Sciences*, no 104, pp4814–4819, Washington, DC, March.

33   Ibid.

CHAPTER 4

1   By 1975 DuPont had financed studies by J. E. Lovelock (Reading University UK), J. N.

Pitts and J. A. Taylor (University of California), C. Sandorfy (University of Montreal) and R. A. Rasmussen (Washington State University) that had confirmed that CFCs were rapidly accumulating in the atmosphere. J. W. Swinnerton (US Naval Research Laboratory) reported similar results. See McCarthy, Ray and Schuyler, Roy L. (for DuPont) (1975) 'Testimony to the Subcommittee on the Upper Atmosphere', US Senate, October.

2   See, for example, R. L. McCarthy's testimony of 11–12 December 1974 before the House of Representatives Subcommittee on Public Health and Environment.

3   See Orfeo, R. S. (1986) 'Response to questions to the panel from Senator John H. Chafee', *Ozone Depletion, the Greenhouse Effect and Climate Change: Hearings before the Senate Subcommittee on Environmental Pollution of the Committee on Environment and Public Works*, US GPO, Washington, DC, pp189–192, June; Strobach, Donald (1986) 'A search for alternatives to the current commercial chlorofluorocarbons', in *Protecting the Ozone Layer: Workshop on Demand and Control Technologies*, US EPA, Washington, DC.

4   Hoppe, Richard (1986) 'Ozone: Industry is getting its head out of the clouds', *Business Week*, 13 October, p110.

5   'Dear Freon™ customer', letter from DuPont to customers, dated 26 September 1986.

6   Masaaki Yamabe is Research Coordinator of the National Institute of Advanced Industrial Science and Technology, Japan, former member of the board of Asahi Glass Company, former member of Solvents TOC and the Co-Chair of Chemicals TOC under UNEP.

7   On 24 March 1988 DuPont notified customers that it was beginning an orderly phase-out of ozone-depleting substances and urged customers to seek new solutions.

8   Carnevale, Mary Lu (1988) 'DuPont plans to phase out CFC output', *The Wall Street Journal*, 25 March.

9   See Andersen, Stephen O. and Durwood, Zaelke (2002) *Industry Genius: Inventions and People Protecting the Climate and Fragile Ozone Layer*, Greenleaf, Sheffield UK; Andersen, Stephen O., Frech, Clayton and Morehouse, E. Thomas (1997) *Champions of the World: Stratospheric Ozone Protection Awards*, US EPA, Washington, DC; Andersen, Stephen O. (1997) *Newest Champions of the World: Winners of the 1997 Stratospheric Ozone Protection Awards*, US EPA, Washington, DC; Le Prestre, Philipe G., Reid, John D. and Morehouse, E. Thomas (1998) *Protecting the Ozone Layer: Lessons, Models and Prospects*, Kluwer Academic Publishers, Boston; Cook, Elizabeth (1996) *Ozone Protection in the United States: Elements of Success*, World Resources Institute, Washington, DC.

10   Osamu Kataoka is General Manager of the Global Environment Department at Daikin Industries.

11   See Andersen, Frech and Morehouse (1997), op cit (Note 9); Andersen (1997), op cit (Note 9); Le Prestre, Reid and Morehouse (1998), op cit (Note 9); Cook, Elizabeth (1996), op cit (Note 9).

12   See Berkman, Barbara N. (1999) 'AT&T's big push to stay ahead of environmental woes' and Kerr, John (1999) 'Smart answers for toxic troubles' and 'Tektronix on toxics: Looking good by using less', *Electronic Business*, 18 September.

13   See EPA (1998) *How Industry is Reducing Dependence on Ozone-Depleting Chemicals*, US EPA, Washington, DC.

14   See Kerr (1999), op cit (Note 12).

15   US law required public reporting of toxic emissions, including CFCs and other ODSs, and in some countries chemical manufacturers and customers voluntarily disclosed similar information.

16   Since 1990 the US EPA has made annual 'Stratospheric Ozone Protection Awards' to recognize outstanding contributions to the protection of the ozone layer. By 1996, 495 individuals, enterprises and organizations from 40 countries had earned this award.

Other awards for ozone layer protection have included the US EPA's 'Best-of-the-Best Stratospheric Ozone Protection Awards'; the Nobel Prize in Chemistry; the UNEP 'Sasakawa Environment Prize', 'Global 500 Awards', 'Global Ozone Award', '1995 Ozone Awards', '1997 Ozone Awards', '1998 Children's Ozone Painting Award' (UNEP OzonAction and the Egyptian Environmental Affairs Agency), 'OzonAction Programme 2001: Global Video Competition on Ozone Layer Protection' and '2005 Ozone Awards'; the 'Better World Society Environmental Award'; 'Japan Nikkann Kogyo Shimbun Awards'; 'Brazil Stratospheric Protection Award'; 'US Department of Defense Award for Ozone Excellence'; 'Vietnam Ozone Awards'; 'International Cooperative for Ozone Layer Protection'; 'World Resources Institute Legislator Ozone Award'; and 'Mobile Air Conditioning Society 20th Century Award for Environmental Leadership'. A complete listing of awards for stratospheric ozone protection from 1990 to 2002 is presented in Appendix 6 of Andersen, S. and Sarma, K. M. (2002) *Protecting the Ozone Layer: The United Nations History*, Earthscan, London.

17 See Andersen and Durwood (2002), op cit; (Note 9) Andersen, Frech and Morehouse (1997), op cit (Note 9); Andersen (1997), op cit (Note 9); Le Prestre, Reid and Morehouse (1998), op cit (Note 9); Cook (1996), op cit (Note 9).

18 Experts frequently use a variety of definitions in describing technical choices under the Montreal Protocol. The Technology and Economic Assessment Panel (TEAP) generally used the following definitions. When one chemical replaces another with the same basic function, it is a 'substitute'. When a product or process change replaces a chemical, it is an 'alternative'. A fluorinated chemical (HCFC or HFC) is an 'in-kind chemical substitute' while a non-fluorinated chemical (hydrocarbon, ammonia, carbon dioxide, etc.) is a 'not-in-kind chemical substitute'. Examples of in-kind substitutes include HFC-134a replacing CFC-12 in vehicle air-conditioning, HCFC-123 replacing CFC in building chillers and HFC-225 replacing halon-1211. A product alternative satisfies the same consumer need with a different product or service. Examples of product alternatives to CFC aerosol deodorants include pumps and sticks. A process change can eliminate the need for an ODS. For example, no-clean solders eliminate solvent use.

19 In May 1986 the EPA Chlorofluorocarbon Chemical Substitutes International Committee (a highly respected panel of experts from Germany, Japan, the UK and the US) concluded that the absence of a market – rather than technical or environmental issues – was the principal barrier to commercialization of alternatives to CFCs, and listed all of the chlorinated chemicals that could replace CFC-11, CFC-12 and CFC-113. The list included HCFC-225, which was invented by committee member Dr Masaaki Yamabe and his Asahi company team during the four months the committee was meeting. See EPA Press Release, 24 May 1986 and EPA (1987) 'Protection of stratospheric ozone: Proposed rule', 40 CFR Part 82, US Environmental Protection Agency, 14 December.

20 McLinden, M. O. and Didion, D. A. (1987) 'Quest for alternatives: A molecular approach demonstrates tradeoffs and limitations are inevitable in seeking refrigerants', *ASHRAE Journal*, December, pp32–42; Nagengast, B. (1988) 'A historical look at CFC refrigerants', *ASHRAE Journal*, November, pp37–39; Kauffman, G. B. (1989) 'Midgley: Saint or serpent?' *Chemtech Magazine*, December, pp717–725; Manzer, L. E. (1990) 'The CFC–ozone issue: Progress on the development of alternatives to CFCs', *DuPont Articles*, vol 249, 6 July, pp31–35.

21 A complete history of the development of CFC refrigerants is found in Cagin, S. and Dray, P. (1993) *Between Earth and Sky: How CFCs Changed our World and Endangered the Ozone Layer*, Pantheon, New York; Nagengast (1988), op cit (Note 20); Midgley, T. Jr (1937) 'From the periodic table to production', *Journal of Industrial Engineering Chemistry*, February; Midgley, T. Jr and Henne, A. L. (1930) 'Organic fluorides as refrigerants', *Industrial and Engineering Chemistry*, vol 22, no 5, pp542–545.

22   Water (as a solvent and foam-blowing agent), hydrocarbons (as refrigerants and foam-blowing agents) and carbon dioxide (as a refrigerant and foam-blowing agent) are the new alternatives to the HFCs that replaced ODSs in large applications. Environmentally superior substitutes and alternatives are identified for all HCFC uses, with the conspicuous exceptions of HCFC-123 used in building air-conditioners (HCFC-123 systems have near-zero emissions and conspicuously higher energy efficiency), HCFC-225 used for cleaning oxygen systems (HCFC-225 cleans systems with complex geometry) and HCFC blends used in medical sterilization (HCFCs safely inert the flammability of ethylene oxide sterilization conducted on-site in medical facilities and field hospitals).

23   It is significant that corporate counsels often prohibited their enterprises from directly disclosing or endorsing the technology selected to replace ODSs but were comfortable and proud to have EPA publish this information. The legal concern was that manufacturers of ODSs would dispute their decisions or that they would be held liable if enterprises relying on their advice to select a technology were later disappointed by the results. EPA membership in these partnerships also offered other advantages. For example, enterprises alone could not satisfy US Federal Trade Commission (FTC) and internationally equivalent requirements requiring proof of advertising claims regarding environmental leadership. An enterprise announcing a goal or reporting success could not say 'first to declare a CFC-113 phaseout' or 'first to halt CFC-113 use' because surely a small enterprise somewhere might have done that first, and how could the enterprise provide any proof to the FTC that it had thoroughly eliminated this possibility? However, the EPA can, and often did, make such claims in its reports, press releases and awards descriptions. Thus without the EPA, there would have been less reward for environmental leadership enterprises driving market transformation.

24   The environmental acceptability of chemical substitutes depends on the direct and indirect environmental effects, and human health and safety. The direct atmospheric effects are analysed in terms of ozone depletion potential, global warming potential, atmospheric fate, and biological and ecosystem effects. The indirect atmospheric effects are the emissions resulting from generating the energy to power the application. Additional environmental concerns include resource consumption, aquatic and terrestrial toxicity, and sustainability. Human health and safety concerns include human toxicity and flammability. Further, technical performance includes measures of materials compatibility, physical properties such as boiling point, chemical properties such as solvency, and other measures depending on the application.

25   Conspicuous exceptions, where SNAP was not the path-finder, include metered-dose inhalers (MDIs) and alternatives to methyl bromide. The US Food and Drug Administration (FDA), not SNAP, has primary regulatory authority over medicine in the US, but did not act as quickly as the EC in approving MDIs, partly due to the complex FDA regulatory procedure and the backlog of pharmaceutical products under review. Alternatives to methyl bromide have also been approved more quickly for use outside the US.

26   Nissan Motor Company (1989) 'Nissan to ban CFCs on cars', news release, 2 August; Volvo Cars of North America (1989) 'Volvo to eliminate CFC use by 1995', news release, 27 October. Fierce competition among vehicle manufacturers to phase out CFCs continued with great pride at every benchmark. In March 1995 Mercedes-Benz announced it had completely transformed its global new vehicle production to HFC-134a (Mercedes-Benz of North America, Inc. (1995) 'Corporate leadership in elimination of CFCs from automobiles', news release, 7 March).

27   The US military was so satisfied with the superiority of performance standards developed for ODS replacement that it expanded the programme to replace all

prescriptive standards with performance standards as part of its 'pollution prevention' initiative.

28 The UNEP scientific assessment has always included detailed consideration of the impacts of substitutes on climate change, including presentation of the global warming potentials of proposed alternatives and substitutes. TEAP and its Technical Options Committees featured comprehensive sections in their reports on toxicity, flammability, global warming potential, energy efficiency and other factors. They repeated this information prominently in executive summaries. Assessment Synthesis Reports highlighted climate change, global warming potential, energy efficiency and other measures of environmental acceptability for substitutes and alternatives.

29 Akira Okawa is Deputy Secretary General, Japan Industrial Conference for Ozone Layer and Climate Protection (JICOP).

30 The most comprehensive comparisons of ODS regulations and related measures are found in UNEP, MLF, SEI (1996) *Regulations to Control Ozone-Depleting Substances: A Guidebook*, updated 2000 edition, published by the Division of Trade, Industry and Economics (DTIE), UNEP, Paris. Additional descriptions of national programmes with comparisons with other countries are found on national websites.

31 Countries with taxes or fees on ODSs include Australia, Belarus, Bulgaria, the Czech Republic, Denmark, Hungary, Republic of (South) Korea, the Seychelles, Singapore, South Africa, Sweden, Thailand, Vietnam and the US.

32 Naohiro Murayama is Deputy Director of the Ozone Layer Protection Office, Ministry of Economy, Trade and Industry (METI), Japan.

33 Yuichi Fujimoto is a consultant and former director of the Japan Electrical Manufacturers' Association (JEMA), former advisor to the Japan Industrial Conference for Ozone Layer and Climate Protection (JICOP), former Senior Expert Advisor of TEAP and former member of the Solvents TOC.

34 Australia, Bulgaria, Canada, China, the Czech Republic, Gambia, Germany, Jamaica, Jordan, Malaysia, New Zealand, Romania, Spain, Sweden, Syria and the US require special environmental labelling for products containing ODSs; Malaysia, Syria and the US require labels for products made using ODSs. Some countries also require or encourage labelling of products not containing ODSs.

35 Tsutomu Odagiri is Secretary General, Japan Industrial Conference on Cleaning (JICC).

36 Shigehiro Uemura is Secretary General, Japan Industrial Conference for Ozone Layer and Climate Protection (JICOP).

37 Simple economic analysis fails to account for the investment that would have to occur without the phaseout (overstating the net investment required) and improvements in operating costs, reliability and performance (understating the benefits). Furthermore, many calculations improperly count tax revenue as a cost when it is properly considered as a transfer payment. See Vogelsberg, F. A. (1996) 'An industry perspective: Lessons learned and the cost of the CFC phaseout', paper presented at the International Conference on Ozone Protection Technologies, Washington, DC.

CHAPTER 5

1 This chapter was written with the substantial contributions of Rick Golde, Brian Howack, Bella Maranion, Dave Koehler, John Manzione, Tom Morehouse and Pete Mullenhard and with the assistance of David Liddy.

2 UNEP/DTIE (1999) 'Maintaining military readiness by managing ozone-depleting substances', United Nations Environment Programme/Division of Trade, Industry and Economics, September.

3   United Nations Environment Programme (1989) *Final Report of the Halons Technical Options Committee*, Ozone Secretariat, UNEP, Nairobi, p4.
4   US National Safety Council, www.nsc.org/ehc/chemical/Bromomet.htm, retrieved 17 March 2006.
5   National Academy of Sciences (1997) *Fire Suppression Substitutes and Alternatives to Halon for US Navy Applications*, National Academy of Sciences, p65 (quoted in United Nations Environment Programme (1989), op cit, Note 3).
6   UNEP (2001) 'More action needed to guarantee recovery of ozone layer: New substances may damage Earth's protective shield', press release, 14 September, retrieved 17 March 2006 from www.grida.no/newsroom.cfm?pressReleaseItemID=164.
7   United Nations Environment Programme (1989), op cit (Note 3), Executive Summary.
8   Ibid.
9   United Nations Environment Programme (1989), op cit (Note 3), p5.
10   Holder, B. (1993) 'Call for halon replacement has lab in firefighting work', *Skywriter*, 19 November 1993.
11   With the exception of halon-2402, which was normally limited to outdoors applications due to its high toxicity.
12   A notable exception is methyl bromide, which was occasionally used as a degreasing solvent for wool. Unlike many other ODSs, methyl bromide is highly toxic.
13   UNEP (1989a) *Electronics, Degreasing and Dry Cleaning Solvents Technical Options Report*, Ozone Secretariat, UNEP, Nairobi, p13.
14   Holder (1993), op cit.
15   Gilbert F. Decker and Robert L. Walker (1995) 'Endorsement of the strategic guidance and planning for eliminating ozone-depleting chemicals from US Army applications', October.
16   Gary Vest, US Principal Assistant Deputy Under Secretary of Defense (Environmental Security), speech on global climate change and the military.
17   In March 1971 the US House of Representatives voted not to continue funding development of the American SST. In 1973 Japan Air Lines, Pan Am, Qantas and TWA cancelled their orders for Concorde SSTs. Only British Airways and Air France were then flying Concordes across the Atlantic Ocean. See Andersen and Sarma (2002) *Protecting the Ozone Layer: The United Nations History*, Earthscan, London.
18   In 1974 Stephen O. Andersen had been a member of the team that assessed the consequences of hypothesized climate change and ozone depletion from the SST; Gary Vest had been a US Air Force officer organizing environmental programmes.
19   Andersen, Stephen O., Frech, Clayton and Morehouse, E. Thomas (1997) *Champions of the World: Stratospheric Ozone Protection Awards*, US EPA, Washington, DC, p35.
20   Andersen et al (1997), op cit (Note 19), p36.
21   UNEP (1998) 'Report of the Halons Technical Options Committee', Ozone Secretariat, UNEP, Nairobi, p6.
22   Andersen et al (1997), op cit (Note 19), p36.
23   UNEP/DTIE (1999), op cit (note 2).
24   Proceedings of 'The importance of military organisations in stratospheric ozone and climate protection', 6–8 February 2001, Brussels.
25   Stephen O. Andersen et al (1988) 'Stratospheric ozone and the US Air Force', *The Military Engineer*, vol 80, no 523, August. See also Stephen O. Andersen et al (1994) 'The military's role in protection of the ozone layer', *Environmental Science and Technology*, vol 28, no 13.

CHAPTER 6

1   The authors are indebted to Mike Jeffs and Sally Rand for significant contributions to this chapter.

2   UNEP Foams Technical Options Committee (1989) *Technical Progress on Protecting the Ozone Layer*, flexible and rigid foams technical options report, United Nations Environment Programme, Nairobi, Kenya, 30 June, p13.

3   Ibid, p14.

4   The OORG developed guidance for implementation and published reports which are available at www.worldbank.org under its Montreal Protocol section. Some of the key reports for foams are:

*   #3 'Reducing ODS use in developing countries in domestic refrigerator/freezer insulating foams' (10/1993);

*   #8 'Foam-blown pre-insulated pipes' (5/1994);

*   #9 'Transitional and zero ODS domestic refrigerator/freezer foam' (5/1994);

*   #14 'Factors which affect operating costs: Foam densities and plastic liner grades' (9/1998); and

*   #15 'Foam density report' (1/2000).

5   Paragraph 32 of the Report of the Sixteenth Meeting of the Executive Committee of the Multilateral Fund, UNEP/Ozl.Pro/Excom 16/20, 17 March 1995, www.multilateralfund.org.

6   UNEP/OzL.Pro/ExCom/20/72, Decision 20/45, Para. 45 (a (iii)); supporting document: UNEP/OzL.Pro/ExCom/20/65.

7   Report of the Twentieth Meeting of the Executive Committee of the Multilateral Fund, UNEP/OzL.Pro/ExCom/20/72,dated 18 October 1996 Decision 20/45, Para. 45 (a (iii)); supporting document: UNEP/OzL.Pro/ExCom/20/65, www.multilateralfund.org.

8   Report of the Twelfth Meeting of the Executive Committee of the Multilateral Fund, UNEP/OzL.Pro/ExCom/12/37 dated 31 March 1994, para. 168; supporting document:UNEP/OzL.Pro/ExCom/12/34, www.multilateralfund.org.

9   Report of the Twentieth Meeting of the Executive Committee of the Multilateral Fund, UNEP/OzL.Pro/ExCom/20/72 dated 18 October 1996, www.multilateralfund.org. Decision 20/48, para. 72 (b, c).

10   Report of the Fifteenth Meeting of the Executive Committee of the Multilateral Fund, UNEP/OzL.Pro/ExCom/15/45 dated 16 December 1994, para. 90. www.multilateralfund.org.

11   Report of the Fifteenth Meeting of the Executive Committee of the Multilateral Fund, UNEP/OzL.Pro/ExCom/15/45 dated 16 December 1994, www.multilateralfund.org; Para 129. Report of the Seventeenth Meeting of the Executive Committee of the Multilateral Fund,UNEP/OzL.Pro/ExCom/17/60, dated 28 July 1995, Decision 17/17 para. 26, www.multilateralfund.org.

12   Report of the Nineteenth Meeting of the Executive Committee of the Multilateral Fund, UNEP/OzL.Pro/ExCom/19/64 dated 10 May 1996, Decision 19/2, para. 17, www.multilateralfund.org.

13   Report of the Twenty-seventh Meeting of the Executive Committee of the Multilateral Fund, UNEP/OzL.Pro/ExCom/27/48, dated 12 April 1999 Decision 27/13, para. 35, www.multilateralfund.org.

14   Report of the Thirty-eighth Meeting of the Executive Committee of the Multilateral Fund, UNEP/OzL.Pro/ExCom/38/70/Rev.1, dated 20 January 2003, Decision 38/38, para. 74, www.multilateralfund.org.

15 UK Department of Trade and Industry (2000) 'Guidance on the new EC Regulation No 2037/2000 on Substances that Deplete the Ozone Layer', retrieved 5 October 2006 from www.dti.gov.uk/files/file29100.pdf.

16 EC 2037/2000, http://ec.europa.eu/environment/ozone/pdf/reg2037_2000_en.pdf.

17 EC 2037/2000, Article 5.1d and Article 16.

18 The safety risks after conversion to HCFC-141b are comparable to those from the previously used CFC-11, although HCFC-141b is flammable under certain conditions. HFC-152a is also flammable, but it is not often used as a foam-blowing agent.

19 UNEP Foams Technical Options Committee (1989), op cit (Note 2).

20 Report of the Seventeenth Meeting of the Executive Committee of the Multilateral Fund,UNEP/OzL.Pro/ExCom/17/60, dated 28 July 1995, Decision 17/14, para. 23, www.multilateralfund.org; Report of the Twentieth Meeting of the Executive Committee of the Multilateral Fund, UNEP/OzL.Pro/ExCom/20/72 dated 18 October 1996, www.multilateralfund.org, Decision 20/45, para. 45, Decision 20/45, para. 45 (a(iii))).

21 Many refrigeration projects included a foam-blowing element. Refrigerators require insulation, and many enterprises had been using CFC-blown foam. It was possible to convert this CFC-blown foam to hydrocarbon-blown foam; however, most enterprises selected HCFC-141b.

22 The projects implemented together in this group were: IND/FOA/17/INV/45, IND/FOA/17/INV/42, IND/FOA/17/INV/41, IND/FOA/17/INV/40 and IND/FOA/17/INV/37.

23 IND/FOA/17/INV/ 35, IND/FOA/17/INV/ 52 and IND/FOA/17/INV/17.

24 IND/FOA/17/INV/52.

25 Iceberg Technology in Russia settled for HCFC-141b in part because it has fewer installation and operating costs; see World Bank (2004) 'Russia implementation completion report', Document No 30981-RU, World Bank, Washington, DC.

26 JSC Stroidetal could have used either the CO_2 technology or the butane technology alone in its foam-blowing operations, but the enterprise found that the combination provided a better quality product and was therefore willing to co-finance the combination technology rather than settle for one or the other.

27 For example by BHL in the Czech Republic and Iceberg Technology in Russia; see World Bank (2004), op cit (Note 25).

28 This was the case for Iceberg Technology in Russia; see World Bank (2004), op cit (Note 25).

29 UNEP/ozL.Pro/Excom/33/6 dated 1 March 2001, Final Report on the evaluation of foam sector projects, Multilateral Fund, www.multilateralfund.org.

30 This was the Atlant plant in Belarus; see World Bank (2001) 'Belarus implementation completion report', Report No 22229.

31 For example, required safety equipment for the Thai hydrocarbon conversion included exhaust and ventilation systems, fire sprinkler systems, high pressure pumping/metering systems, ionized air blowers, gas sensing systems, and modifications and retrofitting to the process equipment and manufacturing plant.

32 Used in refrigerators and freezers at Frigolux and in the production of sandwich panels for building and cold store construction at Metalucon.

33 See Note 31.

34 World Bank (2001), op cit (Note 30).

35 World Bank (2001a) 'Poland implementation completion report', Report No 22341; Polish commercial practices were used as an alternative to World Bank procurement rules.

36 UNEP/ozL.Pro/Excom/33/6 dated 1 March 2001, Final Report on the evaluation of foam sector projects, Multilateral Fund, www.multilateralfund.org.

37 One Hungarian company – Metisol – that selected liquid carbon dioxide foam-blowing technology was unhappy with the outcome. Metisol planned to use carbon dioxide, but foamed panels it produced had an unacceptable degree of dimensional instability. Metisol was forced to continue using HCFC-141b until a better alternative could be found. World Bank (1999) 'Hungary implementation completion report', Report No 19368, World Bank, Washington, DC.

38 Trimo, an enterprise in Slovenia, found that worker safety was improved by reducing worker exposure to CFC fumes. World Bank (1998) 'Slovenia implementation completion report', Report No 18822, World Bank, Washington, DC.

39 World Bank (1999), op cit (Note 37).

40 UNEP/ozL.Pro/Excom/33/6 dated 1 March 2001, Final Report on the evaluation of foam sector projects, Multilateral Fund, www.multilateralfund.org.

41 United Nations Development Programme (2003) 'Project implementation report: Kazakhstan Country Programme', UNDP.

CHAPTER 7

1 The authors are grateful to Radhey Agarwal for his peer review.

2 Other refrigerants included ammonia, butane, carbon bisulphide, carbon dioxide, dichloroethylene, ethane, ethylamine, ethyl bromide, gasoline, isobutane, methyl formate, methylene chloride, methylamine, methyl chloride, naphtha, nitrous oxide, propane, sulphur dioxide, trichloroethylene and trimethylamine.

3 In 1803 Thomas Moore coined the word 'refrigerator' to describe his ice-cooled, rabbit-fur-insulated wooden butter-transport box; in 1824 Nicholas Leonhard and Sadi Carnot invented the Carnot Refrigeration Cycle; in 1834 Jacob Perkins received British patent 6662 for closed-cycle refrigeration; in 1850 A. Twining received a British refrigeration patent that mentions CO_2 refrigerants; in 1866 Thaddeus S. C. Lowe made the first recorded use of CO_2 as a refrigerant; and in 1884 William Whiteley invented air-conditioning for horse carriages consisting of a wheel-driven fan blowing air across blocks of ice. In the late 1890s F. Swarts pioneered fluorocarbon chemistry.

4 For a complete history, see Stephen O. Andersen and K. Madhava Sarma (2002) *Protecting the Ozone Layer: The United Nations History*, Earthscan, London.

5 See Andersen and Sarma (2002), op cit (Note 4). For reviews of refrigeration history see Midgley, T. J. and Henne, A. L. (1930) 'Organic fluorides as refrigerants', *Industrial and Engineering Chemistry*, vol. 22, no 5, pp542–545; McLinden, M. O. and Didion, D. A. (1987) 'Quest for alternatives: A molecular approach demonstrates tradeoffs and limitations are inevitable in seeking refrigerants', *ASHRAE Journal*, December, pp32–42; Nagengast, B. (1988) 'A historical look at CFC refrigerants', *ASHRAE Journal*, November, pp37–39; Kauffman, G. B. (1989) 'Midgley: Saint or serpent?' *Chemtech Magazine*, December, pp717–725; Manzer, L. E. (1990) 'The CFC–ozone issue: Progress on the development of alternatives to CFCs', *DuPont Articles*, 6 July, vol 249, pp31–35; Cagin, S. and Dray, P. (1993) *Between Earth and Sky: How CFCs Changed Our World and Endangered the Ozone Layer*, Pantheon, New York.

6 Midgley and Henne (1930), op cit (Note 5); McLinden and Didion (1987), op cit (Note 5); Nagengast, B. (1988), op cit (Note 5); Kauffman (1989), op cit (Note 5); Manzer (1990), op cit (Note 5).

7 With the exception that ammonia kept the market for gas-powered refrigeration on recreational vehicles, boats and cabins not connected to power grids.

8 Andersen and Sarma (2002), op cit (Note 4).

9   For a review of the commercial history of ODSs in all their applications see Andersen and Sarma (2002), op cit (Note 4), pp188–196.

10   See Cagin and Dray (1993), op cit (Note 5); Andersen and Sarma (2002), op cit (Note 4).

11   Molina, Mario and Rowland, F Sherwood (1974) 'Stratospheric sink for chlorofluoromethanes: Chlorine atom-catalyzed destruction of ozone', *Nature*, vol 249, 28 June.

12   Andersen and Sarma (2002), op cit (Note 4). The original 1987 Montreal Protocol envisioned only a 50 per cent decrease in CFC production and a freeze on halon production with controls only for developed countries. However, subsequent scientific assessments indicated that more stringency and scope were required, and technology assessments simultaneously confirmed that replacement technology would be available. Acting on this information, Parties adopted the phaseout strategy, increased the list of controlled substances, and expanded the control obligations to include all developed and developing nations.

13   See Andersen and Sarma (2002), op cit (Note 4). Thomas Midgley and his research team identified both HFC-134a and HCFC-123, but without ozone layer concerns these chemicals could not compete against CFC-12 and CFC-11, which were less expensive and safer to produce.

14   It is noteworthy that in 1987 fluorocarbon manufacturers worked aggressively to prevent the use of HFC-152a as a refrigerant, even though they were marketing HFC-152a for aerosol and feedstock uses.

15   Lorentzen, G. and Pettersen, J. (1993) 'A new, efficient and environmentally benign system for car air-conditioning', *International Journal of Refrigeration*, vol 16, no 1, pp4–12.

16   See Andersen and Sarma, (2002), op cit (Note 4). HFC-134a (ODP=0, GWP=1300) is ozone safe and has a GWP drastically lower than CFC-12 (ODP=1, GWP=10,700).

17   Andersen, Stephen O, Frech, Clayton and Morehouse, E Thomas (1997) *Champions of the World: Stratospheric Ozone Protection Awards*, US EPA, Washington, DC.

18   For a complete chronology, see Andersen and Sarma (2002), op cit (Note 4). The timeline included at the end of the book is particularly useful.

19   The first Greenpeace 'Greenfreeze' was an energy-inefficient model with propane/butane refrigerant and energy-inferior insulation. The German industry made it a success by switching to isobutane refrigerant and to improved cyclopentane blown insulation. See Andersen and Sarma (2002), op cit (Note 4).

20   See UNEP (1991) *Report of the Refrigeration, Air Conditioning and Heat Pumps Technical Options Committee Technical Options Committee*, United Nations Environment Programme, Nairobi, Kenya, p19. The first part of this evolving story is documented in 'Daimler-Chrysler – The Champagne of natural refrigerants', Chapter 3 in Stephen O. Andersen and Durwood Zaelke (2003) *Industry Genius: Inventions and People Protecting the Climate and Fragile Ozone Layer*, Greenleaf Publishing, Sheffield, UK, pp53–70. See also Intergovernmental Panel on Climate Change (2005) *Safeguarding the Ozone Layer and the Global Climate System*, Cambridge University Press, Cambridge, UK, p305.

21   Intergovernmental Panel on Climate Change (2005), ibid, p310.

22   Each refrigerant is designated by industry with a unique 'R' number and also commonly described with a variation more precisely disclosing the chemical composition. For example, R-12 is CFC-12; R-600 (isobutane) is HC-600. This book chooses to use the more descriptive numbering that implies chemical composition.

23   Hill, B., Papasavva, S. and Major, G. (2004) 'A comparison of R134a, R134a Enhanced, R744 and R744 Enhanced automotive refrigerant systems based on lifecycle', presented at the 2004 MAC Summit, 13–14 April, Washington, DC. See also Intergovernmental Panel on Climate Change (2005), op cit (Note 20).

24 Refrigeration, Air Conditioning and Heat Pumps Technical Options Committee (1991) *Report of the Refrigeration, Air Conditioning and Heat Pumps Technical Options Committee*, United Nations Environment Programme, Nairobi, Kenya, p19.

25 Refrigeration, Air-Conditioning and Heat Pumps Technical Options Committee (2002) *Montreal Protocol on Substances that Deplete the Ozone Layer: 2002 Report of the Refrigeration, Air-conditioning and Heat Pumps Technical Options Committee – 2002 Assessment*, http://ozone.unep.org/teap/Reports/RTOC/RTOC2002.pdf

26 Ball, J. (2005) 'Companies try keeping ice cream frozen, emissions down', *Wall Street Journal*, 4 May.

27 Prolonged exposure to carbon dioxide is generally considered harmful to human health at concentrations above 4 per cent.

28 Sub Paragraph (c) of Article 2F (7) of the Montreal Protocol on Substances that 29the Ozone Layer, http://hq.unep.org/ozone/Montreal-Protocol/Montreal-Protocol2000.shtml.

29 Paragraph (a) of Decision VI/13 of the Report of the Sixth MOP, UNEP/Ozl.Pro/6/7 dated 10 October 1994, available at http://ozone.UNEP.Org.

30 Report of the Tenth MOP, UNEP/Ozl.Pro/10/9 dated 5 December 1998, available at http://ozone.UNEP.Org.

31 IPCC (2005) *Special Report on Safeguarding the Ozone Layer and the Global Climate System: Issues related to Hydrofluorocarbons and Perfluorocarbons*, Cambridge University Press, available for download at www.ipcc.ch

32 Agenda 21 is the outcome of the United Nations Conference on Environment and Development held in Rio De Janeiro, Brazil in 1992. The agenda sets out the tasks before the world to promote sustainable development. Chapter 9 of the Agenda deals with protection of the atmosphere and Section C deals with the protection of the ozone layer. Agenda 21 can be downloaded from www.un.org/esa/sustdev/documents/agenda21/english/Agenda21.pdf

33 Alliance for Responsible Atmospheric Policy (ARAP) (2002, updated 2006) *Responsible Use Principles for HFCs*, Partnership statement of ARAP, US EPA, UNEP and Japan Ministry of Economy, Trade and Industry (METI), available at www.arap.org/responsible.html

34 'Trane, it's about making green', Chapter 9 in Andersen and Zaelke (2003), op cit (Note 20), pp159–175.

35 World Bank (1998) 'Czech Republic implementation completion report', Report No 18384, World Bank, Washington, DC.

36 United Nations Development Programme (2003) 'Project implementation report: Lithuania country programme', UNEP, New York. A new owner agreed to the terms of the GEF project document, and manufacturing trials were successfully completed in December 2002 with full-scale manufacturing commencing shortly thereafter.

37 World Bank (2001) 'Poland implementation completion report', Report No 22341.

38 World Bank (2005) 'Ukraine implementation completion report', Report No UA-32788.

39 World Bank (2004) 'Russia implementation completion report', Document No 30981-RU.

40 Global Environment Facility and World Bank (1996) 'Russian Federation: Ozone depleting substance consumption phase-out project', Document No 15326; World Bank (2004) 'Russia implementation completion report', Document No 30981-RU, retrieved 10 December 2006 from www.gefweb.org/wprogram/May99/ozone/Russia2.doc.

41 World Bank (2001), op cit (Note 37).

42 Multilateral Fund 'Evaluation of compressor projects in China', retrieved 15 December 2006 from www.multilateralfund.org/files/evaluation/3417.pdf.

43 'Conversion of CFC-12 small and medium open-type refrigerating compressor at Zhejiang Commercial Machinery Factory', MLF Project CPR/REF/28/INV/300.
44 Ibid.
45 Regulation 2037:2000, retrieved November 2006 from http://ec.europa.eu/environment/ozone/pdf/reg2037_2000_en.pdf.
46 Global Environment Facility and the World Bank (1996), op cit (Note 40).
47 Trane (2005) 'TRANE extends energy efficiency advantage of EarthWise™ CenTraVac™ chillers: Technology breakthrough enables energy efficiency of 0.448 KW/tonne, 13.5 per cent better than the next best chiller?, press release, 16 December, www.trane.com/Commercial/.
48 World Bank (2000) Bulgaria Implementation Completion Report, Report No 20679.
49 Ibid.
50 'Production of small and medium sized ammonia refrigerating compressors at Yantai Moon Group Co. Ltd (Yantai Refrigerating Machinery Works)', MLF Project CPR/REF/16/INV/114. This project phased out 240 ODP tonnes.

CHAPTER 8

1 Geno Nardini significantly contributed to this chapter, with the assistance of Montfort Johnson and Jorge Corona. The authors are also indebted to Drs Helen Tope and Ashley Woodcock for their substantial contribution to the discussion of MDIs. Additional technical information came from the Reports of the Aerosols Technical Options Committee (2002, 2006 and 2007), United Nations Environment Programme, Nairobi, Kenya.
2 The authors based part of the analysis of technology transfer on presentations at the meeting of the Network of the National Ozone Officers of South Asia in Colombo in December 2006; it was then reviewed by Ashley Woodcock and Helen Tope.
3 Report of the Forty-ninth Meeting of the Executive Committee, UNEP/OzL.Pro/ExCom/49/39, 14 July 2006.
4 The company was Krka in Slovenia; see World Bank (1998), 'Slovenia implementation completion report', Report No 18822, World Bank, Washington, DC.
5 World Bank (1998), op cit (Note 4).
6 World Bank (2004) 'Russia implementation completion report', Report No 30981-RU.
7 World Bank (2001) 'Poland implementation completion report', Report No 22341; World Bank (2004), op cit (Note 6); World Bank (1998), op cit (Note 4).
8 Information drawn from Global Environment Facility and the World Bank (1996) 'Russian Federation: Ozone depleting substance consumption phase-out project', Document No 15326.
9 World Bank (2005) 'Ukraine implementation completion report', Report No UA-32788.
10 Ibid.
11 Ibid.

CHAPTER 9

1 The authors are grateful to Bella Maranion (US EPA and the Halons Technical Options Committee – HTOC) and to Dave Catchpole and Dan Verdonik (Co-Chairs of the HTOC) for their substantial contributions to this chapter.
2 Verdonik, Daniel P. (1992) *Strategic Plan for Replacing Ozone Depleting Chemicals in US Army Tactical Weapon System Applications*, US Army Acquisition Pollution Prevention Support

Office, Alexandria, VA.

3 Fire protection is defined as 'fixed' when the fire extinguishing equipment is fixed in place (for example tanks, pipes and nozzles protecting a building) and is defined as 'portable' when the extinguisher can be moved to the location of the fire. Portable fire extinguishers are further defined as 'hand-held' if they are carried or 'wheeled' if they are part of a cart. 'Total flooding' halon systems function by increasing the concentration of halon to a level that prevents fire from burning anywhere in the protected space, which can be contrasted with 'partial flooding' or 'local application' halon systems, which increase the concentration of halon to a level that prevents fire from burning only in a small area in front of the nozzle.

4 Carbon dioxide is a naturally occurring gas that can be inhaled safely and breathed safely at low concentrations, but is toxic at higher concentrations.

5 NFPA (2006) 'NFPA 2010 Standard for Fixed Aerosol Fire-Extinguishing Systems'.

6 UNEP/ozL.Prol/Excon/40/8 dated 19 June, 2003. 'Final report on the evaluation of the halon sector', Executive Committee of the Multilateral Fund, UNEP.

7 Ibid.

8 UNEP (2001) 'Standards and codes of practice to eliminate dependency on halons: Handbook of good practices in the halon sector', UNEP, available at www.uneptie.org/ozonaction/library/policy/main.html#halstand.

9 World Bank (2001) 'Belarus implementation completion report', Report No 22229, World Bank, Washington, DC; World Bank (2001) 'Poland implementation completion report', Report No 22341, World Bank, Washington, DC; World Bank (2005) 'Ukraine implementation completion report', Report No UA-32788, World Bank, Washington, DC.

10 World Bank (1999) 'Hungary implementation completion report', Report No 19368, World Bank, Washington, DC.

11 World Bank (2004) 'Russia implementation completion report', Report No 30981-RU, World Bank, Washington, DC.

12 Ibid.

13 In 1990 Parties to the Montreal Protocol agreed to phase out halons, provided that there were exceptions for critical uses to be identified and recommended by the Technology and Economic Assessment Panel (TEAP) and its Halons Technical Options Committee (HTOC). In their report of 1992, the TEAP and the HTOC reported that it was not possible to define essential uses as they varied from country to country, that many alternatives existed for most of the halon applications, that there was an adequate available quantity of halon recoverable from equipment in non-essential uses for use in applications still requiring halon, and that Parties could choose to halt halon production in 1994, two years earlier than envisaged for CFCs, methyl chloroform and carbon tetrachloride. Their findings and recommendations were incorporated in the Adjustments to the Protocol agreed in Copenhagen in 1992.

CHAPTER 10

1 The authors are grateful to Brian Ellis, Darrel Staley and William Kenyon for significant contributions to this chapter.

2 See Andersen and Sarma (2002) *Protecting the Ozone Layer: The United Nations History*, Earthscan, London, Chapter 5.

3 This chapter lists each ODS and ozone-safe solvent by both common generic trade names and chemical formula in order to clearly distinguish solvents. This is particularly important because, in some cases, different regions of the world use the same acronym

for different solvents.

4   US National Library of Medicine Toxicology Data Network website, http://toxnet.nlm.nih.gov.

5   UNEP (1989) *Reports of the Electronics, Degreasing and Dry-Cleaning Solvents Technical Options Committee*, Ozone Secretariat, United Nations Environment Programme, Nairobi, Kenya, June 30, p153.

6   Ibid, p37.

7   Ibid, pp123–124.

8   Ibid, p102.

9   Joint Assessment of Commodity Chemicals (1988) 'Technical report no 32', Brussels, Belgium, May: 'The low risks calculated by this procedure [measurement of in vivo kinetic constants in mice and calculated in vitro and validated to rats, hamsters and man] indicate that man is adequately protected from the risk of methylene chloride induced cancer by the current hygiene standards which are based on the formation of carboxyhaemoglobin from this chemical'; Joint Assessment of Commodity Chemicals (1989) 'Technical report no 34', March: 'On the basis of these data it concluded that man is not susceptible to the carcinogenic effect of methylene chloride at the doses encountered in occupational and other settings.'

10   Joint Assessment of Commodity Chemicals (1994) 'Technical report no 60', May: 'Taking all of this information [cohort studies of >18,000 workers] into account, it is concluded that exposure to TCE does not present a carcinogenic hazard to man at levels of current occupational exposure standards.'

11   Joint Assessment of Commodity Chemicals (1990) 'Technical report no 37', May: 'It is concluded that, overall, the design and outcome of epidemiological studies failed to demonstrate a relationship between tetrachloroethylene and the occurrence of cancer in man'; Joint Assessment of Commodity Chemicals (1999) 'Technical report no 30', December: 'Overall, the epidemiological studies of greatest relevance are insufficient in both their design and outcome to demonstrate a relationship between exposure to PCE and the occurrence of cancer in humans.'

12   HCFC-225 is a blend of isomers ca and cb, which are respectively non-toxic and toxic. Asahi Glass has successfully separated the ca from the cb and markets the ca as a non-toxic solvent.

13   US National Toxicology Program (2005) *Report on Carcinogens*, Eleventh Edition, US Department of Health and Human Services, Public Health Service, National Toxicology Program, Washington, DC.

14   Andersen, Stephen O. and Durwood, Zaelke (2003) *Industry Genius: Inventions and People Protecting the Climate and Fragile Ozone Layer*, Greenleaf Publishing, Sheffield, UK, p180.

15   The ODP of n-propyl bromide depends on the geographic location of emissions because it has a short atmospheric lifetime and because all ODSs travel from point of emission to the equator, where upwelling of air currents transports chemical substances to the stratosphere. When emitted near the equator, the ODP is highest, because a larger portion reaches the stratosphere than when it is emitted in northern or southern latitudes, where a large portion of the chemical substance is degraded on the route to the equator and then on to the stratosphere.

16   HCFC-225 was invented by Masaaki Yamabe and colleagues (Asahi Glass) while he was serving on the US EPA experts committee that was first to scientifically confirm that chemical substitutes were available for most CFC-113 uses, albeit at higher costs. See Andersen and Durwood (2003), op cit (Note 14), pp84–103.

17   For a complete description of the world's first research centre devoted to 'sustainable chemistry' see Chapter 5, 'Japan's F-Center for developing greenhouse gas alternatives: Sustainable living through better chemistry', in Andersen and Durwood (2003), op cit (Note 14), pp84–103.

18 This is an extract from the 2001 United Nations Environment Programme OzonAction (2001) 'Technical brochure on cleaning without ODSs', United Nations Environment Programme, DTIE, Paris, www.unep.fr/ozonaction.

19 Table 4 from the *Report on the Intermediate Evaluation of the Solvent Sector Phaseout Plan in China*, UNEP/OzL.Pro/ExCom/42/13, dated 2 March 2004, available on the Multilateral Fund website at www.multilateralfund.org/evalution_document_library_.htm; data on projects approved for China are from inventory data of the MLF Secretariat and from the China State Environmental Protection Administration, presented to the Executive Committee of the MLF for the Implementation of the Montreal Protocol, Forty-second Meeting (UNEP/OzL.Pro/ExCom/42/13), Montreal, 29 March–2 April 2004.

20 In the Democratic Peoples' Republic of Korea, trichloroethylene was decreed to have a maximum time-weighted average operator exposure level of 5 ppm, as opposed to 50 ppm in many developed countries. This made this low-cost and useful solvent unusable in this country, with the result that the cost of projects – requiring more sophisticated techniques – often soared.

21 GEF Project Completion Report for Minsk Instrument Building Company, Belarus; World Bank (2001) 'Belarus implementation completion report', Report No 22229, World Bank, Washington, DC.

22 GEF Project Completion Report for Minsk Instrument Building Company, Belarus.

23 GEF Project Completion Report on the phaseout of the 1,1,1-trichloroethane photo-resistant developer at Hitelap using an alkaline solution; World Bank (1999) 'Hungary implementation completion report', Report No 19368, World Bank, Washington, DC.

24 World Bank (2001) op cit (Note 21).

25 The explosive concentration of vapours in air has a lower and upper limit. The lower explosive limit (LEL) is the lowest concentration that will ignite. The upper explosive limit (UEL) is the highest concentration that will ignite. At concentrations in air below the LEL there is not enough fuel to ignite; at concentrations above the UEL there is not enough oxygen to begin a reaction. There is a serious risk of fire or explosion if the vapour concentration is between the LEL and the UEL.

26 World Bank (1999) op cit (Note 23).

27 Ibid.

CHAPTER 11

1   The authors thank Stephen O. Andersen, K. Madhava Sarma and Kristen N. Taddonio for providing editorial comments.

2   Methyl Bromide Global Coalition (1994) 'Methyl bromide annual production and sales for the years 1984–1992', Methyl Bromide Global Coalition, Washington, DC.

3   Ibid.

4   United Nations Environment Programme (1992) 'Methyl bromide: Its atmospheric science, technology and economics', Montreal Protocol Assessment Supplement, UNEP, Nairobi.

5   World Meteorological Organization (1992) 'Scientific assessment of ozone depletion: 1991', World Meteorological Organization Global Ozone Research and Monitoring Project, Report No 25, WMO, Geneva.

6   United Nations Environment Programme (1992), op cit (Note 4).

7   Ozone Secretariat (2006) *Handbook for the Montreal Protocol on Substances that Deplete the Ozone Layer*, seventh edition, Ozone Secretariat, UNEP, Nairobi.

8   MBTOC (Methyl Bromide Technical Options Committee) (2007) *Report of the Methyl Bromide Technical Options Committee: 2006 Assessment*, UNEP, Nairobi.

9   Butler, J. H. and Montzka, S. A. (2003) 'Methyl bromide in the atmosphere – An update', National Oceanic and Atmospheric Administration, Annual International Research Conference on Methyl Bromide Alternatives, San Diego, 3 November, www.mbao.org.

10   International Plant Protection Convention (2002) 'International Standards for Phytosanitary Measures (ISPM) No 15: Guidelines for regulating wood packaging material in international trade', Secretariat of the International Plant Protection Convention, FAO, Rome.

11   US Department of Agriculture (1998) *Treatment Manual*, Animal and Plant Health Inspection Service, US Department of Agriculture, Washington, DC.

12   Knol, T. et al (2005) 'The release of pesticides from container goods', RIVM Report 609021033/2005, RIVM (Dutch National Institute for Public Health and the Environment), Bilthoven.

13   Paull, R. E. and Armstrong, J. W. (1994) 'Insect pests and fresh horticultural products: Treatments and responses', CAB International, Wallingford, UK; MBTOC (1998) 'Report of the Methyl Bromide Technical Options Committee. 1998 Assessment', UNEP, Nairobi.

14   Lower House (1981) Report of The Netherlands Parliamentary Session 1980–1981, 16 400, Chapter XIV, 50, 1–23, The Hague; Ketzis, J. (1992) 'Case studies of the virtual elimination of methyl bromide soil fumigation in Germany and Switzerland and the alternatives employed', Proceedings of International Workshop on Alternatives to Methyl Bromide for Soil Fumigation, 19–23 October 1992, Rotterdam, UNEP, Nairobi.

15   Examples are found in IPCS (International Programme on Chemical Safety) (1995) 'Environmental health criteria: 166 methyl bromide', IPCS, World Health Organization, Geneva; Maddy, K. T. et al (1990) 'Illness, injuries, and deaths from pesticide exposures in California, 1949–1988', *Review of Environmental Contamination and Toxicology*, vol 114, pp57–123; Hustinx, W. N., van de Laar, R. T., van Huffelen, A. C., Verwey, J. C., Meulenbelt, J. and Savelkoul, T. J. (1993) 'Systemic effects of inhalational methyl bromide poisoning: A study of nine cases occupationally exposed due to inadvertent spread during fumigation', *British Journal of Industrial Medicine*, vol 50, no 2, pp155–159; Moses, M. (2002) *Chronic Neurological Effects of Pesticides*, Pesticide Education Center, San Francisco; Herzstein, J. and Cullen, M. R. (1990) 'Methyl bromide intoxication in four field-workers during removal of soil fumigation sheets', *American Journal of Industrial Medicine*, vol 17, no 3, pp321–326.

16   Maddy et al (1990), op cit (Note 15).

17   Alavanja, M. C. R. et al (2003) 'Use of agricultural pesticides and prostate cancer risk in the Agricultural Health Study Cohort', *American Journal of Epidemiology*, vol 157, no 9, pp800–814.

18   MBTOC (1995) 'Report of the Methyl Bromide Technical Options Committee. 1994 Assessment', UNEP, Nairobi.

19   MBTOC (2002) 'Report of the Methyl Bromide Technical Options Committee. 2002 Assessment', UNEP, Nairobi.

20   MBTOC (2002), op cit (Note 19); UNEP (2000) *Case Studies on Alternatives to Methyl Bromide*, UNEP/DTIE, Paris; UNEP (2002) *Case Studies on Alternatives to Methyl Bromide*, Volume 2, UNEP/DTIE, Paris; Miller, M. K. (2001). *Sourcebook of Technologies for Protecting the Ozone Layer: Alternatives to Methyl Bromide*, UNEP/DTIE, Paris; Pizano, M. (2001) *Floriculture and the Environment. Growing Flowers without Methyl Bromide*, UNEP/DTIE, Paris; Batchelor, T. A. and Bolivar, J. (2002) *Proceedings of International Conference on Alternatives to Methyl Bromide*, 5–8 March 2002, Seville, European Commission, Brussels.

21 MBTOC (2007), op cit (Note 8).

22 MBTOC (2002), op cit (Note 19); MBTOC (2007), op cit (Note 8); Vos, J. and Bridge, J. (eds) (2006) *Cases of Methyl Bromide Alternatives Used in Commercial Practice*, CAB International, Wallingford, UK.

23 The barrier sheets are Virtually Impermeable Film (VIF) and Low Permeability Barrier Film (LBPF). These films are spread on the ground to minimise the emissions of methyl bromide and reduce the dosage of methyl bromide needed.

24 TEAP (Technology and Economic Assessment Panel) (2005) 'Report of the Technology and Economic Assessment Panel: May 2005 Progress Report', UNEP, Nairobi; TEAP (2006a) 'Report of the Technology and Economic Assessment Panel: May 2006 Progress Report', UNEP, Nairobi; TEAP (2006b) 'Special Report: Validating the performance of alternatives to methyl bromide for pre-plant fumigation', UNEP, Nairobi; MBTOC (2007), op cit (Note 8).

25 MBTOC (2007), op cit (Note 8). Compiled from Ozone Secretariat database (November 2006), Meeting of the Parties reports, and national licensing and authorization documents relating to authorized methyl bromide production and imports. The UNEP data understates recent US methyl bromide use and emissions because companies maintained stockpiles of methyl bromide produced before the phaseout; this was only recently disclosed by US EPA.

26 Mueller, D. K. (1998) *Stored Product Protection: A Period of Transition*, Insects Limited Inc, Indianapolis.

27 Spotti, C. (2004) 'The use of fumigants and grafted plants as alternatives to methyl bromide for the production of tomatoes and vegetables in Italy', Proceedings of Fifth International Conference on Alternatives to Methyl Bromide, 27–30 September, Lisbon, European Commission, Brussels.

28 Carrera, T., Carrera, A. and Pedros, V. (2004) 'Use of 1,3-dichloropropene/chloropicrin for the production of strawberries in Spain', Proceedings of Fifth International Conference on Alternatives to Methyl Bromide, 27–30 September, Lisbon, European Commission, Brussels

29 Marcotte, M. and Tibelius, C. (1998) *Improving Food and Agriculture Productivity – and the Environment. Canadian Initiatives in Methyl Bromide Alternatives and Emission Control Technologies*, Agriculture and Agri-Food Canada, Ottawa.

30 MBTOC (1998), op cit (Note 13).

31 Gullino, M. L. and Camponogara, A. (2002) 'Italian strategy on methyl bromide phase-out', *Methyl Bromide Action in China*, no 6, State Environmental Protection Administration, Beijing, and GTZ, Eschborn.

32 Porter, I. (2002) 'Methyl bromide phaseout strategy in Australia', *Methyl Bromide Action in China*, no 9, State Environmental Protection Administration, Beijing, and GTZ, Eschborn.

33 Vos and Bridge (2006), op cit (Note 22); UNEP (2002), op cit (Note 20).

34 Vos and Bridge (2006), op cit (Note 22).

35 Porter (2002), op cit (Note 32).

36 Methyl Bromide Consultative Group (1998) *National Methyl Bromide Response Strategy. Part 1: Horticultural Uses*, Environment Australia, Canberra, p75.

37 MBTOC (2002), op cit (Note 19).

38 Porter (2002), op cit (Note 32); Methyl Bromide Consultative Group (1998), op cit (Note 36); MBTOC (2002), op cit (Note 19).

39 Moeller, K. (2004) 'Reducing the use of methyl bromide via EUREPGAP', Proceedings of Fifth International Conference on Alternatives to Methyl Bromide, 27–30 September, Lisbon, European Commission, Brussels.

40 Tateya, A. (2007) personal communication from A. Tateya, Technical Adviser, Japan Fumigation Technology Association, Japan; Zhang, J. (2002) 'Policies to control methyl

bromide in Japan', *Methyl Bromide Action in China*. no. 8, State Environmental Protection Administration, Beijing, and GTZ, Eschborn.

41   Fernández, L. M. (2002) 'Commercial policies in Spain influencing the use of methyl bromide by growers', Proceedings of International Conference on Alternatives to Methyl Bromide, 5–8 March 2002, Seville, Spain.

42   Miller M. K. (1999) *Towards Methyl Bromide Phase Out. A Handbook for National Ozone Units*, UNEP, Paris.

43   Environment Canada (1996) *Canada's Ozone Layer Protection Program – A Summary*, Environment Canada, Ottawa.

44   Lower House (1981), op cit (Note 14); UNEP (1992a) Proceedings of International Workshop Alternatives to Methyl Bromide for Soil Fumigation, 19–23 October 1992, Rotterdam, UNEP, Nairobi.

45   Lower House (1981), op cit (Note 14).

46   UNEP (1992a), op cit (Note 44).

47   VROM (1997) 'Good grounds for healthy growth', Ministry of Environment and Spatial Planning, The Hague; De Barro, P. (1995) 'Strawberry production in The Netherlands without methyl bromide', in Banks, H. J. (ed) *Agricultural Production Without Methyl Bromide – Four Case Studies*, CSIRO, Canberra; Prospect (1997) 'Provision of services with regards to the technical aspects of the implementation of EC legislation on ozone-depleting substances. Methyl bromide. Final report', Report B7-8110/95/000178/MAR/D4 commissioned by the European Commission, Prospect C&S, Brussels.

48   Hallas, T. E. et al (1993) 'Methyl bromide in the Nordic Countries – Current use and alternatives', Nord 1993:34, Nordic Council of Ministers, Copenhagen; Nordic Council (1995) 'Alternatives to methyl bromide', TemaNord 1995:574, Nordic Council of Ministers, Copenhagen.

49   Hallas, T. E. et al (1993), op cit (Note 48).

50   Ministry of the Environment (1994) Statutory Order from the Ministry of the Environment No 478 of 3 June 1994 Prohibiting the Use of Certain Ozone Depleting Substances, Ministry of the Environment, Copenhagen.

51   MBTOC (1995), op cit (Note 18); California Department of Pesticide Regulation (2004) Pesticide Use Reports, CDPR, www.cdpr.ca.gov/docs/pur.

52   Marcotte and Tibelius (1998), op cit (Note 29).

53   EC Regulation No 2037/2000.

54   European Commission (2006) 'European Community management strategy for the phase out of the critical uses of methyl bromide', submitted to the Ozone Secretariat under Decision Ex.I/4 of the Montreal Protocol, European Commission, Brussels.

55   Ibid.

56   Prospect (1997), op cit (Note 47).

57   MBTOC (2002), op cit (Note 19).

58   Porter (2002), op cit (Note 32).

59   Parliament of the Czech Republic (1995) Act on Protection of the Ozone Layer of the Earth, 86, 20 April 1995.

60   Slusarski, C. (2001) 'Report on Regional Policy Development Workshop to Assist Methyl Bromide Phaseout in Eastern and Central Europe', UNEP, Paris; Slusarski, C. (2002) 'Policy development in CEIT countries', *Methyl Bromide Action in China*, no 8, GTZ and State Environmental Protection Agency, Beijing.

61   Slusarski (2002), op cit (Note 60).

62   Watts, M. (1994) *Externalities and Economic Instruments in Pesticide Reduction Policies*, University of Auckland, New Zealand.

63   Prospect (1997), op cit (Note 47).

64　Vickers, R. A. (1995) 'Tomato production in Italy without Methyl Bromide' in Banks, H. J. (ed) *Agricultural Production without Methyl Bromide – Four Case Studies*, CSIRO Division of Entomology, Canberra, Australia; Gullino and Camponagara (2002) op cit (Note 31).

65　MBTOC (2002), op cit (Note 19).

66　MBTOC (2007), op cit (Note 8).

67　Multilateral Fund (2005) 'Evaluation of methyl bromide phase-out projects. Sub-sector reports and case studies', MLF.

68　Valeiro, A. (2002) 'Progress report on UNDP methyl bromide phaseout project in Argentina', Instituto Nacional de Tecnología Agropecuaria (INTA) [National Institute of Agriculture Technology], Tucumán, Argentina and UNDP, New York.

69　Valeiro (2002), op cit (Note 68); MBTOC (2007), op cit (Note 8).

70　MBTOC (2007), op cit (Note 8).

71　Ibid.

72　Ibid.

73　Ibid.

74　Slusarski (2002), op cit (Note 60).

75　Ibid.

76　Ibid.

77　Ozone Secretariat (2006), op cit (Note 7).

78　Hasse, V. (2001) 'Jordan's methyl bromide phaseout policy', *Methyl Bromide Action in China*, no 5, GTZ and State Environmental Protection Agency, Beijing.

79　Horn, F. and Horn, P. (2004) 'Fresh fruit fumigation with pure phosphine as alternative for methyl bromide', Annual International Research Conference on Methyl Bromide Alternatives, www.mbao.org.

80　Slusarski (2002), op cit (Note 60).

81　Pizano, M. (2004) 'Alternatives to methyl bromide for producing cut flowers and bulbs in developing countries', Proceedings of Fifth International Conference on Alternatives to Methyl Bromide, 27–30 September, Lisbon, European Commission, Brussels.

82　Cao, A. (2006) personal communication, cited in MBTOC (2007), op cit (Note 8).

83　Horticultural Crops Development Authority (2003) 'Information on crop prices in 2003', Horticultural Crops Development Authority, Nairobi.

84　Van't Hoff, P. (2004) 'International MPS certification system for cut-flower production', Proceedings of Fifth International Conference on Alternatives to Methyl Bromide, 27–30 September, Lisbon, European Commission, Brussels.

85　Miller, M. K. (1997) *International Directory of Enterprises Supplying Alternatives to Methyl Bromide*, Environmental Research & Policy, Napier.

86　MBTOC (2007), op cit (Note 8).

87　TEAP (2006a), op cit (Note 24).

88　MBTOC (2007), op cit (Note 8).

89　MLF (2005) 'Final report on the evaluation of methyl bromide projects', synthesis evaluation report (document UNEP/Ozl.Pro/ExCom/46/7), available at www.multi-lateralfund.org; Ausher, R., Valeiro, A., Pizano, M., Böye, J. and Mück, O., Consultants (2005) Background and support documents (obtained from the authors) for UNEP/OzL.Pro/ExCom/46/ 'Evaluation of methyl bromide phaseout projects. Sub-sector reports and case studies', Multilateral Fund Secretariat, Montreal, Canada.

90　MBTOC (2007), op cit (Note 8).

91　MLF (2005). Report of 46th ExCom meeting, UNEP/OzL.Pro/ExCom/46/47, p34, paragraph 118, Multilateral Fund Secretariat for the Implementation of the Montreal Protocol, Montreal, Canada.

92   TEAP (2006a), op cit (Note 24).
93   MLF (2006) 'Final evaluation report on case studies on non-compliance (Follow-up to decision 46/6)', UNEP/Ozl.Pro/Excom/50/9, Multilateral fund for the Implementation of the Montreal Protocol, Montreal, Canada.
94   Ibid.
95   International Plant Protection Convention (2002), op cit (Note 10).

CHAPTER 12

1   See reports of the Intergovernmental Panel on Climate Change, the Technology and Economic Assessment Panel, and the World Trade Organization for detailed discussion of barriers to technology transfer, for example IPCC (2001) 'Special report on technology transfer', Cambridge University Press, Cambridge, UK; World Trade Organization (2002) 'Working Group on Trade and the Transfer of Technology: A taxonomy on country experiences on international technology transfers', WT/GGTTT/W/3, 11 November, WTO, Geneva, Switzerland.
2   United Nations 'Agenda 21', retrieved 3 March 2006 from www.un.org/esa/sustdev/documents/agenda21/index.htm.
3   UNCTAD (United Nations Conference on Trade and Development) (2001) *Transfer of Technology*, United Nations, Geneva, p63, available at www.unctad.org/en/docs/psiteiitd28.en.pdf.
4   Maskus, Keith E. (2004) *Encouraging International Technology Transfer*, International Centre for Trade and Sustainable Development (ICTSD) and United Nations Conference on Trade and Development (UNCTAD), Geneva; Intergovernmental Panel on Climate Change (2001) *Methodological and Technological Issues in Technology Transfer: Special Report of the Intergovernmental Panel on Climate Change*, Cambridge University Press, Cambridge, UK; UNCTAD (2001), op cit (Note 3).
5   Blackman, Allen (1999) 'The economics of technology diffusion: Implications for climate policy in developing countries', Discussion Paper 99-42, Resources for the Future, Washington, DC.
6   For example, in Ukraine, the project completion report of the World Bank states that 'a significant number of potential beneficiaries that would not have survived transition to the market economy were eliminated before GEF financing was committed' (World Bank (2005) *Ukraine Implementation Completion Report*, Report No UA-32788, World Bank, Washington, DC).
7   UNDP projects 20/12, 13 and 15, 23/21–25, 26/31-39.
8   K. Madhava Sarma, former Executive Secretary of the Ozone Secretariat, personal communication, 22 March 2007.
9   Jayashree Watal (2000) 'The issue of technology transfer in the context of the Montreal Protocol', in Jha, V. and Hoffmann, U. (eds) (2000) *Achieving Objectives of Multilateral Environmental Agreements: A Package of Trade Measures and Positive Measures – Elucidated by Results of Developing Country Case Studies*, UNCTAD, Geneva, UNCTAD/ITCD/TED/6.
10   Jha and Hoffmann (2000), op cit (Note 9), contains a number of useful observations on the difficulties that arose in the transfer of some technologies to developing countries.
11   Information from Jayashree Watal (2000), op cit (Note 9).
12   Project No CPR/REF/20/INV/182: Conversion of CFC-11 and CFC-12 centrifugal refrigerating compressor to centrifugal technology at Chongqing General Machinery Factory.

13 Project No IDS/SOL/20/INV/50: To phase out consumption of TCA used as a solvent in the production of correction fluid at PT Cahaya Biru Sakti.

14 See www.china.org.cn/english/2001/Jan/6434.htm; www.multilateralfund.org/files/evaluation/4212.pdf, p17.

15 Information from Jayashree Watal (2000), op cit (Note 9).

16 Korean Trade Promotion Agency (2000) 'The Republic of Korea and the Montreal Protocol', in Jha and Hoffmann, op cit (Note 9).

17 Detailed discussion of barriers to technology transfer can be found in Intergovernmental Panel on Climate Change (2001), op cit (Note 9); World Trade Organization (2002), op cit (Note 1).

18 These lessons are taken from the UNEP OzonAction Programme (1999) 'Lessons learned and case studies in technology transfer', and a workshop in Bangkok held in March 1999 by the Centre for Science and Technology of the Non-Aligned and Other Developing Countries, UNEP and the Asian and Pacific Centre for Transfer of Technology.

CHAPTER 13

1 The authors are grateful for the contributions of Etienne Gonin, Samira de Gobert, Jim Curlin, Yerzhan Aisabayev and Nirupa Ram.

2 See Agenda 21, Chapter 34, www.un.org/esa/sustdev/documents/agenda21/index.htm

3 www.unep.fr/ozonaction.

4 The United Nations Environment Programme (UNEP), the United Nations Development Programme (UNDP), the United Nations Industrial Development Organization (UNIDO) and the World Bank (2005) *The Montreal Protocol: Partnerships Changing the World*, www.uneptie.org/ozonaction/information/mmc/lib_detail.asp?r=4326.

5 Benedick, R. (1998) *Ozone Diplomacy, New Directions in Safeguarding the Planet*, Harvard University Press, Cambridge, MA.

6 Shende, R. and Gorman, S. (1997) 'Lessons in technology transfer under the Montreal Protocol', Division of Technology Industry and Economics – OzonAction Programme, UNEP, Paris.

7 COWI Consult (1995) 'Study on the financial mechanism of the Montreal Protocol', Ozone Secretariat, UNEP, Nairobi.

8 Information available at www.uneptie.org/Ozonaction/topics/capacity.htm.

9 Swedish International Development Cooperation Agency (SIDA) and UNEP's Division of Technology Industry and Economics – OzonAction Programme (2002) *Networking Counts: Building Bridges for a Better Environment, the Montreal Protocol Experience in Making Multilateral Environmental Agreements Work*, Swedish International Development Cooperation Agency (SIDA) and UNEP's Division of Technology Industry and Economics – OzonAction Programme, www.uneptie.org/ozonAction/information/mmcfiles/3947-e.pdf.

10 Shende, R. (1996) 'Developing a blueprint for the phaseout of ODSs in Middle East countries – The role of the United Nations Environment Programme', Conference on CFC and Halon Replacement Strategies, Dubai, 28–29 February.

11 Andersen, S. O. and Sarma, K. (2002) *Protecting the Ozone Layer: The United Nations History*, Earthscan, London, p272.

12 For an elaboration, see the 2003 UNFCCC technical report on technology, available at transferhttp://unfccc.int/resource/docs/tp/tp0301.pdf.

13   COWI Consult (1995), op cit (Note 7).
14   Information available at www.halontrader.org.
15   See also United Nations Framework Convention on Climate Change (UNFCCC) (2003) *Capacity Building in the Development and Transfer of Technologies*, http://unfccc.int/resource/docs/tp/tp0301.pdf
16   Gerard, D. and Lester, L. (2003) *Implementing Technology Forcing Policies: The 1970 Clean-Air Act Amendments and the Introduction of Advanced Automotive Emissions Control*, Centre for the Study and Improvement of Regulation, Department of Engineering and Public Policy, Carnegie Mellon University Press, Pittsburgh, PA.
17   UNEP Division of Technology, Industry and Economics (DTIE) (2001) *Ten-year Review of Capacity Building Activities of the OzonAction Programme (1992–2001)*, UNEP/DTIE, Paris.
18   See www.unep.fr/ozonaction/information/mmc/list.asp?x=a for further details of all tools mentioned in this box.
19   UNEP/DTIE (2006) 'Activity and performance report 2006', UNEP/DTIE, Paris.
20   Stockholm Environment Institute (2005) *Interlinked ODS Phase-Out Activities – A Handbook for Improved Effectiveness of ODS Phase-Out Activities in the Refrigeration Servicing Sector*, Stockholm Environment Institute, Stockholm, Sweden.
21   Croatia is also very active in networking activities designed to share its experience with others.
22   Stockholm Environment Institute (2005), op cit (Note 20).
23   UNEP/DTIE (2005) 'SolarChill – The vaccine cooler powered by nature', www.solarchill.org/ and www.refrigerantsnaturally.com/.
24   UNEP/DTIE – OzonAction Programme (2001) *Protecting the Ozone Layer*, Volume 2, UNEP/DTIE, Paris; UNEP (1999) 'Report of the Solvents, Coatings and Adhesives, Technical Options Committee (STOC), 1998 Assessment', UNEP, Nairobi; O'Connor, David C. (1991) *Policy and Entrepreneurial Responses to the Montreal Protocol: Some evidence from the Dynamic Asian Economies*, Organisation for Economic Co-operation and Development (OECD), Paris.
25   Metz, B., Davidson, O. R., Martens, J. W., van Rooijen, S. N. M. and Van Wie McGrory, L. (2000) *Methodological and Technological Issues in Technology Transfer, A Special Report by the IPCC Working Group III*, Cambridge University Press; Asian and Pacific Centre for Transfer of Technology (APCTT) (1999) *Louder Lessons in Technology Transfer: Lessons Learned and Case Studies*, APCTT, Centre for Science and Technology of the Non-Aligned and other Developing Countries, UNEP/DTIE.
26   Metz et al (2000), op cit (Note 25).
27   Metz et al (2000), op cit (Note 25); APCTT (1999), op cit (Note 25).
28   Benedick, R. (2007) 'Avoiding gridlock on climate change', *Issues in Science and Technology*, winter, pp37-40, available at www.issues.org/23.2/p_benedick.html.
29   UNEP/DTIE – OzonAction Programme, The MLF for the Implementation of the Montreal Protocol (2006) *Return of the Ozone Layer: Are We There Yet?*, UNEP/DTIE, Paris.
30   Greenpeace (2000) *Green Olympics, Dirty Sponsors*, report, Greenpeace International, April.

CHAPTER 14

1   The views expressed in this chapter, and indeed throughout the book, are those of the authors and do not necessarily reflect the views of the sponsors of this book, the US Environmental Protection Agency and the Montreal Protocol Technology and

Economic Assessment Panel, or other organizations where the authors are employed or serve. The authors are indebted to Alan Miller (International Finance Corporation), Scott Stone and Durwood Zaelke (Institute for Governance and Sustainable Development) for their significant contributions to this chapter.

2   Zaelke, D., Kaniaru, D. and Kružiková, E. (2005) *Making Law Work: Environmental Compliance and Sustainable Development*, Chapter 13, Compliance and competitiveness: The Porter Hypothesis?, Cameron May, London.

3   Velders, G. J. M., Andersen, S. O., Daniel, J. S., Fahey, D. W. and McFarland, M. (2006) 'The importance of the Montreal Protocol in protecting the climate', *The Proceedings of the National Academy of Sciences of the United States of America (PNAS)*, vol 104, no 12, pp4814–4819.

4   Life-cycle climate performance (LCCP) measures the impact of the direct chemical emissions plus the indirect energy consumption emissions over the full life-cycle of the product. It is the only way to properly compare the different options for refrigeration, air-conditioning and thermal insulating foam.

5   There was significant inequality among the Parties to the Montreal Protocol when it was signed in 1989. Developed countries bore most of the responsibility for inventing, irresponsibly marketing, and frivolously using ODSs in emissive uses such as cosmetic and convenience aerosol products.

6   United Nations Environment Programme (2006) '2006 assessment of the Montreal Protocol Science Assessment Panel', retrieved April 2007 from http://ozone.unep.org/ Assessment_Panels/SAP/Scientific_Assessment_2006_Exec_Summary.pdf.

7   Andersen, S. O. and Sarma, K. M. (2002) *Protecting the Ozone Layer: The United Nations History*, Earthscan, London, Chapter 5, pp187–233; Economic Options Committee of the Montreal Protocol Technical and Economics Assessment Panel (1991, 1994 and 1998) Reports of the Economic Options Committee, Ozone Secretariat, UNEP.

8   The Intergovernmental Panel on Climate Change (IPCC) was established in 1990 by the United Nations Environment Programme (UNEP) and the World Meteorological Organization (WMO) to periodically assess the science, impacts and socio-economics of climate change and to identify mitigation and adaptation options. The IPCC provides, on request, advice to UNFCCC, similarly to the Assessment Panels providing advice on request to the Parties to the Montreal Protocol.

9   IPCC (2007) 'Summary for policymakers', in 'Climate change 2007: The physical science basis', contribution of Working Group I to the assessment report of the IPCC, accessed April 2007 from www.IPCC.ch.

10   Bronowski, Jacob (1974) *The Ascent of Man*, Chapter 3, Little Brown, London.

11   Generally attributed to Johann Wolfgang von Goethe, this passage was actually written by William H. Murray in *The Scottish Himalaya Expedition* (1951). The association with Goethe came from Murray's going on to quote, 'with deep respect', the Goethe couplet:
     Whatever you can do or dream you can, begin it.
     Boldness has genius, power and magic in it!

12   Hoffman, A. J. (2006) 'Getting ahead of the curve: Corporate strategies that address climate change', report prepared for and published by the Pew Center on Global Climate Change, Washington, DC.

13   United Nations Environment Programme (2007), UNEP/OzL.Pro.WG.1/27/3, February, retrieved April 2007 from www.unep./ozone.

14   Hoffman (2006), op cit (Note 12).

15   Wells, H. G. (1920) *Outline of History: Being a Plain History of Life and Mankind*, vol 2, chapter 41, George Newnes, London.

16   From www.un.org/millenniumgoals.

17   Intergovernmental Panel on Climate Change (2001) 'Executive Summary of Chapter 3, paragraph 3.9, contribution of Working Group III: Mitigation', retrieved April 2007 from www.grida.no/climate/ipcc_tar/wg3/index.htm.

18   Ibid.

19   Global Environment Facility (2005) 'Third overall performance study of the Global Environment Facility (executive version)', June, retrieved April 2007 from www.gefweb.org.; the GEF was established as a pilot programme of the World Bank in 1991 as a means of dedicated financing for the incremental costs to developing countries of achieving global environmental objectives. A restructured, independently governed GEF was established by international agreement in 1994 and includes financing for ozone phaseout in Russia and Eastern European countries that were not eligible for similar financing from the MLF established under the Montreal Protocol by the London Amendments of 1990.

20   The GEF is currently reviewing and updating its focal area strategies based on independent expert opinion. These views have been prepared partly as a contribution to this process.

21   Report of the meeting, FCCC/CP/2001/5, from UNFCCC website, www.unfccc.int.

22   World Business Council for Sustainable Development (2007) *Policy Directions to 2050: A Business Contribution to the Dialogues on Cooperative Action*, World Business Council for Sustainable Development, March

23   United Nations Framework Convention on Climate Change (2007) 'Background note', retrieved April 2007 from www.unfccc.int.

24   Ibid.

List of Acronyms and Abbreviations

AFC	Australian Fluorocarbon Council (was AFCAM)
AFCAM	Association of Fluorocarbon Consumers and Manufacturers (now AFC)
AFCFCP	Alliance for Responsible CFC Policy (now ARAP)
AFEAS	Alternative Fluorocarbons Environmental Acceptability Study
APC	Aerosol Promotion Council (India)
ARAP	Alliance for Responsible Atmospheric Policy (was AFCFCP)
ASEAN	Association of Southeast Asian Nations
ASHRAE	American Society of Heating, Refrigerating, and Air-Conditioning Engineers
ASTM	American Society of Testing and Materials
ATOC	Aerosols, Sterilants, Miscellaneous Uses and Carbon Tetrachloride Technical Options Committee (TEAP)
CAP	Compliance Assistance Programme
CCOL	Coordinating Committee on the Ozone Layer (UNEP)
CDM	Clean Development Mechanism
CEIT	country with economy in transition
CFC	chlorofluorocarbon
CFM	chlorofluoromethane
CMA	Chemical Manufacturers Association
COAS	Council of Atmospheric Sciences
COP	Conference of the Parties (to the Vienna Convention)
COPD	chronic obstructive pulmonary disease
CT	clean technology
CTC	carbon tetrachloride
CUE	critical use exemption
	Consumption
DLA	Defense Logistics Agency (US)
DME	dimethyl ether
DPI	dry powder inhaler
DTI	Danish Technological Institute
DTIE	Division of Technology, Industry and Economics (of UNEP)
EAP	Economic Assessment Panel
EC	European Commission
ECE	Economic Commission for Europe (UN)
ED	Executive Director
EDF	Environmental Defense Fund (US) (now Environmental Defense)

EEAP	Environmental Effects Assessment Panel
EEC	European Economic Community
EIA	Environmental Impact Assessment
ENGO	environmental non-governmental organization
EO	ethylene oxide
EOC	Economic Options Committee (TEAP)
EPA	Environmental Protection Agency (US)
EU	European Union
EUE	essential use exemption
EUREP-GAP	European standards of good agricultural practice
FAA	Federal Aviation Administration (US)
FAO	Food and Agriculture Organization of the United Nations
FC	fluorocarbon
FCCC	Framework Convention on Climate Change (UN)
FDA	Food and Drug Administration (US)
FDI	foreign direct investment
FOE	Friends of the Earth
FPA	Fire Protection Association (Hungary)
FPI	Foodservice and Packaging Institute
FTOC	Foams Technical Options Committee (TEAP)
GATT	General Agreement on Tariffs and Trade
GAW	Global Atmosphere Watch (established by WMO)
GC	Governing Council (UNEP)
GEF	Global Environment Facility
GEMS	Global Environment Monitoring System
GHG	greenhouse gas
GM	General Motors
Gt	gigatonne
GTZ	Deutsche Gesellschaft für Technische Zusammenarbeit (Bilateral Aid agency of Germany)
GWP	global warming potential
HAP	hydrocarbon aerosol propellant
HARC	Halon Alternatives Research Corporation (US)
HBFC	hydrobromofluorocarbon
HBR&R	halon banking, recovery and recycling
HC	hydrocarbon
HCFC	hydrochlorofluorocarbon
HFC	hydrofluorocarbon
HFE	hydrofluoroether
HTOC	Halons Technical Options Committee (TEAP)
HUNC	Halon Users National Consortium
ICAO	International Civil Aviation Organization (UN)
ICEL	International Cooperative for Environmental Leadership (was ICOLP)
ICI	Imperial Chemical Industries
ICOLP	International Cooperative for Ozone Layer Protection (now ICEL)

ICSU	International Council of Scientific Unions
IGPOL	Industry Group for the Protection of the Ozone Layer (European Community)
IIR	International Institute of Refrigeration
IMO	International Maritime Organization (UN)
IOC	International Ozone Commission
IPAC	International Pharmaceutical Aerosol Consortium
IPCC	Intergovernmental Panel on Climate Change (UN Framework Convention on Climate Change)
IPM	integrated pest management
IPPC	Integrated Pollution Prevention and Control
ISO	International Standards Organization
JACET	Japan Association of Cleaning Engineering and Technology
JAHCS	Japan Association for Hygiene of Chlorinated Solvents
JEMA	Japan Electrical Manufacturers' Association
JI	Joint implementation
JICC	Japan Industrial Conference on Cleaning
JICOP	Japan Industrial Conference for Ozone Layer Protection
K	kelvin (unit of thermodynamic temperature)
LCCP	life-cycle climate performance
LCD	liquid carbon dioxide
LEK	lower explosive limit
LIA	low index/additive (technology)
LPG	liquefied petroleum gas
MAC	mobile air-conditioning
MACS	Mobile Air Conditioning Society Worldwide
MBAN	Methyl Bromide Alternatives Network
MBTOC	Methyl Bromide Technical Options Committee (TEAP)
MC	methylene chloride
MDI	metered-dose inhaler
MEA	multilateral environmental agreement
METI	Ministry of Economy, Trade and Industry (Japan)
MLF	Multilateral Fund for the Implementation of the Montreal Protocol
MOP	Meeting of the Parties (to the Montreal Protocol)
MP	Montreal Protocol
MPS	Milieu Programma Sierteelt certification programme
Mt	megatonne
NASA	National Aeronautics and Space Administration (US)
NATO	North Atlantic Treaty Organization
NCAR	National Center for Atmospheric Research (US)
NDSC	Network for the Detection of Stratospheric Change
NFPA	National Fire Protection Association (US)
NGO	non-governmental organization
NO	nitric oxide
NO_2	nitrogen dioxide
N_2O	nitrous oxide

NOAA	National Oceanic and Atmospheric Administration (US)
NOU	National Ozone Unit
nPB	normal propyl bromide
NRDC	Natural Resources Defense Council (US)
O_3	ozone
ODA	official development assistance
ODP	ozone depletion potential
ODS	ozone-depleting substance
OECD	Organisation for Economic Co-operation and Development
OEWG	Open-Ended Working Group (of the Parties to the Montreal Protocol)
OORG	Ozone Operations Resource Group (World Bank)
PAFT	Program for Alternative Fluorocarbon Toxicity Testing
PATF	Process Agent Task Force (TEAP)
PCE	perchloroethylene
PFC	perfluorocarbon
ppt	parts per trillion
PU	polyurethane
QPS	quarantine and pre-shipment
R&D	research and development
RMP	refrigeration management plan
RTOC	Refrigeration, Air-Conditioning and Heat Pumps Technical Options Committee (TEAP)
SACEP	South Asia Cooperative Environment Programme
SAE	Society of Automobile Engineers
SAGE	tratosphere Aerosol and Gas Experiment
SAP	Scientific Assessment Panel (UNEP)
SBSTA	Subsidiary Body for Scientific and Technological Advice (UN Framework Convention on Climate Change)
SCOPE	Standing Committee on the Problems of the Environment
SEI	Stockholm Environment Institute
SEPA	State Environmental Protection Administration (China)
SF_6	sulphur hexafluoride
SMEs	small and medium-sized enterprises
SNAP	Significant New Alternatives Program (US EPA)
SORG	Stratospheric Ozone Research Group (UK)
SSP	Solvent Sector Plan (China)
SST	supersonic transport aircraft
STOC	Solvents, Coatings and Adhesives Technical Options Committee (TEAP)
SUBSTA	Subsidiary Body for Scientific and Technological Advice (UNFCCC)
TAP	Technical Assistance Programme
TCA	1,1,1-trichloroethane (also methyl chloroform)
TCE	trichloroethylene
TEAP	Technology and Economic Assessment Panel (UNEP Montreal Protocol)

TEWI	total equivalent warming impact
TOC	Technical Options Committee (of TEAP)
TOMS	total ozone-mapping spectrometer
TRP	Technology Review Panel (merged with Economic Panel to become TEAP)
UARS	upper atmosphere research satellite
UEL	upper explosive limit
UN	United Nations
UNCED	United Nations Conference on Environment and Development (known as the Earth Summit)
UNCTAD	United Nations Conference on Trade and Development
UNDESA	United Nations Department of Economic and Social Affairs
UNDP	United Nations Development Programme
UNEP	United Nations Environment Programme
UNESCO	United Nations Educational, Scientific and Cultural Organization
UNFCCC	United Nations Framework Convention on Climate Change
UNIDO	United Nations Industrial Development Organization
UNOPS	United Nations Office for Project Services
US DoD	United States Department of Defense
US EPA	United Stated Environmental Protection Agency
USSR	Union of Soviet Socialist Republics
UV	ultraviolet
UV-A	ultraviolet A radiation
UV-B	ultraviolet B radiation
UV-C	ultraviolet C radiation
VIF	virtually impermeable film
VOC	volatile organic compound
WHO	World Health Organization
WMO	World Meteorological Organization
WRI	World Resources Institute
WTO	World Trade Organization

Glossary

Aerosol products: Containers filled with ingredients and an ODS, hydrocarbon or HFC propellant to release the product in a fine spray.

Amendments and Adjustments: 'Amendments' (Articles 9 and 10 of the Convention) must be ratified by individual Parties and are binding, after entry into force of the Amendment, only on those Parties that ratify the Amendment. 'Adjustments' (Article 2, Paragraph 9 of the Montreal Protocol) to the ozone depletion potential or control measures of substances already listed in the annexes to the Protocol, once duly approved by Meetings of the Parties and notified, will be binding on all the Parties, without any process of ratification, after the expiry of six months after the notification.

Annex A substances: Eight chemicals: Group I, five CFCs; and Group II, three halons.

Annex B substances: Three groups: Group I, ten CFCs; Group II, carbon tetrachloride; and Group III, methyl chloroform.

Annex C substances: Three groups: Group I, 40 HCFCs; Group II, 34 HBFCs; and Group III, bromochloromethane.

Annex E substance: Methyl bromide.

Aqueous cleaning: Cleaning parts with water.

Article 5 Party: A developing country Party to the Montreal Protocol whose annual calculated level of consumption is less than the limits prescribed in Article 5, Paragraph 1 – 0.3kg per capita of the controlled substances in Annex A, and less than 0.2kg per capita of the controlled substances in Annex B, on the date of the entry into force of the Montreal Protocol for that country or any time thereafter. An Article 5 Party is entitled to delay its implementation of the control measures of Article 2 by a period specified in Article 5 of the Protocol, in order to meet its basic domestic needs.

Atmospheric lifetime: A measure of the time a chemical remains without breaking up, once released into the atmosphere.

Basic domestic needs: The Protocol does not define this term. The first and seventh Meetings of the Parties made clarificatory decisions on this term.

Blowing agents: Expanding gases that foam the plastic matrix, creating open or closed foam cells.

Carbon tetrachloride (CCl_4): A solvent with an ODP of approximately 1.1 that is controlled under the Montreal Protocol. It is used primarily as a feedstock material for the production of other chemicals.

Cataract Damage: Damage to the eye, in which the lens is partly or completely clouded, impairing vision and sometimes causing blindness. Exposure to ultraviolet radiation can cause cataracts.

Chlorofluorocarbons (CFCs): A family of organic chemicals composed of chlorine, fluorine and carbon atoms.

Chlorofluoromethanes: A subset of chlorofluorocarbons that contain only one carbon atom.

Command and control: To prescribe regulations to compel implementation of particular policies.

Consumption: As defined by the Montreal Protocol, production plus imports minus exports.

Controlled substance: Any ozone-depleting substance that is subject to control measures under the Montreal Protocol, such as a phase-out requirement. Specifically, it refers to a substance listed in Annexes A, B, C or E of the Protocol, whether alone or in a mixture.

Decommissioning: The physical process of removing a halon system from service.

Developing countries: Countries listed as 'developing' by the first Meeting of the Parties to the Montreal Protocol in 1989 and subsequently modified by Meetings of the Parties.

Dobson unit: A unit to measure the total amount of ozone in a vertical column above the Earth's surface. A typical value for the amount of ozone in a column of the Earth's atmosphere is 300 Dobson units.

Drop-in replacement: The procedure of replacing CFC refrigerants with non-CFC refrigerants in existing refrigerating, air-conditioning and heat pump plants without doing any plant modifications.

Essential use: Decision IV/25 of the Parties to the Montreal Protocol defined criteria for essential uses for which Parties could get exemptions for continued use of ODSs after the phase-out date.

Executive Committee (of the Multilateral Fund): Established by Article 10 of the Montreal Protocol to administer the Multilateral Fund.

Feedstock: A chemical that is entirely transformed or destroyed while being used for manufacture of another chemical.

Financial mechanism: Established by Article 10 of the Montreal Protocol for providing financial and technical cooperation, including the transfer of technologies, to Article 5 Parties. It may also include other means of multilateral, regional and bilateral cooperation.

Freon™: The trade name of CFC and HCFC products marketed by DuPont.

Global Environment Facility (GEF): Provides financing to the eligible developing countries and countries with economies in transition for programmes that achieve global environmental benefits in one or more of four focal areas: biological diversity; climate change; international waters; and the ozone layer.

Global warming: The warming of the Earth due to the heat-trapping action of natural and manufactured greenhouse gases, which leads to climate change.

Global warming potential (GWP): The relative contribution of greenhouse gases, e.g. carbon dioxide, methane, CFCs, HCFCs and halons, to the global warming effect when the substances are released to the atmosphere. The standard measure of GWP is relative to carbon dioxide, whose GWP is 1.0.

Greenhouse gas: A gas, such as water vapour, carbon dioxide, methane, CFCs and HCFCs, that absorbs and re-emits infrared radiation, warming the Earth's surface and contributing to climate change.

Halocarbon: A compound derived from methane (CH_4) and ethane (C_2H_6), in which one or several of the hydrogen atoms are substituted with chlorine (Cl), fluorine (F) and/or bromine (Br). CFCs, HCFCs and HFCs are examples of halocarbons. If all the hydrogen atoms are substituted in a halocarbon, it is 'fully halogenated'. Otherwise it is 'partially halogenated'.

Halon: A halon is a bromochlorofluorocarbon, a chemical consisting of one or more carbon atoms surrounded by fluorine, chlorine and bromine. Halons are fully halogenated hydrocarbons that are used as fire-extinguishing agents and as explosion suppressants.

Halon bank: The total quantity of halon existing at a given moment in a facility, organization, country or region. The halon bank includes the halon in fire protection systems, in portable fire extinguishers, in mobile fire extinguishers and in storage containers.

Halon bank management: A method of managing a supply of banked halon. Bank management consists of keeping track of halon quantities at each stage: initial filling; installation; 'recycling'; and storage. A major goal of a halon bank is to avoid demand for new (virgin) halons by redeploying halons from decommissioned systems or non-essential applications to essential uses.

Harmonized System: A numbering system developed by the World Customs Organization to cover all imported or exported goods in international trade.

HCFC cap: The non-Article 5 Parties must freeze their consumption of HCFCs from 1996 at the level of the HCFC cap, i.e. their 1989 consumption of HCFC plus 2.8 per cent of their 1989 CFC consumption.

Hydrobromofluorocarbons (HBFCs): A family of hydrogenated chemicals related to halons consisting of one or more carbon atoms surrounded by fluorine, bromine, at least one hydrogen atom and sometimes chlorine.

Hydrocarbons (HCs): Chemical compound consisting of one or more carbon atoms surrounded only by hydrogen atoms. Examples of hydrocarbons are propane (C_3H_8, HC-290), propylene (C_3H_6, HC-1270) and butane (C_4H_{10}, HC-600). HCs are commonly used as a substitute for CFCs in aerosol propellants and refrigerant blends and have an ODP of zero. Although they are used as refrigerants, their highly flammable properties normally restrict their use as low concentration components in refrigerant blends.

Hydrochlorofluorocarbons (HCFCs): A family of chemicals related to CFCs that contain hydrogen, chlorine, fluorine and carbon atoms. HCFCs are partly halogenated and have much lower ODP than the CFCs.

Hydrofluorocarbons (HFCs): A family of chemicals related to CFCs that contain one or more carbon atoms surrounded by fluorine and hydrogen atoms. Since no chlorine or bromine is present, HFCs do not deplete the ozone layer. HFCs are widely used as refrigerants. Examples of HFC refrigerants are HFC-134a (CF_3CH_2F) and HFC-152a (CHF_2CH_3). HFCs have a high global warming potential.

Implementation Committee: A committee of the Parties to the Montreal Protocol, established by the non-compliance procedure of the Protocol to investigate and recommend action to the Meetings of the Parties on instances of non-compliance by Parties with the provisions of the Montreal Protocol.

Implementing agencies: United Nations Development Programme (UNDP), United Nations Environment Programme (UNEP), United Nations Industrial Development Organization (UNIDO) and the World Bank have been designated as the implementing agencies of the Multilateral Fund. UNDP, UNEP and the World Bank are the implementing agencies of the GEF.

Incremental costs: The Multilateral Fund meets the agreed incremental costs of Article 5 Parties. An indicative list of such incremental costs was agreed to by the second and fourth Meetings of the Parties to the Montreal Protocol.

Industrial rationalization: An exchange of production quotas between Parties, provided total production is within the limit set by the Protocol.

Langmuir periodic table: A variation of the periodic table that grouped all the known elements by their atomic weight.

Low volume ODS-consuming countries (LVC countries): Defined by the Multilateral Fund's Executive Committee as Article 5 Parties whose calculated level of ODS consumption is less than 360 ODP tonnes annually.

Metered-dose inhalers (MDI): A method of dispensing inhaled pulmonary drugs.

Methyl bromide: A colourless, odourless, highly toxic gas composed of carbon, hydrogen and bromine, used as a broad spectrum fumigant in commodity, structural and soil fumigation. Methyl bromide has an ODP of approximately 0.6.

Methyl chloroform: see 1,1,1-trichloroethane.

Montreal Protocol: An international agreement, under the Vienna Convention to Protect the Ozone Layer, to phase out the production and consumption of ozone-depleting chemicals according to a time schedule.

Multilateral Fund: Part of the financial mechanism under the Montreal Protocol, established by the Parties to provide financial and technical assistance to Article 5 Parties.

National ozone unit: The government unit in an Article 5 country that is responsible for managing the national ODS phase-out strategy as specified in the country programme. These units are responsible for, inter alia, fulfilling data reporting obligations under the Montreal Protocol.

No-clean soldering: A method of soldering which leaves no residues to be cleaned up.

Non-Article 5 Parties: Parties that do not operate under Article 5. They are obliged to implement the control measures of Article 2.

Non-Party: With respect to any controlled substance, the term 'state not party to this Protocol' (non-Party) includes a country that has not agreed to be bound by the control measures for that substance. For instance, a Party to the Montreal Protocol of 1987, which listed and defined the control measures for the Annex A substances, is a Party for the Annex A substances. If the same Party does not ratify the 1990 Amendment to the Protocol, which listed and defined the control measures for the Annex B substances, it is a non-Party for the Annex B substances.

Not-in-kind alternatives/substitutes: Approximately 80 per cent of ozone-depleting compounds that would be used today if there were no Montreal Protocol have been successfully phased out without the use of other fluorocarbons. ODS use was eliminated with a combination of 'not-in-kind' chemical substitutes, product alternatives, manufacturing process changes, conservation and doing without.

ODP tonnes: The number of tonnes of a substance multiplied by its ozone depletion potential.

OzonAction Programme: The programme implemented by UNEP with the assistance of the Multilateral Fund. It provides assistance to developing countries under the Montreal Protocol through information exchange, training, networking, country programmes and institutional strengthening projects.

Ozone: A gas consisting of three oxygen atoms, formed naturally in the atmosphere by the association of molecular oxygen (O_2) and atomic oxygen (O). It has the property of blocking the passage of dangerous wavelengths of ultraviolet radiation in the upper atmosphere.

Ozone-depleting substance (ODS): Any substance that can deplete the stratospheric ozone layer.

Ozone depletion: Accelerated chemical destruction of the stratospheric ozone layer by the presence of ozone-depleting substances produced, for the most part, by human activities.

Ozone depletion potential (ODP): A relative index indicating the extent to which a chemical product may cause ozone depletion. The reference level of 1 is the potential of CFC-11 and CFC-12 to cause ozone depletion. If a product has an ozone depletion potential of 0.5, a given weight of the product in the atmosphere would, in time, deplete half the ozone that the same weight of CFC-11 would deplete.

Ozone layer: An area of the stratosphere, approximately 15 to 60 kilometres (9 to 38 miles) above the Earth, where ozone is found as a trace gas (at higher concentrations than other parts of the atmosphere). This relatively high concentration of ozone filters most ultraviolet radiation, preventing it from reaching the Earth.

Ozone-safe: A substance which has zero ODP (example, HFCs, hydrocarbons) or a product which contains no ozone-depleting substance.

Ozone Secretariat: The Secretariat to the Montreal Protocol and Vienna Convention, provided by UNEP and based in Nairobi, Kenya.

Ozonesonde: A lightweight, balloon-borne instrument that is mated to a conventional meteorological radiosonde. As the instruments ascend through the atmosphere, information on ozone and standard meteorological quantities such as pressure, temperature and humidity are transmitted to ground stations.

Parties to the Montreal Protocol: Countries/regional economic integration organizations that have ratified the Protocol.

Perfluorocarbons (PFCs): A group of synthetically produced compounds in which the hydrogen atoms of a hydrocarbon are replaced with fluorine atoms. The compounds are characterized by extreme stability, non-flammability, low toxicity, zero ozone-depleting potential and high global warming potential.

Phaseout: The ending of all production and consumption of a chemical controlled under the Montreal Protocol, consumption being defined as production plus imports minus exports.

Production: As defined by the Montreal Protocol, the amount produced minus the amount destroyed by technologies approved by the Parties minus the amount entirely used as feedstock in the manufacture of other chemicals.

Quarantine and pre-shipment applications: Quarantine applications are treatments to prevent the introduction, establishment or spread of quarantine pests, including diseases. Pre-shipment applications are those treatments applied directly preceding and in relation to exports to meet the sanitary or phytosanitary requirements of the importing or exporting country.

Reclamation: As defined by the Parties to the Montreal Protocol in their Decision IV/24, 'the re-processing and upgrading of a recovered controlled substance through such mechanisms as filtering, drying, distillation and chemical treatment in order to restore the substance to a specified standard of performance. It often involves processing "off-site" at a central facility'.

Recovery: As defined by the Parties to the Montreal Protocol in their Decision IV/24, 'the collection and storage of controlled substances from machinery, equipment, containment vessels, etc, during servicing or prior to disposal'.

Recycling: As defined by the Parties to the Montreal Protocol in their Decision IV/24, 'the re-use of a recovered controlled substance following a basic cleaning process such as filtering and drying'. Refrigerants are normally cleaned by the on-site recovery equipment and recharged back into the equipment.

Refrigerant: The chemical or mixture of chemicals used in refrigeration equipment.

Refrigerant management plan (RMP): A strategy for cost-effective phaseout of ODS refrigerants, which considers and evaluates all alternative technical and policy options.

Service tail: After the phase-out date for ozone-depleting substances, permission to consume ODSs for the purpose of servicing existing equipment.

Skin cancer: There are three types of skin cancer: basal cell carcinoma; squamous cell carcinoma; and cutaneous melanoma.

Stratosphere: The part of the Earth's atmosphere above the troposphere, at about 15 to 60 kilometres (9 to 38 miles). The stratosphere contains the ozone layer.

Technology and Economic Assessment Panel (TEAP): The TEAP is one of the three assessment panels appointed by the Parties to the Montreal Protocol. The TEAP is responsible for reviewing and reporting to the Parties about the status of options to phase out the use of ODSs, recycling, reuse and destruction techniques, and their technological and economic viability. The TEAP is served by Technical Options Committees on Chemicals; Medical uses; Foams; Halons; Methyl Bromide; and Refrigeration, Air-Conditioning and Heat Pumps. These committees consist of hundreds of experts from around the world.

Transitional substance: A chemical whose use is permitted as a replacement for ozone-depleting substances, but only temporarily due to the substance's ODP. The HCFCs were categorized as transitional by the second Meeting of the Parties in 1990, but were included in the list of controlled substances by the fourth Meeting of the Parties in 1992.

1,1,1-trichloroethane: A hydrochlorocarbon commonly used as a blowing agent and as a solvent in a variety of metal, electronic and precision cleaning applications. It has an ODP of approximately 0.11. It is also known as methyl chloroform.

Troposphere: The lower part of the Earth's atmosphere, below 15 kilometres (9 miles). The troposphere is below the stratosphere.

Ultraviolet radiation (UV radiation): Solar radiation at the top of the atmosphere contains radiation of wavelengths shorter than visible light. The shortest of these wavelengths (UV-C) are completely absorbed by oxygen and ozone in the atmosphere. Wavelengths in the middle (UV-B) range are partially absorbed by the ozone layer. The higher UV-A wavelengths are transmitted to the Earth's surface.

Vienna Convention: The international agreement agreed to in 1985 to set a framework for global action to protect the stratospheric ozone layer. This Convention is implemented through its Montreal Protocol.

About the Contributors

Radhey S. Agarwal PhD is Professor of Mechanical Engineering at the Indian Institute of Technology, Delhi, India. He is a member of the Montreal Protocol Technology and Economic Assessment Panel (TEAP) and is a Co-Chair of the TEAP Refrigeration, Air-conditioning and Heat Pumps Technical Options Committee (RTOC). Dr Agarwal has been actively pursuing research in refrigeration and air-conditioning for over thirty years. He earned the 1998 US EPA Stratospheric Ozone Protection Award and the 2007 Best-of-the-Best Stratospheric Ozone Protection Award.

Atul Bagai is the Regional Officer (Networking) for South Asia under UNEP DTIE's Compliance Assistance Programme. Prior to 1 November 2002, he worked in the OzonAction programme office in Paris. Before joining UNEP, he worked for the Government of India as the Director of the Ozone Cell in the Ministry of Environment and Forests and in a variety of field and policy-level assignments at both federal and provincial levels in India as an Indian Administrative Services Officer. He earned the 2007 US EPA Stratospheric Ozone Protection Award.

Natarajan Balaji (India) is the Methyl Bromide Officer for the Asia and the Pacific Region. He has more than 10 years experience on project activities pertaining to strategy formulation and implementation on the Montreal Protocol in different countries across the globe. He has also worked closely with the Government of India implementing different projects, including the CFC production sector and carbon tetrachloride (CTC) production and consumption phaseout plans.

Tom Batchelor PhD (Belgium) is Director of TouchDown Consulting specializing in ozone layer and climate protection programmes. He worked in the European Commission from 1999 to 2006, where he was responsible for national and international leadership on policies to further protect the ozone layer. He was a member of the TEAP Methyl Bromide Technical Options Committee (MBTOC) from 1992 to 2003, Co-Chair of MBTOC and a member of the TEAP. He earned the 2000 US EPA Stratospheric Ozone Protection Award.

Penelope Canan PhD (US) is Professor of Sociology and Director of the International Institute for Environment and Enterprise at the University of Denver. She served on the Economic Options Committee of the Technology and Economic Assessment Panel of the Montreal Protocol. She is co-author, with Nancy Reichman, of *Ozone Connections: Expert Networks in Global Environmental Governance* (2002, Greenleaf Publishing, Sheffield, UK).

Suely Machado Carvalho PhD (Brazil) is Chief of the Montreal Protocol Unit of the United Nations Development Programme (UNDP) in New York and was Co-Chair of the TEAP from 1993 to 2002. Prior to joining UNDP she was Director of Technology Transfer at Companhia de Tecnologia de Saneamento Ambiental (CETESB) in Sao Paulo, Brazil. She earned the 1996 US EPA Stratospheric Ozone Protection Award and the 1997 Best-of-the-Best Stratospheric Ozone Protection Award.

David Catchpole (UK/US) is a consultant for Petrotechnical Resources Alaska, and provides advice to the Alaskan oil and gas industry on alternatives to halon for fire protection and explosion prevention. He is a member of the TEAP and Co-Chair of its TEAP Halons Technical Options Committee (HTOC). He earned the 1993 US EPA Stratospheric Ozone Protection Award and the 1997 Best-of-the-Best Stratospheric Ozone Protection Award.

Jorge Corona (Mexico) was for 19 years Chair of the Environmental Commission of the National Chamber of Industries of Mexico (CANACIN-TRA), where he promoted industrial commitment with the Montreal Protocol. He was Co-Chair of the Solvents Technical Options Committee from 1991 to 2000, and member of the Montreal Protocol Technology and Economic Assessment Panel from 1999 to 2002. He earned the 1991 US EPA Stratospheric Ozone Protection Award and the 1997 Best-of-the-Best Stratospheric Ozone Protection Award.

Jim Curlin (US/France) is Information Manager of the UNEP OzonAction Programme in the Division of Technology, Industry and Economics in Paris. He manages the global Information Clearinghouse service provided by UNEP under the Multilateral Fund for the Implementation of the Montreal Protocol. He is responsible for information, communication and awareness strategies and products, capacity-building services, programme management support and multi-stakeholder partnerships. In 2005, OzonAction became the first United Nations programme to earn the US EPA Stratospheric Ozone Protection Award.

Samira de Gobert is a member of the United Nations Environment Programme, Division of Technology, Industry and Economics, OzonAction Information team. Since 1998, she actively contributed to raising public awareness of the need for ozone layer protection and the Montreal Protocol. She created the OzoNews E- news service and co-managed the OzonAction Newsletter and the OzonAction Multimedia Collection. She is also a focal point for the UNEP Iraqi Marshlands Information Network, the 'Refrigerants, Naturally' partnership and the 'SolarChill' project. In 2005, OzonAction became the first United Nations programme to earn the US EPA Stratospheric Ozone Protection Award.

Brian Ellis is an expert in cleaning technology for the electronics industry and was Solvents Consultant to the Swiss Federal Office for the Environment, Forests and Landscape. He was a member of the TEAP Solvents Technical Options Committee (STOC) from its inception to its merger into the TEAP Chemicals Technical Options Committee, Co-Chair of the TEAP n-propyl bromide Task Force, and Senior Solvents Consultant to the UN Multilateral Fund. He earned the 1994 US EPA Stratospheric Ozone Protection Award.

Yuichi Fujimoto (Japan) is Consultant for Environment, Technology and International Cooperation and previously Advisor and Director of the Japan Industrial Conference for Ozone Layer Protection (JICOP). Prior to founding JICOP, Mr Fujimoto was Director of the Japan Electrical Manufacturers' Association (JEMA) where he organized Japanese industry to protect the ozone layer. He was Senior Expert Member of the TEAP during 1992–2002. He earned the 1993 US EPA Stratospheric Ozone Protection Award and the 1997 Best-of-the-Best Stratospheric Ozone Protection Award.

Etienne Gonin (France) joined the UNEP OzonAction Branch in the Division of Technology, Industry and Economics in 2003. He has worked on projects related to both ozone layer and climate protection, as well as the MLF Green Customs Capacity Building Initiative for customs officers.

Kiwohide Hata (Japan) has retired from Matsushita Electric Industrial Company.

Richard Helmick (US) is the former Director of the Climate Control Systems Division at the US Naval Sea Systems Command (NAVSEA) where he was responsible for the conversion of existing shipboard cooling and refrigeration systems to ozone-friendly refrigerants and the introduction of advanced design, ozone-safe systems in new ships. Prior to joining NAVSEA, he was Head of the Mechanical Systems Branch at the Naval Surface Warfare Center (NSWC) where he was responsible for the CFC Elimination R&D programme. He earned the 1997 US EPA Stratospheric Ozone Protection Award.

Tilman Hertz (Germany) is PhD student in Economic Sciences at the University of Aix-Marseilles III in Aix-en-Provence, France. His research deals with game theoretic approaches to Multilateral Environmental Agreement-making. He draws extensively on the example of the Montreal Protocol. He collaborated in 2007 with the UNEP OzonAction Branch in the Division of Technology, Industry and Economics.

Tony Hetherington BEng (Australia) was the Deputy Chief Officer for Technical Cooperation in the Secretariat of the Montreal Protocol Multilateral Fund (MLF) for eleven years (1995–2006). In that capacity, he was primarily responsible for the review of projects and funding applications. Prior to his service with the Secretariat he was Director of Ozone Protection with the (then) Environment Protection Agency, Commonwealth of Australia.

Michael Jeffs PhD (UK/Belgium) is the Secretary General of ISOPA, a European Trade Association for diisocyanates and polyols for polyurethanes, a role to which he is seconded by Huntsman. He has facilitated ODS-free foams technology transfer. He has been a member of the TEAP Foams Technical Options Committee (FTOC) since 1992 and was a member on the 1999 TEAP HFC and PFC Task Force. He was the foams representative on The World Bank Ozone Operation Resource Group (OORG) from 1992 to 2006. He earned the 1993 US EPA Stratospheric Ozone Protection Award and the 1997 Best-of-the-Best Stratospheric Ozone Protection Award.

Osami Kataoka (Japan) is a senior manager for technical affairs in the Daikin Industries. He serves on various Japanese committees related to the air-conditioning and refrigeration industry. He was a former chairman of the Climate Change Measures committee of the Japan Refrigeration Air conditioning Industry Association (JRAIA) and former Executive Officer of Japan Society of Refrigeration and Air Conditioning Engineers (JSRAE).

William G. Kenyon PhD (US) is President of Global Centre Consulting. He co-founded the Ad Hoc Solvent Benchmarking Team which promoted the accelerated electronics industry phaseout of CFC and methyl chloroform solvents. He has undertaken ODS phaseout projects in China and Indonesia as a consultant to UNDP, and in Central and South America as a consultant to the World Bank. He was a member of the TEAP STOC and member of World Bank Ozone Operations Resource Group (OORG) for Solvents and Process Agents sectors. Dr Kenyon is currently an expert chapter reviewer for the Intergovernmental Panel on Climate Change (IPCC) technical reports. He earned the 1990 US EPA Stratospheric Ozone Protection Award and led the team that earned the 1997 Best-of-the-Best US EPA Stratospheric Ozone Protection Award.

Dave Koehler (US) is a Principal Engineer for Prospective Technology, seeking solutions to industrial challenges by advancing today's environmentally friendly technology. He has over thirty years of government acquisition experience and has participated on a number of Defense and US EPA teams related to pollution prevention in weapon systems, facilities and industrial processes. He currently manages the US Army's Program for the Elimination of Ozone Depleting Substances. He earned the 1998 US EPA Stratospheric Ozone Protection Award.

David J. Liddy PhD (UK) is a chemist and environmental scientist for the UK Ministry of Defence, where he has had a leading role in the replacement of ODSs for the UK Ministry of Defence since 1992 and has senior responsibilities for setting a wide range of the Ministry's environmental policies. He is a member of the TEAP HTOC. He earned the 2000 US EPA Stratospheric Ozone Protection Award.

Jean Lupinacci (US) is Chief of the ENERGY STAR Commercial and Industrial Branch in the US Environmental Protection Agency's Climate Protection Partnerships Division. She was the Co-Chair of the TEAP Foams Technical Options Committee from 1989–1995. She earned a 1996 US EPA Stratospheric Ozone Protection Award.

Bella Maranion (US) is a Sector Analyst in the US EPA Stratospheric Protection Division. She was International Projects Manager for EPA's Technology Transfer and O3 Partnerships Program from 1992 to 1998, implementing bilateral projects under the Montreal Protocol Multilateral Fund to address developing country phaseout of halons. She serves as a member of the TEAP Halon Technical Options Committee (HTOC).

János Maté MA (Canada) is a Consultant to Greenpeace International's Political and Business Unit. He has worked with Greenpeace in various capacities since 1989. Over the years, he has represented Greenpeace at key meetings of the Montreal Protocol. He is also Director of Ozone and Greenfreeze Projects. He is the author of numerous Greenpeace documents on technical and policy issues related to ozone-layer protection, and is the producer of two videos on ozone and climate-friendly technologies in refrigeration. Since 2000 he has coordinated the SolarChill Cooler Project, which he co-initiated with UNEP, and has helped organize and support the work of the corporations and NGOs that are members the 'Refrigerants, Naturally!' initiative.

Alan Miller (US) is Principal Project Officer and Global Environment Facility (GEF) Coordinator at the International Finance Corporation (IFC), the private sector arm of the World Bank. Prior to joining the IFC he was Team Leader for Climate Change at the Global Environment Facility (GEF), Executive Director of the University of Maryland Center for Global Change, a senior associate in the energy and climate programme at the World Resources Institute (WRI), and Senior Attorney at the Natural Resources Defense Council (NRDC). He helped negotiate the voluntary phaseout of CFCs in food packaging and started the litigation forcing the US EPA to protect the ozone layer. He earned the 1992 US EPA Stratospheric Ozone Protection Award and the 2007 Best-of-the-Best Stratospheric Ozone Protection Award.

Melanie Miller PhD (Belgium) is a consultant and specialist in methyl bromide and alternative technologies and member of the TEAP MBTOC since 1993. She has authored many papers and case studies on viable alternatives to methyl bromide and has assisted projects of GTZ, UNDP, World Bank and GEF. She was a sector expert in World Bank's Ozone Operations Review Group (1999–2006) and adviser to the TEAP Economic Options Committee Task Force on Methyl Bromide and MLF Replenishment Task Force. She earned the 1996 US EPA Stratospheric Ozone Protection Award, the 1997 Best-of-the-Best Stratospheric Ozone Protection Award, and the 1997 UNEP Ozone Award.

Toshiyuki Miyajima (Japan) works for Seiko Epson Corporation and is now General Manager in Environmental Strategy Planning Department.

Hideo Mori (Japan) works for Otsuka Pharmaceutical Company and is Chairman of CFC Committee of Federation of Pharmaceutical Manufacturers' Association of Japan (FPMAJ). He has been a member of the TEAP ATOC/MTOC since 1999.

Peter Mullenhard (US) is a Senior Engineer for Science Applications International Corporation and is the Director of the United States Navy Shipboard Environmental Information Clearinghouse (formerly the CFC and Halon Information Clearinghouse). He also serves as the Secretary for the US Department of Defense Ozone Depleting Substances Services Steering Committee (ODS SSC) which has been meeting continuously since 1991 to exchange information and to manage and coordinate phaseouts of ODS for the US military. He earned the 1996 US EPA Stratospheric Ozone Protection Award and the 2007 Best-of-the-Best Stratospheric Ozone Protection Award.

Naohiro Murayama (Japan) is the Deputy Director of the Ozone Layer Protection Office, Ministry of Economy, Trade and Industry (METI), Japan.

Geno Nardini is the former President of the Mexican Aerosol Institute and General Director, of the International Aerosol Institute (INAAC). He is an author of The Aerosol Conversion Technology Manual and The Aerosol Guide (with Dr Montfort A. Johnsen) and the film Safe Sprays. He has spoken on aerosol technology in over 35 countries during a 48-year career. Currently a consultant and businessman, and publisher of a magazine on aerosols, Mr Nardini has served on the Aerosols Technical Options Committee (ATOC) and was the aerosol specialist on the World Bank's OORG committee. He earned the 1992 US EPA Stratospheric Ozone Protection Award.

Tsutomu Odagiri (Japan) is the Secretary General of Japan Industrial Conference on Cleaning (JICC). He had been involved in control and phase-out of 1,1,1-trichloroethane as a member of Japan Association for Hygiene of Chlorinated Solvents (JAHCS, 1989–1990) and as the Deputy Secretary General of the Japan Industrial Conference for Ozone Layer Protection (JICOP, 1990–1997). He earned the 1994 US EPA Stratospheric Ozone Protection Award and the 2007 Best-of-the-Best Stratospheric Ozone Protection Award.

Akira Okawa (Japan) is the Deputy Secretary General of the Japan Industrial Conference for Ozone Layer and Climate Protection (JICOP).

Marta Pizano (Colombia) is a consultant specializing in methyl bromide alternatives for horticulture and particularly floriculture. She has contributed to methyl bromide phaseout programmes in nearly twenty developing countries, has authored numerous papers on this topic and is a frequent speaker at international methyl bromide conferences. Since 1998, Ms Pizano has been a member of the MBTOC and has served as Co-Chair since 2005, when she became a member of the TEAP. She earned the 2004 US EPA Stratospheric Ozone Protection Award.

Jose Pons (Venezuela) is president of Spray Quimica, an aerosol products filling company that phased out CFCs in the early 1980s. Since 2002 he has been Co-Chair of the TEAP and its Medical TOC. He earned the 1995 US EPA Stratospheric Ozone Protection Award and the 1997 US EPA Best-of-the-Best Stratospheric Ozone Protection Award.

Ian Porter PhD (Australia) is the Principal Research Scientist in Plant Pathology within the Department of Primary Industries in Victoria. He is an international expert on soil disinfestation strategies, has conducted key consultancies in China and in countries with economies in transition (CEIT), has served as a member of TEAP and has been Co-Chair of MBTOC since 2005. He has been the National Research Coordinator for methyl bromide research since 1995 and has led the Australian phaseout strategy for preplant soil uses of methyl bromide in horticulture. He earned the 2003 US EPA Stratospheric Ozone Protection Award.

Gopichandran Ramachandran PhD (India) has been associated with the development and implementation of technical assistance of the OzonAction Programme for nearly a decade. He holds two doctoral degrees in chemical and microbial ecology, is an alumnus of the US Department of State International Visitors Leadership Programme and serves as the Programme Director for Environment Management at the Centre for Environment Education, Ahmadabad, India.

Sally Rand (US) is Director of High-GWP Industry Partnerships in the US EPA Climate Change Division. She was Co-Chair of the Technology and Economic Assessment Panel (TEAP) Foams Technical Options Committee (FTOC) from 1996 to 1999, member of the 1999 HFC and PFC Task Force of the TEAP, and a Lead Author of the IPCC/TEAP Special Report on Safeguarding the Ozone Layer and the Global Climate System. In 1998, Ms Rand received the US EPA Stratospheric Ozone Protection Award.

Shunichi Samejima (Japan) is the Head of Secretariat and Director of Commendation at the Asahi Glass Foundation and has been a member of the TEAP CTOC since 2005, after having served as a member of the TEAP STOC from 1998. He has extensive research experience in organic fluorine chemistry and contributed in developing alternatives to ODSs while being employed by Asahi Glass.

Akira Sekiya PhD (Japan) is Principal Research Scientist of the Research Institute for Innovation in Sustainable Chemistry in National Institute of Advanced Industrial Science and Technology (AIST), Japan. Since 1990, he has collaborated with three national projects to develop alternatives to CFCs including new HFEs. He was a member of a research team that earned the 1997 EPA Stratospheric Protection Award and a member of another research team that earned the 1998 EPA Stratospheric Protection Award.

Rajendra Shende (India/France) is Head of the UNEP OzonAction Branch in the Division of Technology, Industry and Economics in Paris. Prior to joining UNEP he represented Indian Industry at the negotiations of the Montreal Protocol when the Multilateral Fund was being created. At UNEP, Mr Shende leads the Capacity Building and Technology Support programme that enables the 145 developing countries and 17 countries in transition to comply with the provisions of the Montreal Protocol. He was review editor for the IPCC Special Report on Technology Transfer and Coordinating Lead Author for IPCC/TEAP Special Report on Ozone and Climate Change. In 2005, OzonAction became the first United Nations programme to earn the EPA Stratospheric Ozone Protection Award.

Ronald W. Sibley was the Program Manager of the US DOD Ozone Depleting Substances Reserve stockpile (1991–2006) and currently serves as a Consultant to the DOD ODS Reserve Program Office. He is a member of the TEAP HTOC and previously served on the TEAP ODS Destruction Technologies Task Force and the TEAP ODS Collection, Recovery and Storage Task Force. Mr Sibley earned the White House Closing of the Circle Award, the 1994 US EPA Stratospheric Ozone Protection Award and the 1997 US EPA Best-of-the-Best Stratospheric Ozone Protection Award.

A. K. Singh Air Marshal of the Indian Air Force has retired as Air Officer Commanding In Chief of the Western Air Command after 40 years of distinguished service. He has won many Presidential Awards including the highest Indian Military award and Mahatma Gandhi Service Award for organizing relief efforts many natural calamities. He has advanced degrees from India and US. He has contributed greatly to the computerization of inventory and phase out of ODS in the Indian Air Force.

Darrel Staley (US) is Director of Real Property Strategic Planning at Boeing. He was a member of the Solvents Technical Options Committee from 1990 to 2005 and has served as a technical consultant from 2001 to present many projects to phase out the use of ozone depleting solvent. He earned the 2002 US EPA Stratospheric Ozone Protection Award.

John R. Stemniski PhD (US) retired from the Massachusetts Institute of Technology (MIT) Draper Laboratory as principal member of technical staff. He is presently a consultant in the area of ODSs and climate change and is a member of the UNEP CTOC committee. He earned the US EPA Stratospheric Ozone Protection Award.

Scott Stone (US) is a Policy Analyst at the Institute for Governance and Sustainable Development (IGSD), a staff attorney at the Secretariat for the

International Network for Environmental Compliance and Enforcement (INECE), and a research fellow at the Program on Governance and Sustainable Development at the Bren School for Environmental Science and Management, University of California, Santa Barbara. He currently works to maximize the climate benefits of ozone protection under the Montreal Protocol.

Helen Tope PhD (Australia) is Principal Consultant for Energy International Australia and Director of Planet Futures, an independent consulting company providing strategic, policy and technical advice and facilitation services to government, industry and other non-governmental organizations on climate change, ozone layer protection and other environmental issues. She is a member of the TEAP and Co-Chair of its Medical Technical Options Committee (MTOC). She earned the 1997 US EPA Stratospheric Ozone Protection Award and the 1997 Best-of-the-Best Stratospheric Ozone Protection Award.

Shigehiro Uemura (Japan) is the Secretary General, Japan Industrial Conference for Ozone Layer and Climate Protection (JICOP).

Daniel P. Verdonik EngScD (US) is the Director of Environmental Programs with Hughes Associates, a firm specializing in fire science and engineering. Dr Verdonik provides consultancy services to the US Department of Defense, the US Environmental Protection Agency and other organizations on alternatives to halon for fire protection and explosion prevention, and on global estimates of available stocks and emissions of halons and halon alternatives. He is a member of the TEAP and Co-Chair of the HTOC. He earned the 1995 US EPA Stratospheric Ozone Protection Award and the 1997 Best-of-the-Best Stratospheric Ozone Protection Award.

Ashley Woodcock MD (UK) is Professor of Respiratory Medicine at the University of Manchester, UK. He researches the impact of the environment on the initiation, prevalence and severity of asthma and allergy. He is Co-Chair of the Medical TOC and member of TEAP. He earned the 2007 US EPA Stratospheric Ozone Protection Award.

Masaaki Yamabe PhD (Japan) is Research Coordinator for environment and energy at the National Institute of Advanced Industrial Science and Technology (AIST). Dr Yamabe has been Co-Chair of the Chemicals Technical Options Committee since 2005 and previously was a TEAP Senior Expert Advisor and a member of the Solvents TOC. He was the Director of Research Center at Asahi Glass, where he had been deeply involved in the research and development of alternatives to CFCs. He was a Coordinating Lead Author for the IPCC/TEAP joint Report on HFCs and PFCs (April 2005). He earned the 1993 US EPA Stratospheric Ozone Protection Award and the 1997 Best-of-the-Best Stratospheric Ozone Protection Award.

Durwood Zaelke (US) is President and Founder of the Institute for Governance and Sustainable Development (IGSD), the Director of the

Secretariat for the International Network for Environmental Compliance and Enforcement (INECE), and the co-director and co-founder of the Program on Governance and Sustainable Development at the Bren School for Environmental Science and Management, University of California, Santa Barbara. He was involved in the ozone negotiations as the founder and past president of the Center for International Environmental Law (CIEL) and is currently working to maximize the climate benefits of ozone protection under the Montreal Protocol.

Index